Wildlife Conservation and Management

Wildlife Conservation and Management

Edited by **Martin Winter**

SYRAWOOD
PUBLISHING HOUSE

New York

Published by Syrawood Publishing House,
750 Third Avenue, 9th Floor,
New York, NY 10017, USA
www.syrawoodpublishinghouse.com

Wildlife Conservation and Management
Edited by Martin Winter

International Standard Book Number: 978-1-68286-157-8 (Hardback)

Printed in the United States of America.

Contents

Preface

Every book is a source of knowledge and this one is no exception. The idea that led to the conceptualization of this book was the fact that the world is advancing rapidly; which makes it crucial to document the progress in every field. I am aware that a lot of data is already available, yet, there is a lot more to learn. Hence, I accepted the responsibility of editing this book and contributing my knowledge to the community.

Wildlife conservation and management is increasingly becoming a subject of great interest and value. The significance of wildlife conservation came into light few decades back when ecologists noticed the negative impacts of habitat loss and extinction of species. The destruction and degradation of different habitats of wild animals meant for their development and preservation, has led to fall in the population of various species. National parks, wildlife sanctuaries and hotspots are designed to protect wildlife on a large scale. This book focuses upon the relevance of study of conservation biology. Some of the significant concepts included in the book are wildlife ecology, morphology, genetics, and conservation practices. It aims to benefit students and academicians who are looking to explore this field.

While editing this book, I had multiple visions for it. Then I finally narrowed down to make every chapter a sole standing text explaining a particular topic, so that they can be used independently. However, the umbrella subject sinews them into a common theme. This makes the book a unique platform of knowledge.

I would like to give the major credit of this book to the experts from every corner of the world, who took the time to share their expertise with us. Also, I owe the completion of this book to the never-ending support of my family, who supported me throughout the project.

Editor

Geographic patterns of vertebrate diversity and identification of relevant areas for conservation in Europe

M. J. T. Assunção-Albuquerque, J. M. Rey Benayas, M. Á. Rodríguez & F. S. Albuquerque

Assunção–Albuquerque, M. J. T., Rey Benayas, J. M. Rodríguez, M. Á. & Albuquerque, F. S., 2012. Geographic patterns of vertebrate diversity and identification of relevant areas for conservation in Europe. *Animal Biodiversity and Conservation*, 35.1: 1–11.

Abstract

Geographic patterns of vertebrate diversity and identification of relevant areas for conservation in Europe.— The 'EU Council conclusions on biodiversity post–2010' re–enforced Europe's commitment to halt biodiversity loss by 2020. Identifying areas of high–value for biodiversity conservation is an important issue to meet this target. We investigated the geographic pattern of terrestrial vertebrate diversity status in Europe by assessing the species richness, rarity, vulnerability (according to IUCN criteria), and a combined index of the three former for the amphibians, reptiles, bird and mammals of this region. We also correlated the value of all indices with climate and human influence variables. Overall, clear geographic gradients of species diversity were found. The combined biodiversity index indicated that high–value biodiversity areas were mostly located in the Mediterranean basin and the highest vulnerability was found in the Iberian peninsula for most taxa. Across all indexes, the proportion of variance explained by climate and human influence factors was moderate to low. The results obtained in this study have the potential to provide valuable support for nature conservation policies in Europe and, consequently, might contribute to mitigate biodiversity decline in this region.

Key words: High–value biodiversity areas, Human influence, Richness, Rarity, Vulnerability.

Resumen

Patrones geográficos de diversidad de vertebrados e identificación de áreas relevantes para su conservación en Europa.— Las conclusiones del 'Consejo de la UE sobre la biodiversidad post–2010' reforzaron el compromiso europeo de detener la pérdida de la misma para el año 2020. La identificación de áreas de alto valor para la conservación de la biodiversidad resulta importante para alcanzar esta meta. En el presente estudio investigamos la distribución geográfica del estatus de la diversidad de vertebrados en Europa evaluando la riqueza de especies, rareza, vulnerabilidad (según criterios de la UICN) y un índice combinado de los tres anteriores para anfibios, reptiles, aves y mamíferos de esta región. Además, se correlacionó el valor de estos cuatro índices con variables climáticas e influencia humana. En general, se identificaron gradientes geográficos claros de diversidad de las especies. El índice combinado de biodiversidad indicó que, para la mayoría de los taxones, las áreas de alto valor de biodiversidad se encuentran principalmente en la cuenca mediterránea y la mayor vulnerabilidad en la península Ibérica. La proporción de variación explicada por el clima y la influencia humana fue de moderada a baja para todos los índices. Los resultados de este estudio tienen el potencial de proporcionar un valioso soporte científico para las políticas europeas de conservación de la naturaleza y, consecuentemente, pueden contribuir a mitigar la pérdida de biodiversidad en esta región.

Palabras clave: Áreas de alto valor de biodiversidad, Influencia humana, Riqueza, Rareza, Vulnerabilidad.

M. J. T. Assunção–Albuquerque, J. M. Rey Benayas & M. Á. Rodríguez, Dept. of Ecology, Univ. of Alcalá, 28871 Alcalá de Henares, Madrid, España (Spain).– F. S. Albuquerque, Grupo de Ecología Terrestre, Centro Andaluz de Medio Ambiente, Univ. Granada–Junta de Andalucía, Av. del Mediterráneo s/n., 18006 Granada, España (Spain).

Corresponding author: M. J. T. Assunção–Albuquerque. E–mail: mariajose.teixeira@edu.uah.es

Introduction

Assessing broad geographical patterns of species distribution is crucial to identify areas with highest species richness, rarity or vulnerability that are relevant for species conservation (Davies et al., 2006; Kati et al., 2004; Mittemeier, 2005; Myers et al., 2000; Orme et al., 2005). Myers (1988) used the term 'hotspots' to refer to those areas with relevant biodiversity characteristics that are threatened with destruction. These areas usually harbour high species richness and a high number of endemic species (Myers et al., 2000). The identification of biodiversity hotspots has been mostly based on the amount of biodiversity per land unit area (Veech, 2000), although some efforts have also considered the distribution of biodiversity threats (Balmford et al., 2000; Fleishman et al., 2006; Rey Benayas & de la Montaña, 2003; Sierra et al., 2002).

Metrics that take biodiversity and the risk of species loss into account in a particular region are important for conservation efforts and allow the identification of areas that need urgent protection (Didier et al., 2010; Margules & Pressey, 2000; Rey Benayas & de la Montaña, 2003). Identifying factors that affect species threats in a particular area may provide the bases for protection and inspire prevention measures to mitigate such threats and thus extinction risk. The relationships between human factors and biodiversity are important to assess such risk of extinction as human pressures are often related to large changes in biological diversity. However, the literature shows contradictory results. Previous studies report that human influence may affect species' spatial distribution both negatively and positively (Young et al., 2005). On the one hand, human factors, such as human activities (Araújo et al., 2002; Cincotta et al., 2000; Clergeau et al., 2006; Donald et al., 2001) and, in particular, the alteration of habitats (Kiesecker et al., 2001; Peres et al., 2010) are major causes of biodiversity loss (Brooks et al., 2002; Cardillo et al., 2004; Gaston, 2006; McKee et al., 2003; McKinney, 2001; Singh, 2002; Van Rensburg et al., 2004). On the other hand, several studies have even shown a positive relationship between human density and biodiversity, indicating that species–rich areas and human settlements often co–occur (Albuquerque & Rueda, 2010; Luck, 2007; Maffi, 2005; Sutherland 2003). However, this might be a purely correlative effect in many instances, particularly for species that are associated with farming and human habitation such as aphids (Pautasso & Powel, 2009) or ants (Schlick-Steiner et al., 2008) that may behave as invasive pests causing an absolute loss of diversity by displacing other species.

The present study joins previous conservation biogeography efforts to identify critical areas to protect European vertebrate diversity (Araújo & Pearson, 2005; Jelaska et al., 2010); it aimed to document geographic patterns of species richness, rarity, vulnerability, and a combined index of the three former measures at the 50–km grain resolution for each major taxa. We also analyzed relationships between human influence and these biodiversity indices, highlighting key areas for vertebrate conservation. Our analysis provides insights into how to address anthropogenically–derived conservation issues.

Material and methods

Distribution data

Distribution data from atlas maps for amphibian, reptile, bird, and mammal species in Europe were obtained from Gasc et al. (1997); Hagemeijer & Blair (1997) and Mitchell–Jones et al. (1999). These maps were digitalized and processed in Arc GIS 9.3 in a grid comprising 2,194 UTM cells of 50 x 50 km each. All islands, except Great Britain, and cells with less than 50% land cover were excluded from the analyses. Preliminary data analyses identified some cells with abnormally low amphibian and reptile richness compared with nearby cells. We identified these cells as outliers and they were excluded from analysis.

Criteria for identifying areas of high–value diversity

We followed Rey Benayas & de la Montaña (2003) to identify areas of high–value diversity of the various taxonomic groups. The following biodiversity criteria were assessed in all cells: a) species richness, b) rarity, c) vulnerability, and d) a combined index of biodiversity that integrates the three former criteria.

Rarity (R) was computed for each cell r as:

$$R = \sum_{i=1}^{s} (1 / n_{ri}) / S_r$$

where n is the number of cells in which species i is present, and S_r is the cell's species richness.

For vulnerability (V), we first ranked the five threat categories defined by the International Union for Nature Conservation (IUCN, 2006) as: (1) non–threatened, (2) insufficiently known, (3) rare, (4) undetermined or vulnerable, and (5) endangered, and then computed the index for each cell as:

$$V = \sum_{i=1}^{s} (v_{ri}) / S_r$$

where v_{ri} is the vulnerability rank of species i, and S_r is the richness of cell r. Initially, we also computed this index using the similar categories defined by the European Nature Information System (EUNIS, 2005) but obtained similar results (not shown) which led us to omit this index from the study.

Then, we calculated the combined index of biodiversity (C), which jointly evaluates the species richness, rarity and vulnerability for each cell:

$$C = \sum_{i=1}^{s} (1/n_{ri}) v_{ri}$$

in which species richness is implicit in the expression $\sum_{i=1}^{s}$, rarity is represented by $1/n_{ri}$, and vulnerability by v_{ri}.

Finally, we calculated a standardized biodiversity index (SBI) by dividing the combined index of biodiversity of each taxonomic group in every cell by its mean across all cells. Next, we summed the four standardized combined indices. The *SBI* formula is:

$$SBI = \sum_{j=1}^{4} 1 / m_j \sum_{i=1}^{jS} (1 / n_{ji}) V_{ji}$$

where m_j refers to the mean combined index of biodiversity of the taxonomic group j across cells.

Climate and human influence variables

We generated 21 variables to explain geographic patterns of vertebrate richness, rarity and vulnerability. These comprised the 19 climate variables of the WorldClim database (annual mean temperature, mean diurnal range, isothermality, temperature seasonality, maximum temperature of warmest month, minimum temperature of coldest month, temperature annual range, mean temperature of wettest quarter, mean temperature of driest quarter, mean temperature of warmest quarter and mean temperature of coldest quarter, annual precipitation, precipitation of wettest month, precipitation of driest month, precipitation seasonality, precipitation of wettest quarter, precipitation of driest quarter, precipitation of warmest quarter, and precipitation of coldest quarter; Hijmans et al., 2005), and two surrogates of human influence, namely, human population density and a habitat fragmentation index. Human density was obtained from the Gridded Population of the World [urban mapping project, version 3 produced by the Center for International Earth Science Information Network (CIESIN) and available at: http://sedac.ciesin.columbia.edu/gpw/ (last accessed February 2012)]. The habitat fragmentation index measures the fragmentation of land by urbanization, transport infrastructure and agriculture. It calculates how many natural complexes are found within each cell and the compactness of these complexes (average size of complex in a cell versus total area of complexes in the cell). This index was produced by the European Environment Agency and is available at http://www.eea.europa.eu/data-and-maps/figures/fragmentation–by–urbanisation–infrastructure–and–agriculture (last accessed February 2012).

Data analysis

Initially, relationships among the four biodiversity variables (species richness, rarity, vulnerability and the combined index of biodiversity) within taxonomic groups were examined by means of Spearman rank correlation using Bonferroni correction for multiple comparisons. We also performed a principal component analysis (PCA) including all biodiversity variables (species richness, rarity, vulnerability, and the combined index of biodiversity) for each taxonomic group as well as the combined biodiversity index to highlight relationships among multiple and highly correlated variables. Additionally, relationships of each biodiversity index with climate and human influence variables were investigated by means of a redundancy analysis–based variation partitioning (Borcard et al., 1992; Legendre & Legendre, 1998;

Péres–Neto et al., 2006). This analysis provides a synthetic view of the relationships by partitioning the variation of a response variable in the study area (*i.e.* a biodiversity index of a particular vertebrate group) into components independently and jointly explained by groups of explanatory variables (*i.e.* climate variables and human factors in this study). Finally, we also took into account the results of Whittaker et al. (2007) who found that relationships of amphibian, bird, and mammal (but not reptile) species richness with solar radiation (a measure of the amount of energy available in the environment) shifted from positive in northern Europe to negative in the south of this region, and that the line separating these two zones was different for each group. Thus, we repeated the above–mentioned analyses separately for each of these regions and species groups. All analyses were performed in R (R Development Core Team, 2009) using the 'vegan' package (Oksanen et al., 2009).

Results

Geographical patterns of vertebrate diversity

There are 817 terrestrial vertebrate species in our study area, of which 52 are amphibians, 108 reptiles, 515 birds, and 142 mammals. Except for birds, which showed higher species richness in central European regions, there was a tendency of the richness of the other three vertebrate groups to increase southwards, with picks of highest richness values occurring in central Europe for amphibians and mammals, and in Mediterranean areas (Iberian peninsula and Greece) for reptiles (figs. 1A–1D). The overall geographic pattern of rarity (*R*) was similar for the four taxonomic groups, with rarity generally increasing southwards, although for birds and mammals it also showed secondary peaks in the north (Norway, Sweden and Finland; figs. 1E–1H).

Higher values of the vulnerability index (*V*) based on the IUCN threat categories for amphibians were recorded in north–eastern Portugal and west–central Spain; for reptiles in France and Germany primarily, and Norway, Sweden and Romania secondarily; and for birds and mammals across the Iberian Peninsula, Poland, Ukraine and Romania, with mammals also picking in north–eastern Europe (figs. 1I–1L).

Amphibians and reptiles showed a clear north–to–south gradient of increasing values of the combined index of biodiversity (*C*), mammals did the same albeit, with a more patchy distribution, and birds showed no clear trend, with high values occurring in localized areas of southern (Iberian and Greek peninsulas), central (*e.g.* Great Britain and Hungary) and northern (Norway, Sweden and Finland) Europe (figs. 1M–1P).

Highest values of the standardized biodiversity index (*SBI*) that integrates all biodiversity criteria for the four taxonomic groups were mainly observed in the Mediterranean basin, especially in Portugal, Spain, Greece and Bulgaria, with a secondary peak in Northern Europe (fig. 2).

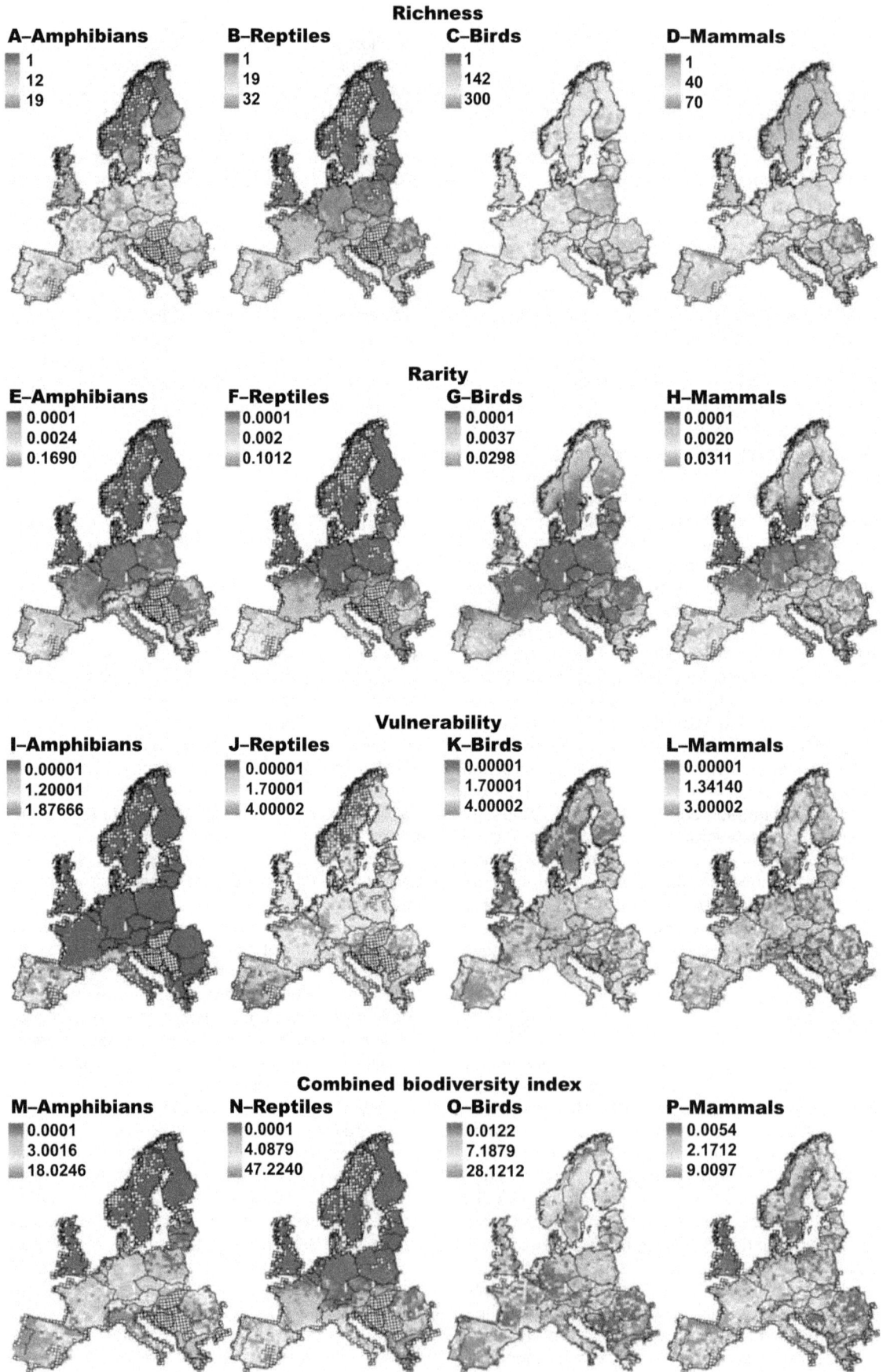

Richness

A–Amphibians
- 1
- 12
- 19

B–Reptiles
- 1
- 19
- 32

C–Birds
- 1
- 142
- 300

D–Mammals
- 1
- 40
- 70

Rarity

E–Amphibians
- 0.0001
- 0.0024
- 0.1690

F–Reptiles
- 0.0001
- 0.002
- 0.1012

G–Birds
- 0.0001
- 0.0037
- 0.0298

H–Mammals
- 0.0001
- 0.0020
- 0.0311

Vulnerability

I–Amphibians
- 0.00001
- 1.20001
- 1.87666

J–Reptiles
- 0.00001
- 1.70001
- 4.00002

K–Birds
- 0.00001
- 1.70001
- 4.00002

L–Mammals
- 0.00001
- 1.34140
- 3.00002

Combined biodiversity index

M–Amphibians
- 0.0001
- 3.0016
- 18.0246

N–Reptiles
- 0.0001
- 4.0879
- 47.2240

O–Birds
- 0.0122
- 7.1879
- 28.1212

P–Mammals
- 0.0054
- 2.1712
- 9.0097

Relationships among biodiversity criteria

Correlation analyses between species richness, rarity, vulnerability, and the combined biodiversity index within each vertebrate group indicated that almost all these biodiversity estimates were significantly and positively correlated (table 1). The combined biodiversity index was positively correlated with all estimates and especially with rarity for all groups.

The two first axes of the PCA performed on all biodiversity criteria absorbed 36.8% and 18.1% of the variation, respectively. The visual inspection of this graph revealed association of rarity, the combined index and the standardized biodiversity index on one side, and of vulnerability and species richness on the other side (fig. 3). Taxonomic groups were spread throughout the PCA bi–plot; however, it is noticeable the fact that the bird diversity criteria are relatively independent from those of all remaining taxa (fig. 3).

Variation of vertebrate diversity explained by climate and human influence

The proportion of variation explained by climate and human influence variables was highest for richness, especially for amphibians (41%) and reptiles (42%) (table 2). Rarity, vulnerability and combined biodiversity indexes were, in general, less associated with climate and human influence variables. In all cases, climate contributed more than human influences to explain these biodiversity variables. This was also reflected in the results of the variation partitioning analyses conducted separately for north and south Europe for amphibians, birds and mammals, although more variation was explained by the models for the north (table 2).

Discussion

This study identified high–value diversity areas for amphibians, reptiles, birds, and mammals in Europe by documenting the geographic distribution of five biodiversity criteria and analysing their relationships with climate and human influence factors. For most groups (amphibians, reptiles and mammals) we observed a general north–to–south gradient of increasing richness, whereas for birds, the patterns were more complex and richness picked at central European regions. Still, climate was more important than human

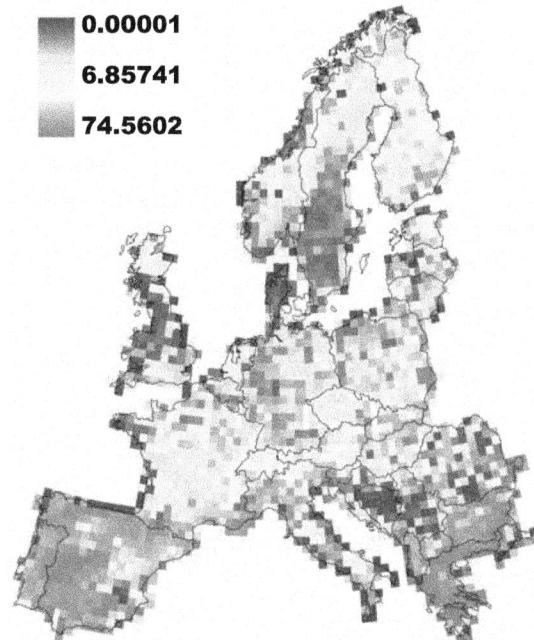

Fig. 2. Geographical patterns of the standardized biodiversity index (SBI) in Europe, which integrates all biodiversity criteria for the four vertebrate groups.

Fig. 2. Patrones geográficos del índice de biodiversidad estandarizado (SBI) en Europa, que integra todos los criterios de diversidad para los cuatro grupos de vertebrados.

influences in driving the patterns in all cases. Similar richness gradients and relationships with climate have been reported by previous studies for these taxa across Europe (Araújo & Pearson, 2005; Carrascal & Díaz, 2003; Nogués–Bravo & Martínez–Rica, 2004; Olalla–Tárraga et al, 2006; Qian & Xiao, 2012; Rodríguez et al, 2005; Rojas et al., 2001).

We also found a strong and positive correlation between rarity and the combined index of biodiversity for all groups, which highlights rarity as a key criterion to identify high–value biodiversity areas over broad geographical extents. This supports

Fig. 1. Geographical pattern of species richness (A–D), rarity (E–H), vulnerability (I–L), and combined biodiversity index (M–P) for amphibians, reptiles, birds, and mammals in Europe based on UTM grid cells with a grain of 50 x 50 km. White cells lack reliable data (see the text).

Fig. 1. Patrón geográfico de la riqueza de species (A–D), su rareza (E–H), su vulnerabilidad (I–L) y el índice de biodiversidad combinada (M–P) para anfibios, reptiles, aves y mamíferos en Europa, basándose en celdas de una cuadrícula de coordenadas UTM de 50 x 50 km. Las celdas blancas carecen de datos fiables (véase el texto).

Table 1. Spearman rank correlation coefficients between criteria used to identify areas of high–value diversity within taxonomic groups in Europe. Coefficients in bold are significant at $p < 0.05$ after applying Bonferroni corrections for multiple comparisons: S. Richness; R. Rarity; V. Vulnerabilty; C. Combined index.

Tabla 1. Coeficientes de correlación de rango de Spearman entre los criterios utilizados en la identificación de las áreas de gran valor en cuanto a diversidad de los grupos taxonómicos de Europa. Los coeficientes en negrita son significativos para p < 0,05 tras aplicar las correcciones de Bonferroni para comparaciones múltiples: S. Riqueza; R. Rareza; V. Vulnerabilidad; C. Índice combinado.

		Amphibians (n = 1,674)			Reptiles (n = 1,648)			Birds (n = 2,144)			Mammals (n = 1,875)		
		S	R	V	S	R	V	S	R	V	S	R	V
Amphibians													
	S	0.25											
	V	0.68	0.34										
	C	0.28	0.81	0.31									
Reptiles													
	S				0.65								
	V				0.49	0.25							
	C				0.57	0.92	0.14						
Birds													
	S							0.03					
	V							0.54	0.26				
	C							0.25	0.73	0.26			
Mammals													
	S										0.22		
	V										0.69	0.51	
	C										0.52	0.69	0.45

previous claims pointing out that rarity is likely to be more effective than richness to identify priority areas for conservation (Williams et al., 1996). This result is important, since richness is the conservation criterion that is used by decision makers most often (Médail & Quézel, 1997; Reyers et al., 2000; Rodrigues et al., 2004).

In general, for the four biodiversity criteria analysed, the proportion of variation explained by climate and human influence factors was moderate to low, suggesting that other factors might be important for the described geographical pattern of vertebrate diversity in Europe. Thus, the patterns found for amphibians and reptiles may be related to the lower dispersal capacity of these groups compared to that exhibited by other vertebrates, as species with low dispersal rates need a longer time to colonize sites away from their origin (Aragón et al., 2010; Araújo & Pearson, 2005), which in turn might be associated with their higher levels of endemism (Williams et al., 2000). In agreement, Araújo & Pearson (2005) reported low levels of equilibrium (*i.e.* the time needed to reach saturated communities) between the distribution of reptile and amphibian species in Europe and current climate, whereas they found that major ice–age refugia (Iberia, Italy and the Balkans) were key determinants of the current distributions of these species across this region (see also Whittaker et al., 2007). The contrasted geographical patterns found for bird richness in Europe (see fig. 1C), and the relatively independent location of this taxon with respect to all other taxa in the ordination of biodiversity criteria (see figure 3), may be related with the location of speciation centres, dispersal capacity and environmental preferences of the species of this taxon (Covas & Blondel, 1998). Also, bird and mammals appear to have been under a strong selective pressure by human disturbance in the northern hemisphere since the last glaciation, which may have also played a relevant role in driving the diversity patterns of these groups (Nogués–Bravo & Matrínez–Rica, 2004; Walther et al., 2002).Previous results have indicated that areas with high species rarity and vulnerability are usually associated with

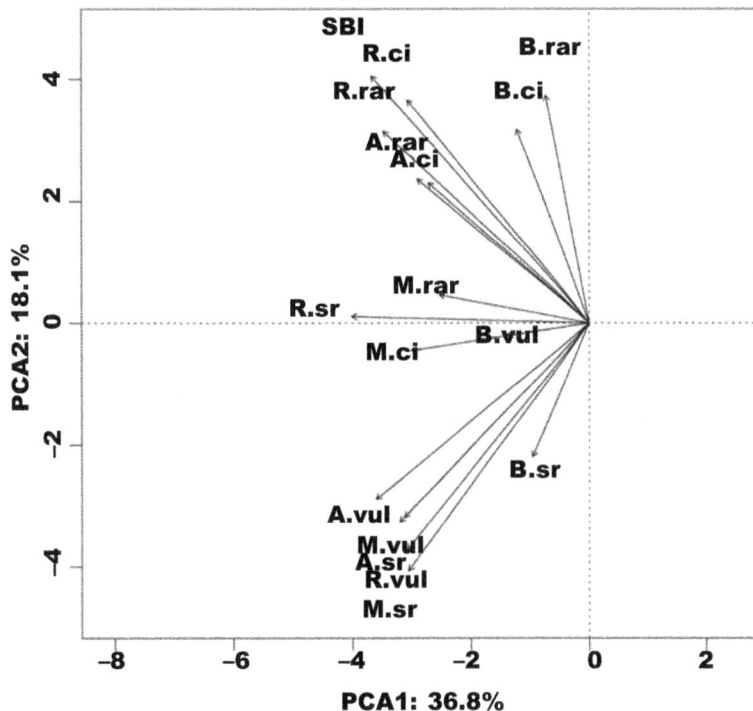

Fig. 3. Principal component analysis of vertebrate diversity criteria in Europe. Labels are the following: A.sr. Amphibian species richness; A.rar. Amphibian rarity; A.vul. Amphibian vulnerability; A.ci. Amphibian combined index; R.sr. Reptile species richness; R.rar. Reptile rarity; R.vul. Reptile vulnerability; R.ci. Reptile combined index; B.sr. Bird species richness; B.rar. Bird rarity; B.vul. Bird vulnerability; B.ci. Bird combined index; M.sr. Mammal species richness; M.rar. Mammal rarity; M.vul. Mammal vulnerability; M.ci. Mammal combined index; SBI. Standardized biodiversity index.

Fig. 3. Análisis de componentes principales de los criterios de la diversidad de vertebrados en Europa. Las etiquetas son las siguientes: A.sr. Riqueza de especies de anfibios; A.rar. Rareza de anfibios; A.vul. Vulnerabilidad de anfibios; A.ci. Índice combinado de anfibios; R.sr. Riqueza de especies de reptiles; R.rar. Rareza de reptiles; R.vul. Vulnerabilidad de reptiles; R.ci. Índice combinado de reptiles; B.sr. Riqueza de especies de aves; B.rar. Rareza de aves; B.vul. Vulnerabilidad de aves; B.ci. Índice combinado de aves; M.sr. Riqueza de especies de mamíferos; M.rar. Rareza de mamíferos; M.vul. Vulnerabilidad de mamíferos; M.ci. Índice combinado de mamíferos; SBI. Índice estandarizado de biodiversidad.

low habitat variety, forest loss, human impacts and climate change (Carrascal & Palomino, 2006; Mainka & Howard, 2010; Nuñeza et al., 2010; Vié et al., 2009), in agreement with theory and empirical evidence that relate population declines with disturbance and habitat homogenization (Echeverrería et al., 2004; Rey Benayas et al., 1999). However, our results show a weak association between rarity and vulnerability of these taxa with climate and human influence variables. This difference may be related to the coarser grain used in this study, in agreement with suggestions that the relationship between the ecological characteristics of a given species and its rarity and vulnerability value are scale–dependent (Murray & Lepschi, 2004). Even though our results suggest relatively minor effects of climate and hu-

man influence variables on vertebrate rarity and vulnerability, it should be noted that these results were obtained for a particular geographical extent (Europe) and grain (cells of 50 km²), and we cannot discard a stronger role of human influence at smaller scales (e.g. see Derraik & Phillips, 2010; Nuñeza et al., 2010; Rowley et al., 2010). Additionally, the IUCN Red List clearly shows that many vertebrate species are under threat of extinction mainly as a direct or indirect result of human activities and climate change (Vié et al., 2009).

This study identified the Mediterranean basin as one of the richest, rarest and most vulnerable areas of Europe in terms of vertebrate diversity, and supports the tenet that Mediterranean basin biodiversity is under strong threat (see fig. 1M–1P).

Table 2. Proportional amounts of the variation of species richness (*S*), rarity (*R*), vulnerability (*V*) and combined index of biodiversity (*C*) of four vertebrate groups explained independently or in concert (Sh, shared 'effects') by climate (Cl) and human (Hm) influence variables, and unexplained variation (Un) in each case. For amphibians, birds and mammals, results are reported for the entire western European region, as well as for the northern and southern areas for which Whittaker et al. (2007) found contrasted relationships between species richness and climate (see Methods).

Tabla 2. Cantidades proporcionales de la variación de la riqueza de especies (S), su rareza (R), su vulnerabilidad (V) y el índice combinado de biodiversidad en cuatro grupos de vertebrados, explicadas independientemente (C) o en concierto (Sh, 'efectos' compartidos) debidas a las variables del clima (Cl) y la influencia humana (Hm) o a una variación inexplicable (Un), en cada caso. En el caso de anfibios, reptiles, aves y mamíferos los resultados provienen de toda la región europea occidental, así como de las áreas septentrionales y meridionales, para las cuales Whittaker et al. (2007) hallaron relaciones contrastadas entre la riqueza de especies y el clima (véase Métodos).

	Amphibians				Reptiles				Birds				Mammals			
	Cl	Sh	Hm	Un	Cl	Sh	Hm	Un	Cl	Sh	Hm	Un	Cl	Sh	Hm	Un
Europe																
S	0.26	0.13	0.02	0.59	0.39	0.03	0.00	0.58	0.25	0.00	0.01	0.74	0.33	0.03	0.00	0.64
R	0.09	0.01	0.00	0.90	0.23	0.00	0.00	0.77	0.16	0.02	0.01	0.81	0.19	0.04	0.00	0.77
V	0.16	0.03	0.01	0.80	0.12	0.02	0.01	0.85	0.09	0.02	0.02	0.87	0.16	0.09	0.02	0.73
C	0.13	0.01	0.00	0.86	0.18	0.01	0.00	0.81	0.14	0.00	0.01	0.85	0.17	0.00	0.00	0.83
Northern areas																
S	0.35	0.32	0.02	0.31	–	–	–	–	–	0.01	0.00	0.65	0.39	0.09	0.01	0.51
R	0.32	0.09	0.00	0.59	–	–	–	–	–	0.07	0.00	0.86	0.11	0.40	0.02	0.47
V	0.20	0.02	0.00	0.78	–	–	–	–	–	0.02	0.02	0.83	0.16	0.15	0.02	0.67
C	0.37	0.33	0.02	0.28	–	–	–	–	–	0.04	0.01	0.88	0.22	0.05	0.00	0.73
Southern areas																
S	0.23	0.00	0.01	0.76	–	–	–	–	–	0.02	0.01	0.69	0.35	0.01	0.01	0.63
R	0.06	0.00	0.00	0.94	–	–	–	–	–	0.00	0.01	0.59	0.13	0.02	0.00	0.85
V	0.18	0.03	0.03	0.76	–	–	–	–	–	0.04	0.01	0.88	0.14	0.08	0.00	0.78
C	0.13	0.02	0.00	0.85	–	–	–	–	–	0.00	0.00	0.75	0.06	0.03	0.00	0.91

This agrees with the findings of Myers et al. (2000), who recognized the Mediterranean as one of the 25 Global Biodiversity Hotspots. These threats are often attributed to human disturbance, natural disasters, habitat loss and degradation, pollution, or invasive alien species (Vié et al., 2009). However, human influence factors explained a small proportion of the variance of each of the four biodiversity criteria that we investigated. Further research might establish to what extent detection of human influence on diversity patterns are dependent on grain in studies conducted in large areas. Irrespectively, our data allow us to conclude that using a range of biodiversity criteria is necessary to accurately identify high–value diversity areas on a large geographic scale.

Knowing the spatial distribution of species richness, rarity and vulnerability is necessary to mitigate biodiversity decline and accomplish the goal of the EU Council on biodiversity post–2010 to re–enforce Europe's commitment to halt biodiversity loss by 2020. The results of this study may be relevant to policy makers to target critical areas in order to strengthen conservation.

Acknowledgments

This research was supported by the Spanish Ministry of Science and Education (grants CGL2010–18312 to JMRB and CGL2010–22119 to MAR) and the Madrid Government REMEDINAL project (S2009AMB–1783). M. J. T. Assunção–Albuquerque was supported by the Brazilian Ministry of Education, through CAPES (Coordenação de Aperfeicçoamento de Pessoal de

Nível Superior) Doctorate scholarship and FSA was supported by BIOTREE–net –project funded by BBVA Foundation. We thank M. A. Olalla–Tárraga for his friendly review on a former version of this manuscript.

References

Albuquerque F. S., Rueda, M., 2010. Forest loss and fragmentation effects on woody plant species richness in Great Britain. *Forest Ecology and Management*, 260: 472–479.

Aragón, P., Rodríguez, M. Á., Olalla–Tárraga, M. Á. & Lobo, J. M., 2010. Predicted impact of climate change on threatened terrestrial vertebrates in central Spain highlights differences between endotherms and ectotherms. *Animal Conservation*, 13: 363–373.

Araújo, M. B. & Pearson, R. G., 2005. Equilibrium of species' distributions with climate. *Ecography*, 28: 693–695.

Araújo, M. B., Williams, P. H. & Turner, A., 2002b. A sequential approach to minimize threats within selected conservation areas. *Biodiversity and Conservation*, 11: 1011–1024.

Balmford, A., Gaston, K. J., Rodrigues, A. L. & James, A. N., 2000. Integrating costs of conservation into international priority setting. *Conservation Biology*, 14: 597–605.

Borcard, D., Legendre, P. & Drapeau, P., 1992. Partialling out the spatial component of ecological variation. *Ecology*, 73: 1045–1055.

Brooks, T. M., Mittermeier, R. A., Mittermeier, C. G., da Fonseca, G. A. B., Rylands, A. B., Konstant, W. R., Flick, P., Pilgrim, J., Oldfield, S., Magin, G. & Hilton–Taylor, C., 2002. Habitat Loss and Extinction in the hotspots of Biodiversity. *Conservation Biology*, 16: 909–923.

Cardillo, M., Purvis, A., Sechrest, W., Gittleman, J. L., Bielby, J., Mace, G. M., 2004. Human Population Density and Extinction Risk in the World's Carnivores. *PLoS Biology*, 2: 0909–0914.

Carrascal, L. M. & Díaz, L., 2003. Asociación entre distribución continental y regional. Análisis con la avifauna forestal y de medios arbolados de la Península Ibérica. *Graellsia*, 59: 179–207.

Carrascal, L. M. & Palomino, D., 2006. Rareza, estatus de conservación y sus determinantes ecológicos. Revisión de su aplicación a escala regional. *Graellsia*, 62: 523–538.

CIESIN, 2007. Center for International Earth Science Information Network., 2000. Gridded population of the world, version 2. http://sedac.ciesin.columbia.edu/plue/gpw.

Cincotta, R. P., Wisnewski, J. & Engelmn, R., 2000. Human Population in the biodiversity hotspots. *Letters to Nature*, 404: 990–992.

Clergeau, P., Croci, S., Jokimaki, J., Kaisanlahti–Jokimaki, M. L. & Dinetti, M., 2006. Avifauna homogenisation by urbanisation: analysis at different European latitudes. *Biollogical Conservation*, 127: 336–344.

Covas, R. & Blondel, J., 1998. Biogeography and history of the Mediterranean bird fauna. *Ibis*, 140: 395–407.

Davies, R. G., Orme, C. D. L., Olson, V., Thomas, G. H., Ross, S. G., Ding, T.–S., Rasmussen, P. C., Stattersfield, A. J., Bennett, P. M., Blackburn, T. M., Owens, I. P. F. & Gaston, K. J., 2006. Human impacts and the global distribution of extinction risk. *Proceedings of the Royal Society of London*, Series B 273: 2127–2133.

Derraik, J. G. B. & Phillips S., 2010. Online trade poses a threat to biosecurity in New Zealand. *Biological Invasions,* 12: 1477–1480.

Didier, K. A., Wilkie, D., Douglas–Hamilton, I., Frank, L., Georgiadis N., Graham, M., Ihwagi, F., King, A., Cotterill, A., Rubenstein D. & Woodroffe, R., 2010. Conservation planning on a budget: a 'resource light' method for mapping priorities at a landscape scale? *Biodiversity and Conservation*, 18: 1979–2000.

Donald, P. F., Green, R. E. & Heath, M. F., 2001. Agricultural intensification and collapse of Europe's farmland bird populations. *Proceedings of the Royal Society London*, B 268: 25–29.

Echeverría, C., Lara, A., Newton, A., Rey Benayas, J. M. & Coomes, D. 2007. Impacts of forest fragmentation on species composition and forest structure in the temperate landscape in southern Chile. *Global Ecology and Biogeography*, 16: 426–439.

EUNIS, 2005. European Nature Information System. http://eunis.eea.europa.eu.

Fleishman, E., Noon, B. R. & Noss, R. F., 2006. Utility and limitations of species richness metrics for conservation planning. *Ecological Indicators*, 6: 543–553.

Gasc, J. P., Cabela, A., Crnobrnja–Isailovic, J., Dolmen, D., Grossenbacher, K., Haffner, P., Lescure, J., Martens, H., Martínez Rica, J. P., Maurin, H., Oliveira, M. E., Sofianidou, T. S., Veith, M. & Zuiderwijk, A. (Eds), 1997. Atlas of amphibians and reptiles in Europe. Collection Patrimoines Naturels 29, Societas Europaea Herpetologica, Muséum National d'Histoire Naturelle & Service du Petrimone Naturel, Paris.

Gaston, K. J., 2006. Biodiversity and extinction: Macroecological patterns and people. *Progress in Physical Geography*, 30: 258–269.

Hagemeijer, W. J. M. & Blair, M. J., 1997. *The EBCC atlas of European breeding birds, their distribution and abundance*. Published for European Bird Census Council. Poyser, London.

Hijmans, R. J., Cameron, S. E., Parra, J. L., Jones, P. G. & Jarvis, A., 2005. Very high resolution interpolated climate surfaces for global land areas. *International Journal of Climatology*, 25: 1965–1978.

IUCN, 2006. Red list of the threatened species. http://www.redlist.org.

Jelaska, S. D., Nikolić, T., Serić Jelaska, L., Kusan, V., Peternel, H., Guzvica, G., Major, Z., 2010. Terrestrial Biodiversity Analyses in Dalmatia (Croatia): A Complementary Approach Using Diversity and Rarity. *Environmental Management*, 45: 616–625.

Kati, V., Devillers, P., Dufrêne, M., Legakis, A., Vokou, D. & Lebrun, P., 2004. hotspots, complementarity or representativeness? Designing optimal small–scale reserves for biodiversity conservation. *Biological Conservation*, 120: 471–480.

Kiesecker, J. M., Blaustein, A. R. & Belden, L. K., 2001. Complex causes of amphibian population declines. *Nature*, 410: 681–684.

Legendre, P. & Legendre, L., 1998. Numerical Ecology, 2nd English edn. Amsterdam: Elsevier Science BV.

Luck, G. W., 2007. A review of the relationships between human population density and biodiversity. *Biological Reviews*, 82: 607–645.

Maffi, L., 2005. Linguistic, cultural, and biological diversity. *Annual Review of Anthropology*, 34: 599–617.

Mainka, S. A. & Howard, G. W., 2010. Climate change and invasive species: double jeopardy. *Integrative Zoology*, 5: 101–111.

Margules, C. R. & Pressey, R. L., 2000. Systematic conservation planning. *Nature*, 405: 243–253.

McKee, J. K., Sciulli, P. W., Fooce, C. D. & Waite, T. A., 2003: Forecasting global biodiversity threats associated with human population growth. *Biological Conservation*, 115: 161–164.

McKinney, M. L., 2001. Role of human population size in raising bird and mammal threat among nations. *Animal Conservation*, 4: 45–57.

Medail, F. & Quézel, P., 1997. Hot–Spots analysis for Conservation of plant biodiversity in the Mediterranean Basin. *Annales Missouri Botanical Garden*, 84: 112–127.

Mitchell–Jones, A. J., Amori, G., Bodgdanowicz, W., Krystufek, B., Reijnders, P. J. H., Spitzenberger, F., Stubbe, M., Thissen, J. B. M., Vohralík, V. & Zima, J., 1999. *The Atlas of European mammals*. Academic Press, London.

Mittermeier, R. A., 2005. *Hotspots Revisited: Earth's Biologically Richest and Most Endangered Terrestrial Ecoregions*. Cemex, Conservation International and Agrupación Sierra Madre, Monterrey, Mexico.

Murray, B. R. & Lepschi, B. J., 2004. Are locally rare species abundant elsewhere in their geographical range? *Austral Ecology*, 29: 287–293.

Myers, N., 1988. Threatened biotas: hotspots in tropical forests. *Environmentalist*, 8: 178–208.

Myers, N., Mittermeier, R. A., Mittermeier, C. G., da Fonseca, G. A. B. & Kents, J., 2000. Biodiversity Hotspots for Conservation Priorities. *Nature*, 403: 853–858.

Nogués–Bravo, D. & Martínez–Rica, J. P., 2004. Factors controlling the spatial species richness pattern of four groups of terrestrial vertebrates in an area between two different biogeographic regions in northern Spain. *Journal of Biogeography*, 31: 629–641.

Nuñeza, O. M., Ates, F. B. & Alicante, A. A., 2010. Distribution of endemic and threatened herpetofauna in Mt. Malindang, Mindanao, Philippines. *Biodiversity and Conservation*, 19: 503–518.

Oksanen, J., Kindt, R., Legendre, P., O'Hara, B., Simpson, G. L., Solymos, P., Stevens, M. H. H. & Wagner, H., 2009. vegan: community ecology package. Rpackage version 1.15–4. http://CRAN.R–project.org/package=vegan.

Olalla–Tárraga, M. A., Rodríguez, M. A. & Hawkins, B. A., 2006. Broad–scale body size patterns in squamate reptiles of Europe and North America. *Journal of Biogeography*, 33: 781–793.

Orme, C. D. L., Davies, R. G., Burgess, M., Eigenbrod, F., Pickup, N., Oslon, V. A., Webster, A. J., Ding, T., Rasmussen, P. C., Ridgely, R. S., Stattersfield, A. J., Bennett, P. M., Blackburn, T. M., Gaston, K. J. & Owens, I. P. F., 2005. Global hotspots of species richness are not congruent with endemism or threat. *Nature*, 436: 1016–1019.

Pautasso, M. & Powell, G., 2009. Aphid biodiversity is positively correlated with human population in European countries. *Oecologia*, 160: 839–846.

Peres, C. A., Gardner, T. A., Barlow, J., Zuanon, J., Michalski, F., Lees, A. C., Vieira, I. C. G., Moreira, F. M. S. & Feeley, K. J., 2010. Biodiversity conservation in human–modified Amazonian forest landscapes. *Biological Conservation*, 143: 2314–2327.

Peres–Neto, P. R., Legendre, P., Dray, S. & Borcard, D., 2006. Variation partitioning of species data matrices: estimation and comparison of fractions. *Ecology*, 87:2614–2625.

Qian, H. & Xiao, M., 2012. Global patterns of the beta diversity–energy relationship in terrestrial vertebrates. *Acta Oecologica*, 39: 67–71.

R Development Core Team, 2009. R: A Language and Environment for Statistical Computing. R Foundation for Statistical Computing, Vienna, Austria, ISBN 3– 900051–07–0. http://www.R–project.org.

Rey Benayas, J. M., Scheiner., M., García S., Colomer., M., & Levassor, C., 1999. Commonness and rarity: theory and application of a new model to Mediterranean montane grasslands. *Conservation Ecology*, 3(1): 5 (on line).

Rey Benayas, J. M. & De la Montaña, E., 2003. Identifying areas of high–value vertebrate diversity for strengthening conservation. *Biological Conservation*, 114: 357–370.

Reyers, B., Van Jaarsveld, A. S. & Krüger, M., 2000. Complementarity as a biodiversity indicator strategy. *The Royal Society*, 267: 505–513.

Rodrigues, A. S. L, Andelman, S. J., Bakarr, M. I., Boitani, L., Brooks, T. M., Cowling, R. M., Fishpool, L. D., Da Fonseca, G. A., Gaston, K. J., Hoffmann, M., Long, J.S., Marquet, P. A., Pilgrim, J. D., Pressey, R. L., Schipper, J., Sechrest, W., Stuart, S. N., Underhill, L. G., Waller, R. W., Watts, M. E. & Yan, X., 2004. Effectiveness of the global protected area network in representing species diversity. *Nature*, 428: 640–643.

Rodríguez, M. Á., Belmontes, J. A. & Hawkins, B. A., 2005. Energy, water and large–scale patterns of reptile and amphibian species richness in Europe. *Acta Oecologica*, 28: 65–70.

Rojas, A. B., Cotilla, I., Real, R. & Palomo, L. J., 2001. Determinación de las áreas probables de distribución de los mamíferos terrestres en la provincia de Málaga. *Galemys*, 13: 217–229.

Rowley, J., Brown, R., Bain, R., Kusrini, M., Inger, R., Stuart, B., Wogan, G., Thy, N., Chan–ard, T., Trung, C. T., Diesmos, A., Iskandar, D. T., Lau, M., Ming, L. T., Makchai, S., Truong, N. Q. & Phimmachak, S., 2010. Impending conservation crisis for Southeast

Asian amphibians. *Biology Letters*, 6: 336–338.

Schlick–Steiner, B. C., Steiner, F. M. & Pautasso, M., 2008. Ants and people: a test of two mechanisms potentially responsible for the large–scale human population–biodiversity correlation for Formicidae in Europe. *Journal of Biogeography*, 35: 2195–2206.

Sierra, R., Campos, F. & Chamberlin, J., 2002. Assessing biodiversity conservation priorities: ecosystem risk and representativeness in continental Ecuador. *Landscape and Urban Planning*, 59: 95–110

Singh, J. S., 2002. The biodiversity crisis: A multifaceted review. *Current Science*, 82: 638–647.

Sutherland, W. J., 2003. Parallel extinction risk and global distribution of languages and species. *Nature*, 423: 276–279.

Van Rensburg, B. J., Erasmus, B. F. N., Van Jaarsveld, A. S., Gaston, K. J. & Chown, S. L., 2004: Conservation during times of change: correlations between birds, climate and people in South Africa. *South African Journal of Science*, 100: 266–72.

Veech, J. A., 2000. Choice of Species–Area Function Affects Identification of Hotspots. *Conservation Biology*, 14: 140–147.

Vié, J.–C., Hilton–Taylor, C. & Stuart, S. N. (Eds.), 2009. Wildlife in a Changing World–An Analysis of the 2008 IUCN Red List of Threatened Species. Gland, Switzerland: IUCN.

Walther, G. R., Post, E, Convey, P., Menzel, A., Parmesank, C., Beebee, T. J. C., Fromentin, J.–M., Hoegh–Guldberg, O. & Bairlein, F., 2002. Ecological responses to recent climate change. *Nature*, 416: 389–395.

Whittaker, R. J., Nogués–Bravo, D. & Araújo, M. B., 2007. Geographical gradients of species richness: a test of the water–energy conjecture of Hawkins et al. (2003) using European data for five taxa. *Global Ecology and Biogeography*, 16: 76–89.

Williams, P. H., Gibbons, D., Margules, C., Rebelo, A., Humphries, C. & Pressey, R., 1996. A Comparison of Richness Hotspots, Rarity Hotspots and Complementary Areas for Conserving Diversity of British Birds. *Conservation Biology*, 10: 155–174.

Williams, P. H., Humphries, C., Araújo, M. B., Lampinen, R., Hagemeijer, W., Gasc, J.–P. & Mitchell–Jones, T., 2000. Endemism and important areas for representing European biodiversity: a preliminary exploration of atlas data for plants and terrestrial vertebrates. *Belgian Journal of Entomology*, 2: 21–46.

Young, J., Watt, A., Nowicki, P., Alard, D., Clitherow, J., Henle, K., Johnson, R., Laczko, E., Mccracken, D., Matouch, S., Niemela, J. & Richards, C., 2005. Towards sustainable land use: identifying and managing the conflicts between human activities and biodiversity conservation in Europe. *Biodiversity and Conservation*, 14: 1641–1661.

2

The competitor release effect applied to carnivore species: how red foxes can increase in numbers when persecuted

J. Lozano, J. G. Casanovas, E. Virgós & J. M. Zorrilla

Lozano, J., Casanovas, J. G., Virgós, E. & Zorrilla, J. M., 2013. The competitor release effect applied to carnivore species: how red foxes can increase in numbers when persecuted. *Animal Biodiversity and Conservation*, 36.1: 37–46.

Abstract

The competitor release effect applied to carnivore species: how red foxes can increase in numbers when persecuted.— The objective of our study was to numerically simulate the population dynamics of a hypothetical community of three species of small to medium–sized carnivores subjected to non–selective control within the context of the competitor release effect (CRE). We applied the CRE to three carnivore species, linking interspecific competition with predator control efforts. We predicted the population response of European badger, the red fox and the pine marten to this wildlife management tool by means of numerical simulations. The theoretical responses differed depending on the intrinsic rate of growth (r), although modulated by the competition coefficients. The red fox, showing the highest r value, can increase its populations despite predator control efforts if control intensity is moderate. Populations of the other two species, however, decreased with control efforts, even reaching extinction. Three additional theoretical predictions were obtained. The conclusions from the simulations were: 1) predator control can play a role in altering the carnivore communities; 2) red fox numbers can increase due to control; and 3) predator control programs should evaluate the potential of unintended effects on ecosystems.

Key words: Predator control, Wildlife management, Competition, Generalist predator, Population dynamics, Population growth.

Resumen

El efecto liberador de competidores aplicado a las especies de carnívoros: cómo puede aumentar el número de zorros cuando son perseguidos.— El objetivo de nuestro estudio consistió en simular numéricamente la dinámica de poblaciones de una comunidad hipotética de tres especies de carnívoros de talla pequeña y mediana sometidas a un control no selectivo en el contexto del efecto liberador de competidores. Aplicamos el modelo del efecto liberador de competidores, que relaciona la competencia interespecífica con el control de predadores, a tres especies de carnívoro. Así, pudimos predecir la respuesta de las poblaciones de tejón, zorro y marta frente a este mecanismo de gestión de la fauna silvestre por medio de simulaciones numéricas. Las respuestas teóricas fueron distintas en función de la tasa intrínseca de crecimiento (r), si bien estuvieron reguladas por los coeficientes de competencia. El zorro, con el valor de r más elevado, puede aumentar sus poblaciones a pesar del control de predadores si este es moderado. Por el contrario, las poblaciones de las otras dos especies disminuyeron con el control hasta extinguirse. Obtuvimos también tres predicciones teóricas. Las conclusiones de las simulaciones fueron: 1) el control de predadores puede alterar las comunidades de carnívoros; 2) la población de zorros puede aumentar debido al control y 3) los programas de control de predadores deberían evaluar los efectos indeseados que podrían producirse en los ecosistemas.

Palabras clave: Control de predadores, Gestión ambiental, Competencia, Predadores generalistas, Dinámica de poblaciones, Crecimiento poblacional.

Jorge Lozano, Dept. de Ecología, Univ. Autónoma de Madrid, c/ Darwin 2, Edificio de Biología, E–28049 Cantoblanco, Madrid, España (Spain).– Jorge Lozano, Jorge G. Casanovas & Juan M. Zorrilla, Dept. de Ecología, Univ. Complutense de Madrid, c/ José Antonio Novais 12, Ciudad Universitaria, E–28040 Madrid, España (Spain).– Emilio Virgós, Dept. de Biología y Geología, Univ. Rey Juan Carlos, c/ Tulipán s/n., E–28933 Móstoles, Madrid, España (Spain).

Corresponding author: J. Lozano

Introduction

Populations of various taxonomic groups are declining sharply due to human activities in the environment (Groombridge, 1992). In particular, predator control has produced a negative and strong impact on populations of many species of large carnivores (*e.g.* Schaller, 1996; Breteinmoser, 1998; Rodríguez & Delibes, 2002). A lot of smaller carnivore species, such as the marten species (*Martes* spp.), the European wildcat (*Felis silvestris* Schreber, 1775) or the European badger (*Meles meles* L.), are also affected (Langley & Yalden, 1977; Lankester et al., 1991; Ruggiero et al., 1994; Caro & Stoner, 2003; Lozano et al., 2007). In contrast, adaptable species of carnivores, such as the red fox (*Vulpes vulpes* L.), can become more abundant in some places as the result of such activities (Baker & Harris, 2006; Beja et al., 2009).

The main goal of predator control, in both natural reserves interested in protecting sensitive species and hunting lands with an interest in harvest management, is to reduce the incidence of predation (Tapper et al., 1991; Harris & Saunders, 1993; Reynolds & Tapper, 1996; Côte & Sutherland, 1997). Predator control techniques vary greatly both in their degree of selectivity and effectiveness with regard to the persecuted species (*e.g.* Calver et al., 1989; Windberg & Knowlton, 1990; Tapper et al., 1991; Hein & Andelt, 1994; Reynolds & Tapper, 1996; Harding et al., 2001; Rushton et al., 2006; Beja et al., 2009). Unfortunately, many of these methods are non–selective (*e.g.* snares, traps, poisoned baits), and negatively affect both the species considered a pest and others of conservation concern (Herranz, 2000; Duarte & Vargas, 2001; Whitfield et al., 2003; Rodríguez & Delibes, 2004; Virgós & Travaini, 2005; Beja et al., 2009; Cabezas–Díaz et al., 2009; Estes & Terborgh, 2010; Lozano et al., 2010).

Many studies have been carried out around the world dealing with the effects of predator control on prey populations (Reynolds & Tapper, 1996; Côte & Sutherland, 1997; Valkama et al., 2005; Oro & Martínez–Abraín, 2007). Nevertheless, the effects of such control on the population parameters of the targeted predators, or on the structure of their natural communities, have received much less attention (*e.g.* Yoneda & Maekawa, 1982; Harris & Saunders, 1993; Reynolds et al., 1993; Estes & Terborgh, 2010).

It is well established that predation, competition and their interaction are important factors shaping natural communities (Chase et al., 2002; Caro & Stoner, 2003). Interspecific interactions within the communities of carnivores can cause the extinction or exhaustion of specialist species or larger predators (*e.g.* lynxes, wolves, coyotes). For example, such interactions may be interguild predation or competition through exploitation (Erlinge & Sandell, 1988; Polis & Holt, 1992; Palomares & Caro, 1999; Müller & Brodeur, 2002). After the disappearance of these top predators, numbers of smaller species could increase, a pattern also observed in more generalist species such as the Iberian mongoose (*Herpestes ichneumon* L.) or the red fox (see Palomares et al., 1995; Creel & Creel, 1996; Palomares & Caro, 1999).

These data imply that different predator management systems could have different effects on the communities of predators and indirectly affect the levels of predation on the prey species (Estes & Terborgh, 2010; Levi et al., 2012). This could even produce paradoxical effects, such as a reduction in the diversity and/or abundance of game species or those of conservation interest, as a result of the increase in the abundance of generalist carnivores (Soulé et al., 1988; Courchamp et al., 1999a; Crooks & Soulé, 2000; Caut et al., 2007).

However, given the lack of field data in relation to this issue, and the difficulty involved in obtaining such information, an alternative to study the possible effects of the applied control techniques on the predator populations consists of developing simple mathematical models. With a minimum number of assumptions, it is possible analyse the population dynamics of these species, and the interspecific interactions on community composition (see the application of this type of procedure in the works by Shorrocks & Begon, 1975; Courchamp et al., 1999a, 1999b; Caut et al., 2007; Fenton & Brockhurst, 2007).

Caut et al. (2007) used this theoretical approach to describe a new ecological mechanism named the Competitor Release Effect (hereafter CRE). According to this mechanism, an inferior competitor can increase in numbers if the superior competitor is controlled, due to the competitive interactions between them. This occurs even though the inferior competitor is also being killed. Moreover, at the same time, theoretical results show negative effects on the population of a shared prey. Shared prey can decline because while numbers of a superior competitor are decreasing, there may be an unwanted and unexpected increase in numbers of the inferior competitor. Caut et al used empirical data from an eradication program of rodents living on islands to test and support their competitor release hypothesis. They also suggested that the same effect could be found in communities of carnivore mammals, where the population of a competitor such as the red fox could increase if the community of predators is being managed using predator control.

The objective of our study was to numerically simulate the population dynamics of a hypothetical community of three species of small to medium–sized carnivores subjected to non–selective control (*i.e.* where all the individuals are being eliminating with similar probability), within the context of the proposed CRE (Caut et al., 2007). The selected species for the simulations were the European badger, the red fox and the pine marten (*Martes martes* L.). Reasons for this choice were: (i) these species are sympatric in a wide range of Europe (Mitchell–Jones et al., 1999); (ii) there is evidence of competition among them (Lindström et al., 1995; Palomares & Caro, 1999; Trewby et al., 2008); (iii) all three species are often controlled (*e.g.* Côté & Sutherland, 1997; Virgós & Travaini, 2005; Trewby et al., 2008) and; (iv) information about their populational parameters can be found in the scientific literature (Bright, 1993).

Our study differs from that of Caut et al. (2007) in that we evaluated a third species in the mathematical model. Moreover, known (*i.e.* real) values for the population growth rates were used, so the model should be more realistic. We specifically wanted to know whether the CRE could increase the population of red fox when the three species are being controlled, and if so, under what conditions such an increase occurs. Furthermore, we used the results of numerical simulations from the theoretical model to obtain a set of predictions which could be tested with empirical data when available.

Material and methods

We modified the CRE model from Caut et al. (2007) by adding an equation to simulate the populational dynamics of a third competitor. This model was based on the classical Lotka–Volterra equations (Powell & Zielinski, 1983; Begon et al., 1996), which were modified to incorporate an additional factor of linear mortality (in a similar way that in Shorrocks & Begon, 1975; Fenton & Brockhurst, 2007). The slope of this new factor is independent from the density and represents the mortality caused by the predator control involved (degree of non–selectivity). Thus, this modification of the classical model, which only includes density–dependent mortality implicitly in the population growth rate, is analogous to those previously proposed by Gause (1934, 1935). In these studies, the author considered mortality independent from density, possibly caused by factors such as parasites, non–specific diseases, or other mortality factors (*e.g.* Caut et al., 2007; Fenton & Brockhurst, 2007).

Our model is defined by three equations that govern the coupled dynamics of three species of competing carnivores:

$$A_{t+1} = A_t + r_A A_t \left(1 - \frac{A_t + \alpha_{AB} B_t + \alpha_{AC} C_t}{K_A} \right) - \omega A_t$$

$$B_{t+1} = B_t + r_B B_t \left(1 - \frac{B_t + \alpha_{BA} A_t + \alpha_{BC} C_t}{K_B} \right) - \omega B_t$$

$$C_{t+1} = C_t + r_C C_t \left(1 - \frac{C_t + \alpha_{CA} A_t + \alpha_{CB} B_t}{K_C} \right) - \omega C_t$$

where A, B and C represent the number of individuals of each particular species at time t. The intrinsic growth rates of each population are r_A, r_B and r_C. The effect of the interspecific competition of one species against another is represented by α (which is the competition coefficient), and the carrying capacity of the environment for each population is K. Finally, included in each equation is the parameter ω (control coefficient) which represents the extraction rate of each species as a result of the non–selective control applied. This can be interpreted as the proportion of the population of a given species which dies during a period t as a consequence of the control. Because the model attempts to determine the effect of non–selective control, this ω parameter was fixed with the same

value for all the species of the community, although more complex scenarios could be developed within the premise of non–selective control.

In this scenario, the set of values of the parameters was chosen to take into consideration that each species of the model represented one for which the intrinsic growth rate (r) was available (Bright, 1993). These values are from British populations of each species, but we assumed that the growth rates were similar for other regions of Europe (Turchin, 2003). The three carnivore species were the European badger (species A), the pine marten (species B), and the red fox (species C). The carrying capacities (K) were set as constant and identical to facilitate the interpretation of results from the numerical analysis. The selected value is theoretical but also realistic in function of the considered spatial scale (K = 30, approximately equivalent to 25–30 km² considering mean density values of the species in Europe; Wilson & Mittermeier, 2009), and it allows a sufficient range of variation to compare the populational dynamics among species. Likewise, the competitive interactions among the species were considered symmetrical (*i.e.* AB = AC = BA = BC = CA = CB). Although asymmetries can be expected in the wild (*e.g.* Palomares & Caro, 1999), unfortunately no quantitative data are available to create more realistic scenarios. Table 1 shows the demographic values used for the parameters in the model.

A total of 72 deterministic numerical analyses were performed per species: one for each combination of parameters, varying in equal intervals α from 0.1 to 0.9, and ω from 0 to 0.7. The value reached by the population in the equilibrium for each combination of parameters was graphically represented. Furthermore, to test whether the predictions arising from the theoretical model were sufficiently robust to stand up against small variations in the values for the intrinsic growth rate (r), a sensitivity analysis was conducted. This consisted of creating simulations varying by a single parameter and leaving the other parameters constant. Thus, α = 0.5, ω = 0.3, r_A = 0.46 and r_C = 1.1 were fixed, whereas r_B (the intermediate intrinsic growth rate regarding original values) varied in five intervals from 0.4 to 1.1. This also could be equivalent to simulating the population dynamics of a competing species other than the pine marten (species B), showing different values of intrinsic growth rate. All these analyses were performed using the computer program STELLA v.9.1.4 for Windows (ISEE, 2010).

The basic model used here and the Lotka–Volterra equations have been analytically studied elsewhere using the same assumptions (for more details see Gilpin, 1975; Shorrocks & Begon, 1975; Caut et al., 2007). Therefore, our work is an extension of these basic models, incorporating a level of predator community complexity more commonly found in natural systems.

Results

Simulation results

From our numerical simulations, we found a qualitatively different behaviour among the carnivore popu-

Table 1. Demographic values for parameters of the model: r is the intrinsic growth rate for each species, and K is the carrying capacity of the environment.

Tabla 1. Valores demográficos utilizados para los parámetros del modelo: r es la tasa intrínseca de crecimiento para cada especie y K es la capacidad de carga del medio.

Letter in equation	A	B	C
Species	*M. meles* (badger)	*M. martes* (pine marten)	*V. vulpes* (red fox)
r	0.46	0.57	1.1
K	30	30	30

lations, considering the competitive interactions that can occur in the community as a result of a predator control program. The sensitivity analysis of the system with regard to the value of species B ($r = 0.57$, the pine marten) indicates that when there is intermediate competition (0.5) and moderate control (0.3), the variation in the intrinsic growth rate affects the equilibrium value of species C ($r = 1.1$, the red fox) but does not appear to affect the species A ($r = 0.46$, the European badger). There is a critical value for r_B, at 0.9, above which the red fox population does not rise above K* (the maximum population value in the presence of the other two species, in this scenario $K^* = 15$ individuals). At intermediate levels of competition, the model predicts a 'paradoxical effect' produced by the non–selective control. In other words, when species B shows an $r > 0.9$ (this being a threshold value), the population increases rather than decreases, despite the applied control.

The simulations of the general scenario, where the intrinsic growth rates were fixed (0.46, 0.57 and 1.1), produced a system dynamic showing very clear patterns of population change depending on competition and predator control intensity. Thus, if the intrinsic growth rate (r) of one species was below the threshold value $r' = 0.9$, then the population would change in time, always showing a decrease in its numbers (see figs. 1, 2). The simulated population with the lowest intrinsic growth rate, corresponding to the European badger ($r = 0.46$, fig. 1), showed a linear decrease in its numbers when predator control was applied. This population became completely extinct with an intermediate degree of control intensity (0.5), in conditions of minimum competition (0.1). The increase in competition implies that the population could be destroyed under conditions of even less intense non–selective control.

Likewise, the general pattern observed in the change of the population with a slightly higher growth rate ($r = 0.57$), belonging in this case to the pine marten, was practically the same (see fig. 2). The difference was that the higher growth rate implied that the population of this species needed a slightly higher level of control intensity than the badger to disappear: a value of 0.6, under conditions of minimum competition (0.1).

In strong contrast, the population of the red fox behaved in a very different manner under conditions of non–selective control, and depending on competition (see fig. 3). The pattern presented a very marked non–linearity which could be attributed to their high intrinsic growth rate ($r = 1.1$). Without competition, or with very low levels of competition (up to 0.2), control efforts reduced the population in a linear sense, such as in the previous species, but maintained a large number of individuals even under conditions of very intense control (0.7). The red fox population could reach extinction only with a very high degree of control intensity, the value of control coefficient being near the maximum. At a medium level of control intensity, the population maintained approximately half the individuals of the population maximum (K), regardless of competition.

Furthermore, with a low level of competition (0.3) this population was not affected by the number of individuals when low intensities of non–selective control were applied (< 0.4). Surprisingly, when starting from this low level of competition (0.3), the low control intensities (< 0.4) led to a sharper population increase (the paradoxical effect mentioned above) if there was a higher degree of competition with the other two species. The increase in red fox occurred precisely when there was a decrease in the number of their competing species, whose populations showed lower intrinsic growth rates. The final result of the application of predator control under these conditions could therefore be a red fox population with double the initial number of individuals. When the degree of competition was higher, the intensity of control needed to be lower to reach the maximum level of increase. The absolute population maximum was then also reached when competition was maximal.

Predictions of the theoretical model

Based on the results of the numerical analysis and due to the CRE, we made the following set of three predictions: i) populations of red fox showing maximal abundances (or those of other generalist predators showing a high intrinsic growth rate) will be present in areas subjected to predator control (usually areas devoted to small game hunting); ii) statistically signi-

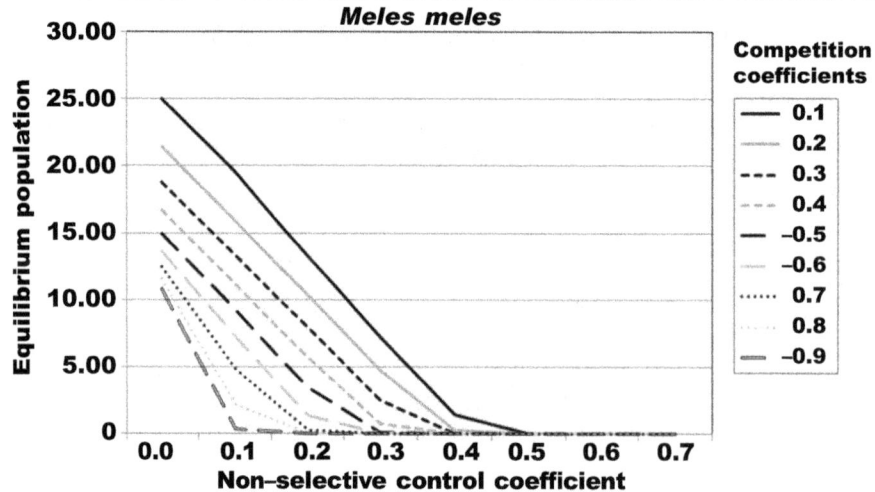

Fig. 1. Functions for European badger (*Meles meles*) populations considering the competition coefficients (α) that relate the value for the population in dynamic equilibrium according to the intensity degree of non–selective control (ω). All populations become extinct at intermediate degrees of control intensity, following a linear pattern.

Fig. 1. Funciones de las poblaciones de tejón (Meles meles) teniendo en cuenta los coeficientes de competencia (α) que relacionan el valor de la población en equilibrio dinámico con el grado de intensidad de control no selectivo (ω). Todas las poblaciones se extinguen con una intensidad de control intermedia siguiendo un patrón lineal.

ficant differences between controlled and uncontrolled areas will not be found in the abundance of red fox (or those of other generalist predators showing a high intrinsic growth rate). But according to prediction 1), if differences appear, the red fox will be more abundant in controlled areas; and iii) the most abundant populations of competing species showing low intrinsic growth rates will be found in areas where predator control programs are not implemented. Thus, statistically significant differences will be found in the abundance of predator populations with low intrinsic growth rate between controlled and uncontrolled areas, with the more abundant populations inhabiting the uncontrolled areas.

Discussion

Many managers of natural areas, gamekeepers and the hunting community in general have the perception that generalist predators (including several species of rodents, corvids, gulls, and carnivores) increase continuously and are so abundant that their populations should be controlled (*e.g.* Herranz, 2000; Garrido, 2008). One of the most persecuted species is the red fox, blamed for reducing populations of game species (Herranz, 2000; Virgós & Travaini, 2005; Rushton et al., 2006; Beja et al., 2009). Although the belief that red fox populations are increasing everywhere and continuously is probably an exageration (see the case of a large Spanish region in Sobrino et al., 2009), it seems true that under certain conditions this species

can increase above normal values (*e.g.* Beja et al., 2009; Trewby et al., 2008). Surprisingly, the paradox is that these real increases in abundance, as in the case of other generalist predators, occur even though their populations are subjected to permanent control campaigns (Herranz, 2000; Virgós & Travaini, 2005; Beja et al., 2009).

Caut et al.´s competitor release effect (CRE) described an ecological mechanism that is applicable to mesocarnivore populations and could theoretically explain this paradox (2007). The CRE framework implies a scenario where the carnivores community is shaped by a number of species presenting negative interspecific interactions (–,–) and experiencing a non–selective predator control program. Under these conditions, our theoretical results predicted that certain changes will occur in the composition and structure of the community. Thus, predator control efforts might eliminate populations with a low intrinsic growth rate (r) and only the population with a high rate of increase might persist (in our case the red fox, whose populations showed a growth rate higher than 0.9), unless the control is extremely intense. Surprisingly, if the control level is moderate, then the populations of these species of generalist predators could increase in a paradoxical way, even surpassing the theoretical value for maximum population in equilibrium in the presence of the remaining species. This is due to both the disappearance of competitors and their higher reproductive capacity. Moreover, the ecological consequences of a mechanism such as the CRE might be similar to those produced by

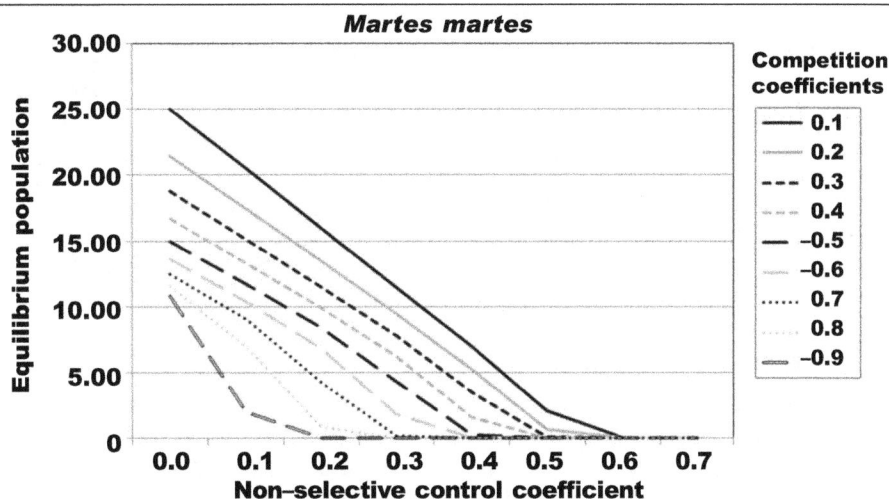

Fig. 2. Functions for pine marten populations (*Martes martes*) follow a similar pattern to those of European badger, although their higher growth rate requires more intense control to completely eliminate the populations.

Fig. 2. Las funciones de las poblaciones de marta (Martes martes) siguen un patrón parecido a las de tejón, aunque su mayor tasa de crecimiento hace necesario intensificar el control para eliminar totalmente las poblaciones.

the mesopredator release effect (see Soulé et al., 1988; Courchamp et al., 1999a), if net predation on certain prey species were to increase as a result of the population increase of control–resistant predators (Caut et al., 2007). Our modelling therefore supports both the CRE hypothesis and the MRE hypothesis.

We found three predictions from our model evaluation that should be specifically tested with empirical data obtained in the field. In general, these predictions are based on the fact that red fox populations (in our case scenario) will increase or maintain their numbers despite the implementation of predator control efforts (Predictions 1 and 2), while other species of carnivores will become less abundant (Prediction 3). It is expected that the most sensitive species disappear over time, so that species richness will decrease in controlled areas (see Estes & Terborgh, 2010). Interestingly, the few data available in the scientific literature seem to support our findings. For example, the results of a study carried out in Portugal showed that in hunting grounds where predator control was practised, the abundance of red fox was almost twice as high as in non–hunting areas (Beja et al., 2009), which appears to bear out Predictions 1 and 2. Moreover, other species of predators tended to be more abundant in uncontrolled areas, supporting Prediction 3 in our study.

Similarly, predator control on badgers in the UK increased the red fox population, again appearing to meet Prediction 1 (Trewby et al., 2008). The results obtained by Virgós & Travaini (2005) in Spain also appear to generally support the CRE predictions. These authors detected the absence of some carnivores

in controlled areas, while the red fox frequency of occurrence in both controlled and uncontrolled areas was similar. However, field data collected following a carefully designed study protocol are needed to reliably test the CRE model predictions found in this study.

In our model, no explicit consideration was given to the effects that spatial heterogeneity, landscape pattern, and structure of a territory could have on the behaviour of population and community dynamics. There is evidence that landscape composition and quality affect interactions among species (Erlinge & Sandell, 1988; Hanski, 1995), the efficacy of predator control programs (Schneider, 2001; Rushton et al., 2006), and therefore the persistence at a regional level of a given pool of species. Given that the model predicts the probabilities of differential extinction of the species in fragmented landscapes and complex environments, the long–term configuration of the communities will also depend on the different probabilities of recolonisation (Hanski, 1994; Schneider, 2001; Rushton et al., 2006). It is possible to speculate about the existence of deterministic processes within a community of carnivores subjected to non–selective control. These processes would occur at the local level, but predictable consequences would result at the landscape scale. These aspects should also be tested independently through further research with empirical data.

The fundamental objective of predator control is an effective increase in the populations of prey species of interest to hunting or conservation (Trout & Tittensor, 1989; Reynolds & Tapper, 1995, 1996; Côte & Sutherland, 1997; Virgós & Travaini, 2005;

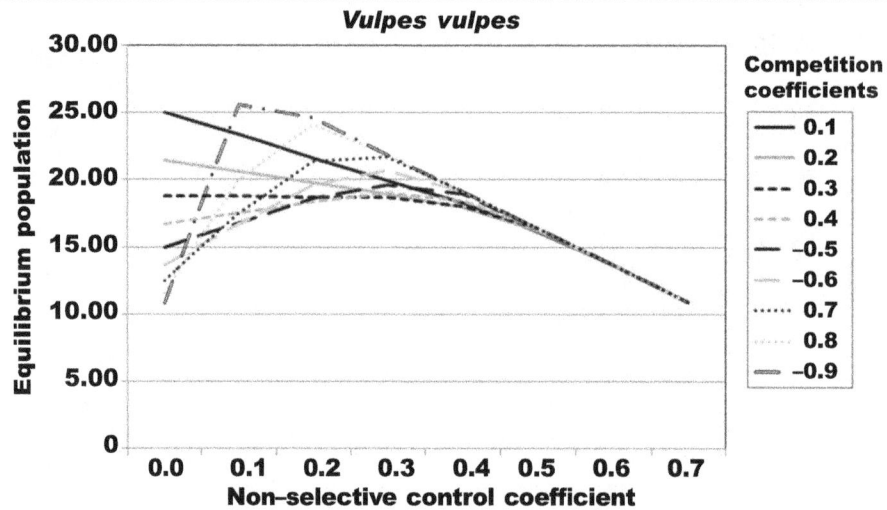

Vulpes vulpes

Fig.3. Functions for red fox populations (*Vulpes vulpes*) are qualitatively different from those of the previous species due to an intrinsic growth rate higher than 0.9. Red fox populations thus show a non–linear response when persecuted depending on the level of competition: populations increase when competition coefficients are greater than 0.3 and control intensity is moderate. Furthermore, red fox populations do not become extinct despite intense predator control.

*Fig. 3. Las funciones de las poblaciones de zorro (*Vulpes vulpes*) son cualitativamente diferentes de las de las especies anteriores debido a una tasa intrínseca de crecimiento mayor que 0,9. Así, las poblaciones de zorro muestran una respuesta no lineal cuando son perseguidas dependiendo del grado de competencia: las poblaciones aumentan cuando los coeficientes de competencia son mayores que 0,3 y la intensidad del control es moderada. Además, las poblaciones de zorro no se extinguen aunque se aplique un control de predadores intenso.*

Reynolds et al., 2010). This is based on studies that found direct effects of control or natural reduction of predators on the abundance and dynamics of prey populations (*e.g.* Marcström et al., 1988, 1989; Small & Keith, 1992; Lindström et al., 1994). However, the success of predator control campaigns is variable and, in general, very expensive (Reynolds & Tapper, 1996; Côte & Sutherland, 1997).

Some studies have thus suggested that these practices are effective (regarding the above indicated objective) when applied at the local level, in conditions of very intense control, but only in the short–term (*e.g.* Reynolds et al., 1993; Harding et al., 2001; Keedwell et al., 2002). Other studies have shown that the predator control was ineffective in meeting management goals (Reynolds & Tapper, 1996; Côte & Sutherland, 1997; Banks, 1999; Herranz, 2000; Kauhala et al., 2000; Keedwell et al., 2002; Martínez–Abraín et al., 2004; Baker & Harris, 2006; King et al., 2009). Moreover, there are many predator species that do not affect game or threatened species. Thus, it has been argued that predator control can not be effective when focused on them (see for the cases of lizards, snakes and large gulls Herranz, 2000; Oro & Martínez–Abraín, 2007). Overall, it has been considered that the unique nature of predator–prey relationships within communities makes it difficult to make generalizations, and that evaluation of the effectiveness of conducting a predator control

program thus requires individual consideration (Sih et al., 1998; Abrams & Ginzburg, 2000; Turchin, 2003; Holt et al., 2008; Valkama et al., 2005).

On the other hand, predator management could have ecological costs that depend on the relative importance of the different uses and intrinsic values of the territory. For example, the ecological consequences of controlling one or various species of predators might be appraised positively or negatively depending on the environmental perception, and on the type of local use of the natural resources (Langley & Yalden, 1977; Banks et al., 1998). Thus, perception might be different if predator control is used to enhance an endangered species rather than a game species (*e.g.* Côte & Sutherland, 1997; Keedwell et al., 2002). Furthermore, predator control also appears to affect different demographic parameters of the target predator species, including density, age structure, and inmigration patterns (see Yoneda & Maekawa, 1982; Rushton et al., 2006).

However, the more notable and more harmful effects of non–selective predator control are related to the conservation of threatened species of predators and the unwanted consequences on ecosystems due to the alteration of natural communities, such as the increase of generalist predators (including target species of the control) and pest species (including rodents), the decline of shared prey species (including those of game interest), and unforeseen effects on vegetation, ecosystem function, and similar owing to chain reactions (*e.g.* Herranz, 2000; Martínez–Abraín et al., 2004; Rodríguez & Delibes, 2004; Virgós & Travaini, 2005; Caut et al., 2007; Cabezas–Díaz et al., 2009; Estes & Terborgh, 2010).

The theoretical results obtained in this study highlight the importance of competitive ecological interactions among predators in the design of an optimum management strategy for their communities (Courchamp et al., 1999a, 1999b; Trewby et al., 2008). Caut et al´s CRE has shown how the complex network of interactions (see also Polis & Holt, 1992; Chase et al., 2002; Caro & Stoner, 2003) among carnivore mammals can also lead to undesired effects, such as a population increase in the target species (the red fox or any predator with high reproductive capacity in our study, or the American mink *Neovison vison* Schreber 1777; see Bright, 1993; King et al., 2009), and the elimination of more sensitive species that might be of conservation interest. The obtained results support the idea that the design of programs to manage predator populations should consider potential consequences to communities and the ecosystem as a whole (Schneider, 2001; Zavaleta et al., 2001; Courchamp & Caut, 2005; Caut et al., 2007), as well as the biological traits of the involved species. To validate our model findings, empirical data should evaluate these responses and not just the individual species' responses of the targeted predator and prey. The development of management strategies for species such as the red fox populations should take the ecological framework into account, and predator control programs should be thoroughly evaluated to determine the potential impact on the community and ecosystem. This is particularly important for predator control programs using non–selective methods (*e.g.* Herranz, 2000; Virgós & Travaini, 2005; Beja et al., 2009), where numbers of red foxes and other generalist predators can increase despite the efforts of managers.

Acknowledgements

We are grateful to Ana Rojas, Raúl Bonal, Sara Cabezas and Cormac de Brun for help in different steps of the study, including the translation to English. We also thank Daniel L. Huertas, Mario Díaz and two anonymous referees who provided useful suggestions and comments to different drafts of this manuscript.

References

Abrams, P. A. & Ginzburg, L. R., 2000. The nature of predation: prey dependent, ratio dependent or neither? *TREE*, 15: 337–341.

Baker, P. J. & Harris, S., 2006. Does culling reduce fox (*Vulpes vulpes*) density in commercial forests in Wales? *European Journal of Wildlife Research*, 52: 99–108.

Banks, P. B., 1999. Predation by introduced foxes on native bush rats in Australia: do foxes take the doomed surplus? *Journal of Applied Ecology*, 36: 1063–1071.

Banks, P. B., Dickman, C. R. & Newsome, A. E., 1998. Ecological costs of feral predator control: foxes and rabbits. *The Journal of Wildlife Management*, 62: 766–772.

Begon, M., Harper, J. L. & Townsend, C. R., 1996. *Ecology: Individuals populations and communities*. Blackwell Science, Oxford.

Beja, P., Gordinho, L., Reino, L., Loureiro, F., Santos–Reis, M. & Borralho, R., 2009. Predator abundance in relation to small game management in southern Portugal: conservation implications. *European Journal of Wildlife Research*, 55: 227–238.

Breitenmoser, U., 1998. Large predators in the Alps: The fall and rise of man's competitors. *Biological Conservation*, 83: 279–289.

Bright, P. W., 1993. Habitat fragmentation–problems and predictions for British mammals. *Mammal Review*, 23: 101–111.

Cabezas–Díaz, S., Lozano, J. & Virgós, E., 2009. The declines of the wild rabbit (*Oryctolagus cuniculus*) and the Iberian lynx (*Lynx pardinus*) in Spain: redirecting conservation efforts. In: *Handbook of nature conservation: global, environmental and economic issues*: 283–310 (J. B. Aronoff, Ed.). Nova Science Publishers, Hauppauge.

Calver, M. C., King, D. R., Bradley, J. S., Gardner, J. L. & Martin, G., 1989. An assessment of the potential target specificity of 1080 predator baiting in western Australia. *Australian Wildlife Research*, 16: 625–638.

Caro, T. M. & Stoner, C., 2003. The potential for interspecific competition among African carnivores. *Biological Conservation*, 110: 67–75.

Caut, S., Casanovas, J. G., Virgós, E., Lozano, J., Witmer, G. W. & Courchamp, F., 2007. Rats dying for mice: Modelling the competitor release effect. *Austral Ecology*, 32: 858–868.

Chase, J. M., Abrams, P. A., Grover, J. P., Diehl, S., Chesson, P., Holt, R. D., Richards, S. A., Nisbet, R. M. & Case, T. J., 2002. The interaction between predation and competition: a review and synthesis. *Ecology Letters*, 5: 302–315.

Côté, I. M. & Sutherland, W. J., 1997. The effectiveness of removing predators to protect bird populations. *Conservation Biology*, 11: 395–405.

Courchamp, F. & Caut, S., 2005. Use of biological invasions and their control to study the dynamics of interacting populations. In: *Conceptual Ecology and Invasions Biology*: 253–279 (M. W. Cadotte, S. M. McMahon & T. Fukami, Eds.). Springer, Dordrecht.

Courchamp, F., Langlais, M. & Sugihara, G., 1999a. Cats protecting birds: modelling the mesopredator release effect. *Journal of Animal Ecology*, 68: 282–292.

– 1999b. Rabbits killing birds: modelling the hyperpredation process. *Journal of Animal Ecology*, 69: 154–165.

Creel, S. & Creel, N. M., 1996. Limitation of African

wild dogs by competition with larger carnivores. *Conservation Biology*, 10: 526–538.

Crooks, K. R. & Soulé, M. E., 2000. Mesopredator release and avifaunal extinctions in a fragmented system. *Nature*, 400: 563–566.

Duarte, J. & Vargas, J. M., 2001. ¿Son selectivos los controles de predadores en los cotos de caza? *Galemys*, 13: 1–9.

Erlinge, S. & Sandell, M., 1988. Co–existence of stoat (*Mustela erminea*) and weasel (*Mustela nivalis*): social dominance, scent communication and reciprocal disrtribution. *Oikos*, 53: 242–246.

Estes, J. A. & Terborgh, J. (Eds.), 2010. *Trophic Cascades: Predators, Prey, and the Changing Dynamics of Nature*. Island Press, Washington.

Fenton, A. & Brockhurst, M. A., 2007. The role of specialist parasites in structuring host communities. *Ecological Research*, 23: 795–804.

Garrido, J. L. (Ed.), 2008. *Especialista en control de predadores*. FEDENCA – Escuela Española de Caza, Madrid.

Gause, G. F., 1934. *The struggle for existence*. Hafner, New York.

– 1935. *Vérifications expérimentales de la théorie mathématique de la lutte pour la vie*. Hermann et Cie Éditeurs, Paris.

Gilpin, M. E., 1975. Limit Cycles in Competition Communities. *The American Naturalist*, 109: 51–60.

Groombridge, B., 1992. *Global Biodiversity: Status of the Earth's Living Resources*. Chapman & Hall, London & New York.

Hanski, I., 1994. Patch–occupancy dynamics in fragmented landscapes. *TREE*, 9: 131–135.

– 1995. Effects of landscape pattern on species interactions. In: *Mosaic landscapes and ecological processes*: 203–224 (L. Hansson, L. Fahrig & G. Merriam, Eds.). Chapman and Hall, London.

Harding, E. K, Doak, D. F. & Albertson, J. D., 2001. Evaluating the effectiveness of predator control: the non–native red fox as a case study. *Conservation Biology*, 15: 1114–1122.

Harris, S. & Saunders, G., 1993. The control of canid populations. *Symposia of the Zoological Society of London*, 65: 441–464.

Hein, E. W. & Andelt, W. F., 1994. Evaluation of coyote attractants and an oral delivery device for chemical agents. *Wildlife Society Bulletin*, 22: 651–655.

Herranz, J., 2000. Efectos de la depredación y del control de predadores sobre la caza menor en Castilla–La Mancha. Ph. D. Tesis, Univ. Autónoma de Madrid.

Holt, A. R., Davies, Z. G., Tyler, C. & Staddon, S., 2008. Meta–Analysis of the Effects of Predation on Animal Prey Abundance: Evidence from UK Vertebrates. *PLoS ONE*, 3(6): e2400.

ISEE, 2010. Stella 9.1.4. System *Thinking for Education and Research. ISEE Systems Inc.* (formerly High Performance Systems). Lebanon, New Hampshire.

Kauhala, K., Helle, P. & Helle, E., 2000. Predator control and the density and reproductive success of grouse populations in Finland. *Ecography*, 23: 161–168.

Keedwell, R. J., Maloney, R. F. & Murray, D. P., 2002. Predator control for protecting kaki (*Himantopus novaezelandiae*)– lessons from 20 years of man-

agement. *Biological Conservation*, 105: 369–374.

King, C. M., McDonald, R. M., Martin, R. D. & Dennis, T., 2009. Why is eradication of invasive mustelids so difficult? *Biological Conservation*, 142: 806–816.

Langley, P. J. W. & Yalden, D. W., 1977. The decline of the rarer carnivores in Great Britain during the nineteenth century. *Mammal Review*, 7: 95–116.

Lankester, K., Van Apeldoorn, H., Meelis, E. & Verboom, J., 1991. Management perspectives for populations of the Eurasian badger (*Meles meles*) in fragmented landscape. *Journal of Applied Ecology*, 28: 561–573.

Levi, T., Kilpatrick, A. M., Mangel, M. & Wilmers, C. C., 2012. Deer, predators, and the emergence of Lyme disease. *PNAS*, 109: 10942–10947.

Lindström, E. R., Andrén, H., Angelstam, P., Cederlund, G., Hornfeldt, B., Jäderberg, L., Lemnell, P. A., Martinsson, B., Sköld, K. & Swenson, J. E., 1994. Disease reveals the predator: sarcoptic mange, red fox predation, and prey populations. *Ecology*, 75: 1042–1049.

Lindström, E. R., Brainerd, S. M., Helldin, J. O. & Overskaug, K., 1995. Pine marten–red fox interactions: a case of intraguild predation? *Annales Zoologici Fennici*, 32: 123–130.

Lozano, J., Virgós, E., Cabezas–Díaz, S. & Mangas, J. G., 2007. Increase of large game species in Mediterranean areas: Is the European wildcat (*Felis silvestris*) facing a new threat? *Biological Conservation*, 138: 321–329.

Lozano, J., Virgós, E. & Mangas, J. G., 2010. Veneno y control de predadores. *Galemys*, 22: 123–132.

Marcström, V., Keith, L. B., Engrén, E. & Cary, J. R., 1989. Demographic responses of arctic hares (*Lepus timidus*) to experimental reductions of red foxes (*Vulpes vulpes*) and martens (*Martes martes*). *Canadian Journal of Zoology*, 67: 658–668.

Marcström, V., Kenward, R. E. & Engrén, E., 1988. The impact of predation on boreal tetraonids during vole cycles: an experimental study. *Journal of Animal Ecology*, 57: 859–872.

Martínez–Abraín, A., Sarzo, B., Villuendas, E., Bartolomé, M. A., Mínguez, E. & Oro, D., 2004. Unforeseen effects of ecosystem restoration on yellow–legged gulls in a small western Mediterranean island. *Environmental Conservation*, 31: 219–224.

Mitchell–Jones, A. J., Amori, G., Bogdanowicz, W., Krystufek, B., Reijnders, P. J. H., Spitzenberger, F., Stubbe, M., Thissen, J. B. M., Vohralík, V. & Zima, J., 1999. *The Atlas of European Mammals*. Academic Press, London.

Müller, C. B. & Brodeur, J., 2002. Intraguild predation in biological control and conservation biology. *Biological Control*, 25: 216–223.

Oro, D. & Martínez–Abraín, A., 2007. Deconstructing myths on large gulls and their impact on threatened sympatric waterbirds. *Animal Conservation*, 10: 117–126.

Palomares, F. & Caro, T. M., 1999. Interspecific killing among mammalian carnivores. *The American Naturalist*, 153: 492–508.

Palomares, F., Gaona, P., Ferreras, P. & Delibes, M., 1995. Positive effects on game species of top predators by controlling smaller predator populations: an example with Lynx, Mongooses and Rabbits. *Conservation Biology*, 9: 295–305.

Polis, G. A. & Holt, R. D., 1992. Intraguild predation: The dynamics of complex trophic interactions. *TREE*, 7: 151–154.

Powell, R. A. & Zielinski, W. J., 1983. Competition and coexistence in mustelid communities. *Acta Zoologica Fennica*, 174: 223–227.

Reynolds, J. C. & Tapper, S. C., 1995. Predation by foxes *Vulpes vulpes* on brown hares *Lepus europaeus* in central southern England, and its potential on population growth. *Wildlife Biology*, 1: 145–158.

– 1996. Control of mammalian predators in game management and conservation. *Mammal Review*, 26: 127–156.

Reynolds, J. C., Goddard, H. N. & Brockless, M. H., 1993. The impact of local fox (*Vulpes vulpes*) removal on fox populations at two sites in Southern England. *Gibier Faune Sauvage*, 10: 319–334.

Reynolds, J. C., Stoate, C., Brockless, M. H., Aebischer, N. J. & Tapper, S. C., 2010. The consequences of predator control for brown hares (*Lepus europaeus*) on UK farmland. *European Journal of Wildlife Research*, 56: 541–549.

Rodríguez, A. & Delibes, M., 2002. Internal structure and patterns of contraction in the geographic range of the Iberian lynx. *Ecography*, 25: 314–328.

– 2004. Patterns and causes of non–natural mortality in the Iberian lynx during a 40–year period of range contraction. *Biological Conservation*, 118: 151–161.

Ruggiero, L. F., Aubry, K. B., Buskirk, S. W., Lyon, L. J. & Zielinski, W. J., 1994. *The scientific basis for conserving forest carnivores: american marten, fisher, lynx and wolverine in the western United States. General Technical Report RM–254*. Forest Service, Rocky Mountain Forest and Range Experiment Station, Fort Collins.

Rushton, S. P., Shirley, M. D. F., Macdonald, D. W. & Reynolds, J. C., 2006. Effects of Culling Fox Populations at the Landscape Scale: A Spatially Explicit Population Modeling Approach. *The Journal of Wildlife Management*, 70: 1102–1110.

Schaller, G. B., 1996. Introduction: carnivores and conservation biology. In: *Carnivore behavior, ecology and evolution*, 2: 1–10 (J. L. Gittleman, Ed.). Cornell Univ. Press, London.

Schneider, M. F., 2001. Habitat loss, fragmentation and predator impact: Spatial implications for prey conservation. *Journal of Applied Ecology*, 38: 720–735.

Shorrocks, B. & Begon, M., 1975. A Model of Competition. *Oecologia*, 20: 363–367.

Sih, A., Englund, G. & Wooster, D., 1998. Emergent impacts of multiple predators on prey. *TREE*, 13: 350–355.

Small, R. J. & Keith, L. B., 1992. An experimental study of red fox predation on arctic and snowshoe hares. *Canadian Journal of Zoology*, 70: 1614–1621.

Sobrino, R., Acevedo, P., Escudero, M. A., Marco, J. & Gortázar, C., 2009. Carnivore population trends in Spanish agrosystems after the reduction in food availability due to rabbit decline by rabbit haemorrhagic disease and improved waste management. *European Journal of Wildlife Research*, 55: 161–165.

Soulé, M. E., Bolger, D. T., Alberts, A. C., Wright, J., Sorice, M. & Hill, S., 1988. Reconstructed dynamics of rapid extinctions of chaparral–requiring birds in urban habitat islands. *Conservation Biology*, 2: 75–92.

Tapper, S. C., Brockless, M. H. & Potts, G. R., 1991. The effect of predator control on populations of grey partridge (*Perdix perdix*). In: *Proceedings of the XXth Congress of the International Union of Game Biologists*: 398–403 (S. Csanyi & J. Ernhaft, Eds.). Godollo, Hungary.

Trewby, L. D., Wilson, G. J., Delahay, R. J., Walker, N., Young, R., Davison, J., Cheeseman, C., Robertson, P. A., Gorman, M. L. & McDonald, R. A., 2008. Experimental evidence of competitive release in sympatric carnivores. *Biology Letters*, 4: 170–172.

Trout, R. C. & Tittensor, A. M., 1989. Can predators regulate wild Rabbit *Oryctolagus cuniculus* population density in England and Wales? *Mammal Review*, 19: 153–173.

Turchin, P., 2003. *Complex population dynamics*. Princeton University Press, Princeton.

Valkama, J., Korpimäki, E., Arroyo, B., Beja, P., Bretagnolle, V., Bro, E., Kenward, R., Mañosa, S., Redpath, S. M., Thirgood, S. & Viñuela, J., 2005. Birds of prey as limiting factors of gamebird populations in Europe: a review. *Biological Reviews*, 80: 171–203.

Virgós, E. & Travaini, A., 2005. Relationship between Small–game Hunting and Carnivore Diversity in Central Spain. *Biodiversity and Conservation*, 14: 3475–3486.

Waechter, A., 1975. Écologie de la Fouine en Alsace. *Revue d'Ecologie (Terre et Vie)*, 29: 399–457.

Whitfield, D. P., McLeod, D. R. A., Watson, J., Fielding, A. H. & Haworth, P. F., 2003. The association of grouse moor in Scotland with the illegal use of poisons to control predators. *Biological Conservation*, 114: 157–163.

Wilson, D. E. & Mittermeier, R. A. (Eds.), 2009. *Handbook of the Mammals of the World. Vol. 1. Carnivores*. Lynx Edicions, Barcelona.

Windberg, L. A. & Knowlton, F. F., 1990. Relative vulnerability of coyotes to some capture procedures. *Wildlife Society Bulletin*, 18: 282–290.

Yoneda, M. & Maekawa, K., 1982. Effects of hunting on age structure and survival rates of red fox in eastern Hokkaido. *The Journal of Wildlife Management*, 46: 781–786.

Zavaleta, E. S., Hobbs, R. J. & Mooney, H. A., 2001. Viewing invasive species removal in a whole–ecosystem context. *TREE*, 16: 454–459.

3

Assessing the extent of occurrence, area of occupancy, territory size, and population size of marsh tapaculo (*Scytalopus iraiensis*)

L. Klemann Jr. & J. S. Vieira

Klemann Jr., L. & Vieira, J. S., 2013. Assessing the extent of occurrence, area of occupancy, territory size, and population size of marsh tapaculo (*Scytalopus iraiensis*). *Animal Biodiversity and Conservation*, 36.1: 47–57.

Abstract

Assessing the extent of occurrence, area of occupancy, territory size, and population size of marsh tapaculo (Scytalopus iraiensis).— First described in 1998, the marsh tapaculo (*Scytalopus iraiensis*) is an endangered bird of the family Rhinocryptidae. It is endemic to Brazil and is restricted to the wet flood plains of rivers and streams. Due to its cryptic habits and environments of occurrence, information available on its biology, natural history and distribution is scarce. We compiled occurrence records (99 records), delimited the extent of occurrences (296,584 km^2), calculated the area of occupancy (84 km^2), estimated territory size (5,313 ± 1,201 m^2 per pair), population density (3.76 ± 0.85 individuals per hectare), and population size (31,584 ± 7,140 mature individuals) of marsh tapaculo. The species was recorded in marshes associated to four types of vegetation and in four ecological zones. This new information is extremely important to support revaluation of the species' threat category and to enhance knowledge about this endemic and little known bird from Brazil.

Key words: *Scytalopus iraiensis*, Distribution, Population, Endangered species, Southeastern Brazil.

Resumen

Evaluación de la extensión de presencia, la superficie de ocupación, el tamaño del territorio y el tamaño de la población del churrín palustre (Scytalopus iraiensis).— Descrito por primera vez en 1998, el churrín palustre (*Scytalopus iraiensis*) es una ave en peligro de extinción de la familia Rhinocryptidae. Es endémica de Brasil y su presencia queda restringida a los planos aluviales de los ríos y los cursos de agua. Debido a sus hábitos crípticos y a los ambientes en los que se halla presente, la información disponible sobre su biología, su historia natural y su distribución es escasa. Compilamos varios registros de presencia (99 registros), delimitamos la extensión de las presencias (296.584 km^2), calculamos la superficie de ocupación (84 km^2) y estimamos el tamaño del territorio (5.313 ± 1.201 m^2 por pareja), la densidad de la población (3,76 ± 0,85 individuos por hectárea) y el tamaño de la población (31.584 ± 7.140 individuos maduros) del churrín palustre. La especie se registró en zonas de marismas asociada a cuatro tipos de vegetación y en cuatro zonas ecológicas. Esta nueva información es fundamental para respaldar la reevaluación de la categoría de situación de peligro de la especie y potenciar el conocimiento de esta ave endémica y poco conocida de Brasil.

Palabras clave: *Scytalopus iraiensis*, Distribución, Población, Especie en peligro de extinción, Sureste de Brasil.

Louri Klemann Júnior, Ecology and Conservation Post–Graduation Program, Univ. Federal do Paraná, P. O. Box 19020, Curitiba, Paraná, Brazil.– Juliana S. Vieira, Entomology Post–Graduation Program, Univ. Federal do Paraná, P. O. Box 19020, Curitiba, Paraná, Brazil.

Corresponding author: L. Klemann Jr. E–mail: klemannjr@yahoo.com.br

Introduction

The marsh tapaculo (*Scytalopus iraiensis*) is a species of the bird family Rhinocryptidae. It was first described in 1998 from specimens taken from two municipalities in the metropolitan region of Curitiba (capital of Paraná state, southern Brazil) and its occurrence is known to be restricted to the wet flood plains of rivers and streams (Bornschein et al., 1998). Its habitat is described as upland marshes and also, at Rio Grande do Sul state (south of Brazil), coastal wetlands associated to grasslands and Atlantic Forest biomes (Machado et al., 2005). According to Bornschein et al. (1998, 2001) this bird occurs in dense humid floodplain watercourses, usually surrounded by alluvial forests, ranging from less than 1 ha to approximately 350 ha, with a vegetation height ranging from 60 to 180 cm.

After the species' description, new records were obtained in Paraná, Santa Catarina and Rio Grande do Sul states (south of Brazil) and in Minas Gerais state (southeast of Brazil) increasing the number of locations, municipalities and states with records of marsh tapaculo (Bornschein et al., 1998, 2001; Accordi et al., 2003; Maurício, 2005; Straube et al., 2005; Bencke et al., 2006; Raposo et al., 2006; Corrêa et al., 2007, 2008; Vasconcelos et al., 2008; Fontana et al., 2008; Rodrigues et al., 2008). Despite the new records, knowledge of this species' real distribution is still very basic. New records are expected even in well–studied regions, an expectation reinforced by the species' description fifteen years ago (Bornschein et al., 1998), based on individuals from a metropolitan region with 1.75 million inhabitants (IBGE, 2007), and from records obtained in well–studied locations of Minas Gerais state (*e.g.* Serra da Canastra and Serra do Cipó) as of 2003 (Vasconcelos et al., 2008).

The cryptic habits, common to all representatives of this genus, and the flooded environment where the species occurs make research of this tapaculo difficult, accounting thereby for the little information produced and available. To date, only two papers have been published on the species' biology and behavior (Hassdenteufel et al., 2006a, 2006b), and three on the species' distribution (Bornschein et al., 2001; Corrêa & Woldan, 2007; Vasconcelos et al., 2008).

Apart from the little information available at the time of the description, a suggestion was made to include marsh tapaculo in the Brazilian list of endangered species (Bornschein et al., 1998, 2001). Nowadays, this tapaculo is considered 'Endangered' at national and international levels (Machado et al., 2005; BirdLife International, 2012).

This lack of information on biological data and distribution details of the marsh tapaculo can lead to erroneous classification of the threat category. It is therefore extremely important to compile records, delimit the extent of occurrence, measure the area of occupancy, and estimate the territory size, population density and population size in order to reevaluate the threat category and to expand knowledge of this endemic and little known bird from Brazil.

Material and methods

Study area

The study area included the extent of occurrence of marsh tapaculo, from Rio Grande do Sul (RG), Santa Catarina (SC) and Paraná state (PR) in Southern Brazil and São Paulo (SP) and Minas Gerais (MG) state in Southeastern Brazil, between coordinates 19° 07'–32° 19' S and 43° 20'–52° 28' W.

Records

Records of the species were obtained from literature, photos and recordings available on the internet (http://www.xeno–canto.org/ and http://www.wikiaves.com.br/), unpublished records from third persons (personal communication) and bird surveys carried out by the first author between 2002 and 2012. For records taken from the literature, internet information and personal communications, the coordinates used were those provided by the author of each record. When this information was not available and the location could not be specified, the municipality centroid was used.

Due to the species' cryptic habits (Bornschein et al., 2001), the records gathered by the first author were obtained by listening to audio communications stimulated by means of song playbacks. The coordinates for the points where the species was recorded were obtained through a global positioning system (GPS) device.

Extent of occurrence, area of occupancy, territory size and population size

The records were represented in a geographical information system (GIS). The extent of occurrence, defined as the smallest area which can be drawn to encompass all the known, inferred or projected sites of present occurrence of a taxon (IUCN, 2001), was delimited by using the minimum convex polygon method (Mohr, 1947; IUCN, 2001); this was defined as the smallest polygon in which none of the internal angles exceed 180 degrees, and containing all the species points of occurrence (IUCN, 2001). This method, usually used to calculate the home range size (Hayne, 1949; White & Garrott, 1990; Harris et al., 1990), was adapted to the delineation of the extent of occurrence (IUCN, 2001).

By overlaying the records and thematic maps, we identified the vegetal formation (IBGE, 1993) and the global ecological zones (ecozones; FRA, 2000) where the species occurs. Considering the minimum convex polygon method, which incorporate large areas that are not used or occupied by the species (Ostro et al., 1999; Powell, 2000; Burgman & Fox, 2003), and aiming to create a realistic extent of occurrence, this polygon was adjusted using the adjusted polygon method (Mills & Gorman, 1987; Li & Rogers, 2005; Grueter et al., 2009) by excluding the vegetal formation without records of the species. We also excluded small separated areas, created by the polygon adjustment, and great water bodies.

The area of occupancy, defined as the area inside the extent of occurrence occupied by the species (IUCN, 2001), was obtained using the grid cell method. To accomplish this goal, the recorded points were overlaid on a grid, the occurrence cells were identified, and their area was added, resulting in a value in square kilometers. This method, which is usually used to calculate home range size (Adams & Davis, 1967; White & Garrott, 1990), was adapted to calculate the area of occupancy according to IUCN (2001).

The size of the cells is a factor that can influence results, overestimating or, more commonly, underestimating the calculated area (Kool & Croft, 1992; IUCN, 2001; Lehmann & Boesch, 2003; Grueter et al., 2009). Although the choice of cell size was usually a decision with no biological basis or known objective procedures (White & Garrott, 1990), the size of cells used in this research (1 km^2) was defined considering the disjunctive distribution of the habitat where the species occurs. We aimed at not producing overestimated values for the area of occupancy, as is preferable from a conservation point of view (IUCN, 2001).

The size of the territory of the marsh tapaculo was estimated by counting individuals in eight different marshes where the species was recorded. The marshes sampled are in two of the four states where the species occur, Paraná and Santa Catarina. These two states contain 74.75% of the compiled records, justifying the implementation of sample sites in these regions. The marshes studied are distributed in the two main ecological zones, subtropical humid forest and tropical mountain system (94.95% of the records have been collected in these zones), with two main vegetation types, araucaria moist forest and grassland (85.86% of the records come from these vegetation types).

Each marsh was sampled twice, in March 2011 and April 2012, using song playback to stimulate vocalization. Considering that the species respond well to the playback and could be heard easily (Bornschein et al., 1998), we followed the point count method (Ferry & Frochot, 1970; Hutto et al., 1986). This is one of the most widely used counting methods in bird population studies (Rosenstock et al., 2002), with an unlimited radius (see Simons et al., 2007). We used an adapted version of the double–observer approach (Nichols et al., 2000; Thompson, 2002), with two observers in each marsh, positioned to visually cover the whole sample area. Due to the species' cryptic habits (Bornschein et al., 1998), we were unable to use other methodologies (such as banding, or observation) to estimate the number of individuals.

For each location sampled we counted the number of individuals vocalizing at the same time, heard in a period of 15 min after playback (done for 10 min). The marshes sampled were vectorized utilizing satellite images, taken in 2010 with 0.5 meter resolution, and the areas were then calculated. Considering the high resolution of the images used and the great facility in marsh identification and delimitation it was possible to vectorize these vegetal formations with great precision, resulting in highly accurate calculations of their areas.

To measure the territory size, the marsh area was divided by the maximum number of individuals vocalizing at the same time after playback, and the average value of this ratio was used. Considering territory as a defended area (Howard, 1920) that provides food, nesting sites and mates (Perrins & Birkhead, 1983), the value obtained would be an estimate of the mean size of the territory occupied by a pair of marsh tapaculo. Population density was then estimated as twice the inverse of territory size, and population size was estimated by multiplying the area of occupancy by the population density value.

Results

Records and occurrence

A total of 99 occurrence records was compiled; 47 from the literature, 22 based on images and recordings from the Internet, six personal communications, and 24 from bird surveys carried out by the first author. The occurrence locations (70) were distributed over 42 municipalities in Paraná, Santa Catarina, Rio Grande do Sul and Minas Gerais states (table 1).

According to the global ecological zones (FRA, 2000), the records were located in four regions: subtropical humid forest (58 records), tropical mountain system (36), tropical moist deciduous forest (3) and tropical rainforest (2).

According to the Brazilian vegetation map (IBGE, 1993) and Brazilian vegetation classification (Veloso et al., 1991), the records were located in marshes within the domains of four types of vegetation: araucaria moist forest (60 records) in its alluvial, montane and cloudy formations (e.g. Rio Grande do Sul, Santa Catarina and Paraná); semi–deciduous seasonal forest (7) and its formations submontane and montane (e.g. Minas Gerais); grassland (25) (e.g. Rio Grande do Sul, Santa Catarina, Paraná and Minas Gerais); pioneer formation areas (5) (e.g. Rio Grande do Sul coastal line); and ecological tension areas (2) between grassland and semi–deciduous seasonal forest (fig. 1).

Extent of occurrence, area of occupancy, territory size and population size

The minimum convex polygon method gave an extent of occurrence of 424,064 km^2. After excluding the vegetal formation without any record of the species, the extent of occurrence obtained was 296,584 km^2 (fig. 1).

We used a 1 km^2 cell grid to calculate the species' area of occupancy (a total of 299,536 cells to cover all of the extent of occurrence), resulting in 84 km^2. This area is distributed over Rio Grande do Sul (6 km^2), Santa Catarina (23 km^2), Parana (44 km^2) and Minas Gerais (11 km^2) states.

The territory size found for the species was 5,313 ± 1,201 m^2 (ranging from 3,589 to 6,990 m^2), resulting in a population density of 3.76 ± 0.85 individuals per hectare. One to five individuals were

Table 1. Records of marsh tapaculo between 1997 and 2013, showing the municipalities, states (MG. Minas Gerais; PR. Parana; RS. Rio Grande do Sul; SC. Santa Catarina), coordinates (*Coordinate of the centroid of the municipality) and source of the records (LKJ. Records collected by the first author; personal communications: PSN. Pedro Scherer Neto; AER. Adrian Eisen Rupp).

*Tabla 1. Registros del churrín palustre entre 1997 y 2013 en los que se muestran los municipios, los estados (MG. Minas Gerais; PR. Parana; RS. Rio Grande do Sul; SC. Santa Catarina), las coordenadas (*Coordenadas del centroide del municipio) y los autores de los registros (LKJ. Registros recopilados por el primer autor; comunicaciones personales: PSN. Pedro Scherer Neto; AER. Adrian Eisen Rupp).*

Municipality	State	Coordinate	Source
Bambuí	MG	20° 14' S, 45° 58' W	Vasconcelos et al., 2008
Catas Altas	MG	20° 07' S, 43° 27' W	Vasconcelos et al., 2008
Itabira	MG	19° 36' 04" S, 43° 18' 03" W*	Silva, 2012
Mariana	MG	20° 19' 46" S, 43° 19' 55" W*	Silva, 2011
Moeda	MG	20° 19' 50" S, 43° 59' 35" W*	Franco, 2013
Morro do Pilar	MG	19° 11' 05" S, 43º 23' 38" W	Faria, 2011
Morro do Pilar	MG	19° 15' S, 43° 31' W	Araújo, 2011
Morro do Pilar	MG	19° 15' S, 43° 31' W	Araújo, 2011
Morro do Pilar	MG	19° 15' S, 43° 31' W	Araújo, 2011
Morro do Pilar	MG	19° 15' S, 43° 31' W	Rodrigues et al., 2008
Ouro Preto	MG	20° 23' 29" S, 43° 36' 39" W*	Silva, 2012
Santa Bárbara	MG	20° 08' S, 43° 31' W	Vasconcelos et al., 2008
Santa Bárbara	MG	20° 00' S, 43° 28' W	Vasconcelos et al., 2008
Santana do Riacho	MG	19° 15' S, 43° 31' W	Vasconcelos et al., 2008
São João Batista do Glória	MG	20° 28' S, 46° 26' W	Vasconcelos et al., 2008
Bituruna	PR	26° 18' 03" S, 51° 30' 12" W	LKJ
Bituruna	PR	26° 18' 09" S, 51° 30' 09" W	LKJ
Bituruna	PR	26° 18' 20" S, 51° 29' 31" W	LKJ
Castro	PR	24° 47' 58" S, 49° 50' 30" W*	IAP, 2009
Cruz Machado	PR	25° 48' S, 51° 05' W	Straube et al., 2005
Cruz Machado	PR	25° 47' S, 51° 07' W	Straube et al., 2005
Cruz Machado	PR	25° 45' S, 51° 05' W	Straube et al., 2005
Curitiba	PR	25° 37' 16" S, 49° 19' 47" W	LKJ
Curitiba	PR	25° 28' 50" S, 49° 20' 02" W	PSN verbally 2011
Curitiba	PR	25° 28' 59" S, 49° 11' 15" W	Straube et al., 2009
Curitiba	PR	25° 35' 58" S, 49° 15' 59" W	Straube et al., 2009
General Carneiro	PR	26° 18' 54" S, 51° 29' 26" W	LKJ
General Carneiro	PR	26° 35' S, 51° 19' W	Straube et al., 2005
Guarapuava	PR	25° 14' 13" S, 51° 17' 04" W	IAP, 2009
Guarapuava	PR	25° 26' 00" S, 51° 13' 00" W	LKJ
Guarapuava	PR	25° 25' 34" S, 51° 13' 40" W	LKJ
Lapa	PR	25° 48' S, 50° 14' W	Bornschein et al., 2001
Palmas	PR	26° 33' 25" S, 51° 34' 09" W	IAP, 2009
Palmeira	PR	25° 26' S, 50° 12' W	Bornschein et al., 2001
Pinhais	PR	25° 26' S, 49° 07' W	Bornschein et al., 2001
Piraí do Sul	PR	24° 27' S, 49° 50' W	Bornschein et al., 2001
Piraquara	PR	25° 27' S, 49° 07' W	Bornschein et al., 2001
Ponta Grossa	PR	25° 14' 51" S, 50° 00' 35" W	PSN verbally 2011

Table 1. (Cont.)

Municipality	State	Coordinate	Source
Quatro Barras	PR	25° 23' S, 49° 05' W	Bornschein et al., 2001
Quatro Barras	PR	25° 23' 38" S, 49° 06' 06" W	PSN verbally 2011
Quatro Barras	PR	25° 23' 38" S, 49° 06' 06" W	PSN verbally 2011
Quatro Barras	PR	25° 23' 38" S, 49° 06' 06" W	PSN verbally 2011
São João do Triunfo	PR	25° 47' S, 50° 13' W	Bornschein et al., 2001
São João do Triunfo	PR	25° 50' S, 50° 14' W	Bornschein et al., 2001
São José dos Pinhais	PR	25° 39' 52" S, 49° 05' 42" W*	Whittaker, 2011
São José dos Pinhais	PR	25° 34' 15" S, 49° 03' 26" W	Athanas, 2008
São José dos Pinhais	PR	25° 34' 12" S, 49° 03' 25" W	Luijendijk, 2010
São José dos Pinhais	PR	25° 34' 59" S, 49° 03' 37" W	By, 1998
São José dos Pinhais	PR	25° 36' S, 49° 09' W	Minns, 2002
São José dos Pinhais	PR	25° 36' S, 49° 09' W	Bornschein et al., 2001
São José dos Pinhais	PR	25° 36' S, 49° 06' W	Bornschein et al., 2001
São José dos Pinhais	PR	25° 37' S, 49° 06' W	Bornschein et al., 2001
São José dos Pinhais	PR	25° 30' S, 49° 09' W	Bornschein et al., 2001
São José dos Pinhais	PR	25° 33' S, 49° 00' W	Bornschein et al., 2001
São José dos Pinhais	PR	25° 34' S, 49° 03' W	Bornschein et al., 2001
São José dos Pinhais	PR	25° 38' S, 49° 09' W	Bornschein et al., 2001
São José dos Pinhais	PR	25° 36' S, 49° 10' W	Bornschein et al., 2001
Teixeira Soares	PR	25° 21' S, 50° 25' W	Bornschein et al., 2001
Tijucas do Sul	PR	25° 48' S, 49° 09' W	Bornschein et al., 2001
Tijucas do Sul	PR	25° 47' S, 49° 09' W	Bornschein et al., 2001
Tijucas do Sul	PR	25° 48' S, 49° 08' W	Bornschein et al., 2001
Tijucas do Sul	PR	25° 51' S, 49° 06' W	Bornschein et al., 2001
Tijucas do Sul	PR	25° 50' S, 49° 12' W	Bornschein et al., 2001
Tijucas do Sul	PR	25° 47' S, 49° 08' W	Bornschein et al., 2001
Tijucas do Sul	PR	25° 52' S, 49° 13' W	Bornschein et al., 2001
Bom Jesus	RS	28° 28' 55" S, 50° 42' 48" W	Fontana et al., 2008
Cambará do Sul	RS	29° 09' 34" S, 50° 07' 35" W	Patrial, 2006
Cambará do Sul	RS	29° 09' 34" S, 50° 07' 35" W	Bencke et al., 2006
Rio Grande	RS	32° 18' 16" S, 52° 25' 19" W	Jacobs, 2008a
Rio Grande	RS	32° 18' 16" S, 52° 25' 19" W	Jacobs, 2008b
Rio Grande	RS	32° 18' 16" S, 52° 25' 19" W	Mauricio, 2005
Rio Grande	RS	32° 18' 57" S, 52° 25' 19" W	Andretti, 2010a
Rio Grande	RS	32° 18' 57" S, 52° 25' 19" W	Andretti, 2010b
Viamão	RS	30° 10' 02" S, 50° 52' 12" W*	Fenalti, 2011
Viamão	RS	30° 06' S, 50° 52' W	Accordi et al., 2003
Água Doce	SC	26° 45' 56" S, 51° 36' 46" W*	Corrêa et al., 2008
Campo Alegre	SC	26° 05' 34" S, 49° 13' 38" W	LKJ
Campo Belo do Sul	SC	27° 50' 51" S, 50° 46' 20" W*	Espínola, 2011
Campo Belo do Sul	SC	27° 59' 45" S, 50° 52' 15" W	Rupp, 2010
Campo Belo do Sul	SC	27° 58' 47" S, 50° 48' 29" W	Rupp, 2010
Jaraguá do Sul	SC	26° 15' 37" S, 49° 13' 47" W	LKJ
Lages	SC	28° 01' 09" S, 50° 20' 21" W*	Corrêa et al., 2008

Table 1. (Cont.)

Municipality	State	Coordinate	Source
Ponte Alta do Norte	SC	27° 07' 39" S, 50° 22' 54" W	LKJ
Ponte Alta do Norte	SC	27° 07' 06" S, 50° 24' 12" W	LKJ
Ponte Alta do Norte	SC	27° 07' 41" S, 50° 22' 53" W	LKJ
Ponte Alta do Norte	SC	27° 07' 15" S, 50° 25' 23" W	LKJ
Ponte Alta do Norte	SC	27° 07' 33" S, 50° 23' 12" W	LKJ
Ponte Alta do Norte	SC	27° 09' 19" S, 50° 22' 28" W	LKJ
Ponte Alta do Norte	SC	27° 07' 30" S, 50° 25' 11" W	LKJ
Santa Cecília	SC	27° 05' 30" S, 50° 23' 42" W	LKJ
Santa Cecília	SC	26° 45' 08" S, 50° 21' 53" W	LKJ
Santa Cecília	SC	26° 46' 00" S, 50° 22' 28" W	LKJ
São Cristóvão do Sul	SC	27° 17' 08" S, 50° 18' 00" W	LKJ
São Cristóvão do Sul	SC	27° 19' 21" S, 50° 16' 34" W	LKJ
São Cristóvão do Sul	SC	27° 18' 38" S, 50° 17' 45" W	LKJ
São Cristóvão do Sul	SC	27° 16' 41" S, 50° 19' 31" W	LKJ
São Cristóvão do Sul	SC	27° 19' 42" S, 50° 16' 17" W	LKJ
Três Barras	SC	26° 12' S, 50° 12' W	Corrêa et al., 2007
Urupema	SC	27° 58' 28" S, 49° 58' 04" W	AER verbally 2011

found vocalizing in the marshes sampled, with an area from 3,589 to 30,422 m², and this number remained constant in the two samples made in each marsh (table 2), reinforcing the efficiency of the method used. The estimated population size, based on the population density and in the area of occupancy, was 31,584 ± 7,140 mature individuals.

Discussion

Occurrence and environment

The expected occurrence in the state of São Paulo, considered by Vasconcelos et al. (2008), was corroborated by the existence of the vegetal formations and ecological zones with known occurrences for the species. The inclusion of São Paulo state in the polygon of the extent of occurrence delimited herein is thus justified. The expected distribution for this state extends from the border with the state of Paraná to the Minas Gerais border, covering an area of approximately 150 km from east to west starting at the Serra do Mar (fig. 1).

The presence of records in the tropical moist deciduous forest (3) and tropical rainforest (2), a few kilometers (11 km maximum) off the boundaries with the mountain tropical system, might be associated with the scale used for mapping the global ecological zones. It is thus convenient to consider only the subtropical humid forest and the mountain tropical system as ecological zones for the species' occurrence.

Occurrence of the marsh tapaculo in the vegetation types where the species were found (araucaria moist forest, semi–deciduous seasonal forest, grassland, pioneer formation zone and ecological tension areas between grassland and semi–deciduous seasonal forest) is directly associated with the presence of pioneer formation areas with fluvial influence, plant communities that occur throughout Brazil in floodplains and flooded depressions (Veloso et al., 1991). The vegetation structure in these communities is quite varied, although the species seems to be restricted to environments dominated by Cyperaceae and/or Poaceae (Bornschein et al., 1998, 2001; Vascocelos et al., 2008).

These vegetal formations suffered greatly from the impact of the Brazilian Federal Government incentive program called Pro–Várzea (established in the 1970s to take advantage of the wetlands for agricultural production, financing the drainage of the wetlands), but are now protected by the Brazilian Forest Code and considered permanent preservation areas. However, the lack of specific data on the alteration of wetlands in Brazil makes it difficult to evaluate and measure the changes that have occurred herein.

On the other hand, we can use the deforestation rates of the Atlantic Forest in the states where the

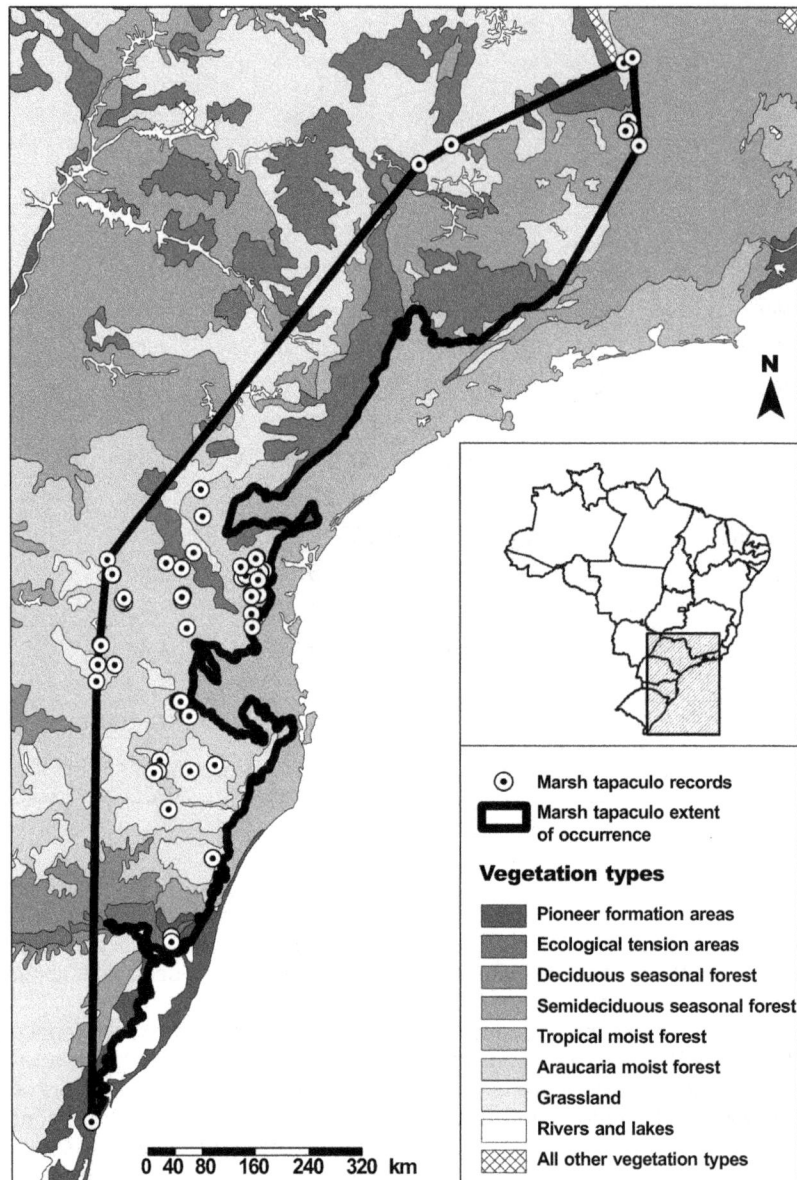

Fig. 1. Map of vegetation types with the location of marsh tapaculo records between 1997 and 2012 and the extent of occurrence delimited through the adjusted polygon method.

Fig. 1. Mapa de los tipos de vegetación con la ubicación de los registros del churrín palustre entre 1997 y 2012 y la extensión de la presencia delimitada mediante el método del polígono ajustado.

marsh tapaculo occur to understand and measure the process of alteration in other vegetal formations associated with these forests. The deforestation rate was 1.23 from 2005–2008 and 0.45% from 2008–2010 in the state of Minas Gerais, 0.51 and 0.17% in Paraná, 0.31 and 0.18% in Rio Grande do Sul, 1.19 and 0.17% in Santa Catarina, and 0.11 and 0.02% in São Paulo (Fundação SOS Mata Atlântica & Instituto Nacional de Pesquisas Espaciais, 2009, 2011).

Considering these rates of deforestation, the reduction in the extent of suitable habitat and consequently the reduction in the species' population cannot be considered significant (≥ 30%). The species' sensitivity to environmental changes, specially caused by fire and by natural or anthropogenic changes in the vegetal structure, is still not fully understood, and therefore does not allow a more significant discussion about the impact of these changes on populations of the species.

Table 2. State of the marshes used to calculate the territory size of marsh tapaculo, with the number of vocalizing individuals at the same time after playback on two sample dates (March 2011 and April 2012), measures of marsh areas (Ma, in m^2) and territory size (Ts, in m^2): S1. Sample 1; S2. Sample 2. (For abbreviations of states, see table 1.)

Tabla 2. Estado de las marismas utilizadas para calcular el tamaño del territorio del churrín palustre, número de individuos que emiten sonidos al mismo tiempo después de reproducir una grabación en las dos fechas del estudio (marzo de 2011 y abril de 2012), mediciones de las superficies de las marismas (Ma, en m^2 y tamaño del territorio (Ts, en m^2): S1. Muestra 1; S2. Muestra 2. (Para las abreviaturas de los estados, ver tabla 1.)

	Vocalizing indiv.			
State	S1	S2	Ma	Ts
PR	1	1	6,990	6,990
PR	1	1	3,589	3,589
PR	2	2	12,614	6,307
PR	4	4	18,121	4,530
SC	4	4	24,203	6,051
SC	2	2	9,535	4,768
SC	2	2	8,363	4,182
SC	5	5	30,422	6,084
Mean				5,313

Extent of occurrence, area of occupancy, territory size and population size

The increasing number of records in the 14 years since the species' description shows a significant growth in the extent of occurrence, going from only one state with records until 2001 (Bornschein et al., 2001) to four states in 2008 (Vasconcelos et al., 2008). The extent of occurrence presented by BirdLife International (2012), 490 km^2, is based on only 20 locations of occurrence, a much smaller number than that presented here: 70 locations and 42 municipalities. This difference in the amount of data used could explain the greater extent of occurrence, area of occupancy and population size presented here.

The territory size found for marsh tapaculo (5,313 ± 1,201 m^2), another factor that contributes to the increase in the estimated population size, is consistent with the value obtained for a species with similar body mass and environment (marsh antwren *Stymphalornis acutirostris*): an average of 2,500 m^2 (Reinert et al., 2007) and 7,000 m^2 (Reinert, 2008) in

tidal marshes and 32,000 m^2 in saw grass marshes (Reinert et al., 2007). There is no available information about the size of territory in relation to other species with similar body size and environment. Averages for territory sizes of forest species with body mass of 10 to 15 g range from 6,000 to 150,000 m^2 (Greenberg & Gradwohl, 1985; Silva, 1988; Terborgh et al., 1990; Skutch, 1996). An unpublished estimate, from Banhado do Maçarico in the state of Rio Grande do Sul (the southernmost locality known for the species), found a population density of 0.5 individuals of marsh tapaculo per hectare (40,000 m^2 territory size) (BirdLife International, 2012). The difference between this value and that obtained here could be explained by the location of marshes sampled in relation to the extent of occurrence of species and/or by differences in vegetation structure of marshes sampled. Such variation can also be observed in the territory size of marsh antwren (2,500 and 32,000 m^2), which varies with the vegetation structure of the marshes (Reinert et al., 2007).

The area of occupancy calculated for marsh tapaculo (84 km^2) represents only 11.24% of the environment considered suitable for this species in one state, Paraná (747 km^2) (Bornschein et al., 2001). Furthermore, the area of occupancy obtained through the grid cell method is influenced by the sampling intensity (Grueter et al., 2009), generating underestimated values when the species is not recorded in points where it occurs. It is therefore expected that the area of occupancy will increase as the number of records increase. This projection is confirmed by the increasing number of new records in recent years and from unpublished information. The growing numbers of searches for the species inside and outside the known range of occurrence will also contribute to increasing the range of occurrence and area of occupancy.

As it is expected that the area of occupancy will extend, the population size for marsh tapaculo is also expected to increase. Thus, given the significant growth of knowledge for the species' distribution presented here, the population size obtained is much larger than the estimates presented previously, 250 to 999 mature individuals (BirdLife International, 2012), and this number tends to increase as new records are made in different localities.

Threat category

The species was classified as 'Endangered' based on the little information available covering aspects of distribution and population. The information compiled and the results presented here show that a revaluation of the threat category is needed. The urgency for this review and for a change in the threat category, due to the fact that the original category is considered misclassified, is recommended by IUCN (2001).

Thus, considering the current extent of occurrence (> 20,000 km^2), area of occupancy (< 100 km^2), population size (> 10,000 individuals) and habitat conditions, the criteria needed to include the marsh tapaculo in the IUCN red list (BirdLife International, 2012) (criteria A3c+4c, B1ab (i, ii, iii, iv, v), C2a(i)) and in the Brazilian red list (Machado et al., 2005) (criteria B2ab) is not obtained for any of the threat categories ('Critically Endangered', 'Endangered' or 'Vulnerable').

Despite the small size of the area of occupancy, caused by the disconnected characteristics of the species' habitat, the number of locations where the marsh tapaculo was recorded was much greater than ten, and no fluctuations in the extent of occurrence, area of occupancy, number of locations or sub–populations, and number of mature individuals were observed. Revaluation of the species' threat category is thus strongly recommended.

Acknowledgements

Thanks to Adrian Eisen Rupp, Pedro Scherer Neto and Raphael Eduardo Fernandes Santos for the marsh tapaculo records kindly provided, and special thanks to Tiago Venâncio Monteiro, Liliane Klemann and Zora Morgenthaler for the translation of the text.

References

Accordi, I. A., Hartz, S. M. & Ohlweiler A., 2003. *O sistema Banhado Grande como uma área úmida de importância internacional.* II Simpósio de Áreas Protegidas – conservação no âmbito do Cone Sul. Pelotas, Brazil.

Adams, L. & Davis, S. D., 1967. The internal anatomy of home range. *Journal of Mammalogy,* 48: 529–536.

Andretti, C. B., 2010a. WA383164, *Scytalopus iraiensis.* Wiki Aves – The Encyclopedia of Brazilian Birds. www.wikiaves.com/383164 (Accessed 3 October 2011).

– 2010b. XC82545, *Scytalopus iraiensis.* Xeno–Canto – Bird sounds around the world. www.xeno–canto.org/82545 (Accessed 3 October 2011).

Araujo, F. M., 2011. WA318746, WA291931, WA300586, WA300597, WA300601, *Scytalopus iraiensis.* Wiki Aves – The Encyclopedia of Brazilian Birds. www.wikiaves.com/318746 (Accessed 3 October 2011).

Athanas, N., 2008. XC22833, *Scytalopus iraiensis.* Xeno–Canto – Bird sounds around the world. www.xeno–canto.org/22833 (Accessed 3 October 2011).

Bencke, G. A., Maurício, G. N., Develey, P. F. & Goerck, J. M., 2006. *Áreas importantes para a conservação das aves no Brasil. Parte I – estados do domínio da Mata Atlântica.* SAVE, São Paulo.

BirdLife International, 2012. Species factsheet: *Scytalopus iraiensis.* www.birdlife.org/datazone/speciesfactsheet.php?id=30032 (Accessed 9 October 2012).

Bornschein, M. R., Reinert, B. L. & Pichorim, M., 1998. Descrição, ecologia e conservação de um novo *Scytalopus* (Rhinocryptidae) do sul do Brasil, com comentários sobre a morfologia da família. *Ararajuba,* 6: 3–36.

– 2001. Novos registros de *Scytalopus iraiensis. Nattereria,* 2: 29–33.

Burgman, M. A. & Fox, J. C., 2003. Bias in species range estimates from minimum convex polygons: implications for conservation and options for improved planning. *Animal Conservation,* 6: 19–28.

By, R. A. de, 1998. XC10687, *Scytalopus iraiensis.* Xeno–Canto – Bird sounds around the world. www.xeno–canto.org/10687 (Accessed 3 October 2011).

Corrêa, L., Bazílio, S., Woldan, D. & Boesing, A. L., 2008. Avifauna da Floresta Nacional de Três Barras (Santa Catarina, Brasil). *Atualidades Ornitológicas,* 143: 38–41.

Corrêa, L. & Woldan, D. R. H., 2007. Registro de *Scytalopus iraiensis* na Floresta Nacional de Três Barras, planalto norte do estado de Santa Catarina, Brasil.In: *Livro de Resumos do XV Congresso Brasileiro de Ornitologia:* 204 (C. S. Fontana, Ed.). EdiPUCRS, Porto Alegre.

Corrêa, L., Woldan, D. R. H. & Bazílio, S., 2007. Registros de aves raras e ameaçadas no planalto norte de Santa Catarina, Brasil. In: *Livro de Resumos do XV Congresso Brasileiro de Ornitologia:* 203 (C. S. Fontana, Ed.). EdiPUCRS, Porto Alegre.

Espínola, C., 2011. WA314543, *Scytalopus iraiensis.* Wiki Aves – The Encyclopedia of Brazilian Birds. www.wikiaves.com/314543 (Accessed 3 October 2011).

Faria, L., 2011. XC76250, *Scytalopus iraiensis.* Xeno–Canto – Bird sounds around the world. www.xeno–canto.org/76250 (Accessed 3 October 2011).

Fenalti, P. R., 2011. WA478969, *Scytalopus iraiensis.* Wiki Aves – The Encyclopedia of Brazilian Birds. www.wikiaves.com/478969 (Accessed 16 January 2012).

Ferry, C. & Frochot, B., 1970. L'avifaune nidificatrice d'une foret de Chenes pedoncules en Bourgogne: Etude de deux successions ecologiques. *Rev Ecol–Terre Vie,* 24: 153–250.

Fontana, C. S., Rovedder, C. E., Repenning, M. & Gonçalves, M. L., 2008. Estado atual do conhecimento e conservação da avifauna dos Campos de Cima da Serra do sul do Brasil, Rio Grande do Sul e Santa Catarina. *Revista Brasileira de Ornitologia,* 16(4): 281–307.

FRA, 2000. *Global Ecological Zones.* Forest Resource Assessment. www.fao.org/geonetwork/srv/en/main.home (Accessed 28 September 2011).

Franco, E. S., 2013. WA844493, *Scytalopus iraiensis.* Wiki Aves – The Encyclopedia of Brazilian Birds. www.wikiaves.com/844493 (Accessed 8 January 2013).

Fundação SOS Mata Atlântica & Instituto Nacional De Pesquisas Espaciais, 2009. *Atlas dos remanescentes florestais da mata atlântica período 2005–2008.* http://mapas.sosma.org.br/dados/ (Accessed 22 December 2011)

– 2011. *Atlas dos remanescentes florestais da mata atlântica período 2008–2010.* http://mapas.sosma.org.br/dados/ (Accessed 22 December 2011)

Greenberg, R. & Gradwohl, J., 1985. A comparative study of the social organization of Antwrens on Barro Colorado Island, Panama. *Ornithological*

Monographs, 36: 845–855.

Grueter, C. C., Li, D., Ren, B. & Wei, F., 2009. Choice of analytical method can have dramatic effects on primate home range estimates. *Primates,* 50: 81–84.

Harris, S., Cresswell, W. J., Forde, P. G., Trewhella, W. J., Woollard, T. & Wray, S., 1990. Home–range analysis using radio–tracking data – a review of problems and techniques particularly as applied to the study of mammals. *Mammal Review,* 20: 97–123.

Hassdenteufel, C. B., Accordi, I. de A. & Hartz, S. M., 2006a. Seleção de micro–hábitat por *Scytalopus iraiensis* em uma fisionomia de área úmida no sul do Brasil. In: *Livro de Resumos (População e Comunidade) do XIV Congresso Brasileiro de Ornitologia:* 41 (R. Ribon, Ed.). Ouro Preto.

Hassdenteufel, C. B., Brandt, C. S., Accordi, I. de A., Hartz, S. M. & Barcellos, A., 2006b. Manifestações sonoras de *Scytalopus iraiensis* em uma fisionomia de área úmida no sul do Brasil. In: *Livro de Resumos (População e Comunidade) do XIV Congresso Brasileiro de Ornitologia:* 55 (R. Ribon, Ed.). Ouro Preto.

Hayne, D., 1949. Calculation of size of home range. *Journal of Mammalogy,* 30:1–18.

Howard, H. E., 1920. *Territory in Bird Life.* E. P. Dutton & Company, New York.

Hutto, R. L., Pletschet, S. M. & Hendricks, P., 1986. A fixed–radius point count method for nonbreeding and breeding season use. *Auk,* 103: 593–602.

IAP, 2009. *Planos de conservação para espécies de aves ameaçadas no Paraná.* IAP/Projeto Paraná Biodiversidade, Curitiba.

IBGE, 1993. Mapa de Vegetação do Brasil. www.ibge.gov.br (Accessed 20 September 2011).

– 2007. Banco de dados–Curitiba (PR). www.ibge.gov.br (Accessed 3 October 2011).

IUCN, 2001. *IUCN Red List Categories and Criteria: Version 3.1.* IUCN Species Survival Commission. IUCN, Gland, Switzerland & Cambridge, UK.

Jacobs, F., 2008a. WA44371, *Scytalopus iraiensis.* Wiki Aves–The Encyclopedia of Brazilian Birds. www.wikiaves.com/44371 (Accessed 3 October 2011).

– 2008b. XC22912, *Scytalopus iraiensis.* Xeno–Canto – Bird sounds around the world. www.xeno–canto.org/22912 (Accessed 3 October 2011).

Kool, K. & Croft, D., 1992. Estimators for home range areas of arboreal colobine monkeys. *Folia Primatologica,* 58: 210–214.

Lehmann, J. & Boesch, C., 2003. Social influences on ranging patterns among chimpanzees (*Pan troglodytes verus*) in the Tai National Park, Cote d'Ivoire. *Behavioral Ecology,* 14: 642–649.

Li, Z. & Rogers, M., 2005. Habitat quality and range use of white–headed langurs in Fusui, China. *Folia Primatologica,* 76: 185–195.

Luijendijk, T., 2010. XC59015, *Scytalopus iraiensis.* Xeno–Canto – Bird sounds around the world. www.xeno–canto.org/59015 (Accessed 3 October 2011).

Machado, A. B. M., Martins, C. S. & Drummond, G. M., 2005. *Lista da fauna brasileira ameaçada de extinção: incluindo as listas de espécies quase ameaçadas e deficientes em dados.* Fundação Biodiversitas, Belo Horizonte.

Maurício, G. N., 2005. Taxonomy of southern populations in the *Scytalopus speluncae* group, with description of a new species and remarks on the systematics and biogeography of the complex (Passeriformes: Rhinocryptidae). *Ararajuba,* 13: 7–28.

Mills, M. G. L. & Gorman, M. L., 1987. The scent–marking behaviour of the Spotted hyaena *Crocuta crocuta* in the southern Kalahari. *Journal of Zoology,* 212: 483–497.

Minns, J., 2002. XC5976, *Scytalopus iraiensis.* Xeno–Canto – Bird sounds around the world. www.xeno–canto.org/5976 (accessed 3 October 2011).

Mohr, C. O., 1947. Table of equivalent populations of North American small mammals. *American Midland Naturalist,* 37: 223–249.

Nichols, J. D., Hines, J. E., Sauer, J. R., Fallon, F. W., Fallon, J. E. & Heglund, P. J., 2000. A double–observer approach for estimating detection probability and abundance from point counts. *Auk,* 117: 393–408.

Ostro, L. E. T., Young, T. P., Silver, S. C. & Koontz, F. W., 1999. A geographic information system (GIS) method for estimating home range size. *Journal of Wildlife Management,* 63: 748–755.

Patrial, E., 2006. XC18297, *Scytalopus iraiensis.* Xeno–Canto – Bird sounds around the world. www.xeno–canto.org/18297 (Accessed 3 October 2011).

Perrins, C. M. & Birkhead, T. R., 1983. *Avian Ecology.* Glasgow & Blackie et Son, London.

Powell, R., 2000. Animal home ranges and territories and home range estimators. In: *Research techniques in animal ecology: controversies and consequences:* 65–110 (L. Boitani, & T. Fuller, Eds.). Columbia Univ. Press, New York.

Raposo, M. A., Stopiglia, R., Loskot, V. & Kirwan, G. M., 2006. The correct use of the name *Scytalopus speluncae* (Ménétriés, 1835), and the description of a new species of Brazilian tapaculo (Aves: Passeriformes: Rhinocryptidae). *Zootaxa,* 1271: 37–56.

Reinert, B. L., 2008. Ecologia e comportamento do bicudinho–do–brejo (*Stymphalornis acutirostris* Bornschein, Reinert e Teixeira, 1995; Aves, Thamnophilidae). Ph. D. Thesis, Univ. Estadual Paulista 'Julio de Mesquita Filho'.

Reinert, B. L., Bornschein, M. R. & Firkowski, C., 2007. Distribuição, tamanho populacional, hábitat e conservação do bicudinho–do–brejo *Stymphalornis acutirostris* Bornschein, Reinert e Teixeira 1995 (Thamnophilidae). *Revista Brasileira de Ornitologia,* 15: 493–519.

Rodrigues, M., Mariana, L. M., Freitas, G. H. S. de, Dias, D. F., Mesquita, E. P., Cavalcanti, M., Rocha, R. P., Rodrigues, L. da C., Ferreira, J. D. & Diniz, F. C., 2008. Campos rupestres úmidos e secos da cadeia do espinhaço: uma abordagem ornitológica. In: *Livro de Resumos do XVI Congresso Brasileiro de Ornitologia:* 194 (T. Dornas & M. de O. Barbosa, Eds.). ECOAVES–UFT, Palmas.

Rosenstock, S. S., Anderson, D. R., Giesen, K. M., Leukering, T. & Carter, M. F., 2002. Landbird counting techniques: Current practices and an alternative. *Auk,* 119: 46–53.

Rupp, A. E., 2010. WA181812, *Scytalopus iraiensis*. Wiki Aves – The Encyclopedia of Brazilian Birds. www.wikiaves.com/181812 (Accessed 3 October 2011).

– 2010. WA244703, *Scytalopus iraiensis*. Wiki Aves – The Encyclopedia of Brazilian Birds. www.wikiaves.com/244703 (Accessed 3 October 2011).

Silva, J. C., 2011. WA487684, *Scytalopus iraiensis*. Wiki Aves – The Encyclopedia of Brazilian Birds. www.wikiaves.com/487684 (Accessed 16 January 2012).

– 2012. WA601765, *Scytalopus iraiensis*. Wiki Aves – The Encyclopedia of Brazilian Birds. www.wikiaves.com/601765 (Accessed 16 May 2012).

– 2012. WA706138, *Scytalopus iraiensis*. Wiki Aves – The Encyclopedia of Brazilian Birds. www.wikiaves.com/706138 (Accessed 8 January 2013).

Silva, J. M. C., 1988. Aspectos da ecologia e comportamento de *Formicivora g. grisea* (Boddaert, 1789) (Aves: Formicariidae) em ambientes amazônicos. *Revista Brasileira de Biologia,* 48(4): 797–805.

Simons, T. R., Alldredge, M. W., Pollock, K. H. & Wettroth, J. M., 2007. Experimental analysis of the auditory detection process on avian point counts. *Auk,* 124(3): 986–999.

Skutch, A. F., 1996. *Antbirds and ovenbirds.* Univ. of Texas Press, Austin.

Straube, F. C., Carrano, E., Santos, R. E. F., Scherer-Neto, P., Ribas, C. F., Meijer, A. A. R. de, Vallejos, M. A. V., Lanzer, M., Klemann–Júnior, L., Aurélio–Silva, M., Urben–Filho, A., Arzua, M., Lima, A. M. X. de, Deconto, L. R., Bispo, A. Â., Jesus, S. de & Abilhôa, V., 2009. *Aves de Curitiba–Coletânea de registros.* Hori Consultoria, Curitiba.

Straube, F. C., Krul, R. & Carrano, E., 2005. Coletânea da avifauna da região sul do estado do Paraná (Brasil). *Atualidades Ornitológicas,* 125: 10–71.

Terborgh, J., Robinson, S. K., Parker Iii, T. A., Munn, C. A. & Pierpont, N., 1990. Structure and organization of an Amazonian forest bird community. *Ecological Monographs,* 60(2): 213–238.

Thompson, W. L., 2002. Towards reliable bird surveys: accounting for Individuals present but not detected. *Auk,* 119(1): 18–25.

Vasconcelos, M. F., Maurício, G. N., Kirwan, G. M. & Silveira, L. F., 2008. Range extension for marsh tapaculo *Scytalopus iraiensis* to the highlands of Minas Gerais, Brazil, with an overview of the species' distribution. *Bulletin BOC,* 128(2): 101–106.

Veloso, H. P., Rangel Filho, A. L. R. & Lima, J. C. A., 1991. *Classificação da vegetação brasileira, adaptada a um sistema universal.* IBGE, Departamento de Recursos Naturais e Estudos Ambientais, Rio de Janeiro.

White, G. & Garrott, R., 1990. *Analysis of wildlife radio tracking data.* Academic, San Diego.

Whittaker, A., 2011. WA511140, *Scytalopus iraiensis*. Wiki Aves – The Encyclopedia of Brazilian Birds. www.wikiaves.com/511140 (Accessed 16 January 2012).

4

Non–invasive sampling of endangered neotropical river otters reveals high levels of dispersion in the Lacantun River System of Chiapas, Mexico

J. Ortega, D. Navarrete & J. E. Maldonado

Ortega, J., Navarrete, D. & Maldonado, J. E., 2012. Non–invasive sampling of endangered neotropical river otters reveals high levels of dispersion in the Lacantun River System of Chiapas, Mexico. *Animal Biodiversity and Conservation*, 35.1: 59–69.

Abstract

Non–invasive sampling of endangered neotropical river otters reveals high levels of dispersion in the Lacantun River System of Chiapas, Mexico.— Patterns of genetic dispersion, levels of population genetic structure, and movement of the neotropical river otter (*Lontra longicaudis*) were investigated by screening eight polymorphic microsatellites from DNA extracted from fecal samples, collected in a hydrologic system of the Lacandon rainforest in Chiapas, Mexico. A total of 34 unique genotypes were detected from our surveys along six different rivers, and the effect of landscape genetic structure was studied. We recovered 16 of the 34 individuals in multiple rivers at multiple times. We found high levels of dispersion and low levels of genetic differentiation among otters from the six surveyed rivers ($P > 0.05$), except for the pairwise comparison among the Lacantún and José rivers ($P < 0.05$). We recommend that conservation management plans for the species consider the entire Lacantún River System and its tributaries as a single management unit to ensure the maintenance of current levels of population genetic diversity, because the population analyzed seems to follow a source–sink dynamic mainly determined by the existence of the major river.

Key words: Assignment test, Chiapas, Lacandon rainforest, Low genetic structure, Microsatellites, Neotropical river otter.

Resumen

El muestreo no invasivo de las nutrias neotropicales, una especie amenazada, revela altos niveles de dispersión en el sistema fluvial del Lacantún, en Chiapas, México.— Se investigaron patrones de dispersión genética, niveles de estructura poblacional genética, y desplazamientos de las nutrias neotropicales (*Lontra longicaudis*) mediante el screening de ocho microsatélites polimórficos tomados del DNA de muestras fecales, que fueron recogidas en el sistema hidrológico del río Lacantún de la Selva Lacandona, Chiapas, México. Se detectó un total de 34 genotipos únicos a lo largo de seis ríos, estudiándose el efecto de la estructura genética según el paisaje. Se identificaron 16 de los 34 individuos estudiados en varios de los ríos y en múltiples ocasiones. Encontramos altos niveles de dispersión y bajos niveles de diferenciación genética entre las nutrias de los seis ríos estudiados ($P > 0,05$), con excepción de las comparaciones por pares entre los ríos José y Lacantún ($P < 0,05$). Se recomienda que en los planes de gestión para esta especie se considere el sistema hidrológico del río Lacantún como una sola entidad, para así mantener los niveles de diversidad genética encontrados, dado que parece que la población analizada sigue una dinámica de fuente-sumidero determinada por el pricipal río de la zona.

Palabras clave: Test de asignación, Chiapas, Selva Lacandona, Estructura genética baja, Microsatélites, Nutria neotropical.

Jorge Ortega & Daya Navarrete, Lab. de Ictiología y Limnología, Depto. de Zoología, Escuela Nacional de Ciencias Biológicas, Inst. Politécnico Nacional, Prolongación de Carpio y Plan de Ayala s/n., Col. Sto. Tomas, 11340, México, D. F.– Jesús E. Maldonado, Center for Conservation and Evolutionary Genetics, Smithsonian Conservation Biology Inst., National Zoological Park, Washington, DC 20008, USA.

Corresponding author: Jorge Ortega E–mail: artibeus2@aol.com

Introduction

The neotropical river otter (*Lontra longicaudis*) is a semi–aquatic carnivore that has been protected by Mexican law since 1994 (NOM, 1994) and is also considered an endangered species according to CITES Appendix I (Larivière 1999). Its current distribution includes Central America, South America, and the island of Trinidad in the Caribbean. It is found in different riverine habitats, including deciduous and evergreen forests, but it is thought to prefer clear rivers and streams. In Mexico, populations of the neotropical river otter currently exist in larger rivers and their tributaries along the northern states of Sinaloa and Jalisco to the southern states of Chiapas, Veracruz, Oaxaca, and Tabasco, including the Yucatan Peninsula. he Lacandon forest is located in the southeast part of the Mexican state of Chiapas and occupies an area of ca. 957, 240 ha (fig. 1). The protected area has an intricate hydrologic system composed of permanent rivers that are well interconnected by smaller tributaries that become desiccated during the dry season. In the Lacantún River System, neotropical river otters subsist in moderate densities, in part because the Lacandon forest has been protected by the Mexican government and has been minimally impacted by anthropogenic activities. The most obvious way of dispersal for otters is the use of water currents to increase their speed and mobility. They also inhabit transformed rivers where the original vegetation has been cleared by anthropogenic activities.

Molecular markers have proven to be an important tool for investigating the patterns of gene flow and levels of genetic structure of otters in different parts of the world. Microsatellite genotype variation in North American river otters (*Lontra canadensis*) revealed that ecological barriers prevented dispersal, and this resulted in the genetic differentiation of three inland and coastal subpopulations in Louisiana (Latch et al., 2008). Another microsatellite study revealed high levels of gene flow among seven different populations of river otters, where gender differences were fundamental to the understanding of genetic diversity and recolonization in the area (Blundell et al., 2002; Reid et al., 1994). In European otters (*Lutra lutra*), microsatellite markers revealed that steep and dry areas prevented proper dispersal in 10 river basins and two genetically differentiated otter populations were described for the area (Janssens et al., 2008). In addition, mitochondrial DNA (mtDNA) data helped discover two well–defined subpopulations of the Southern river otter (*Lontra provocax*) in the southern part of Argentina (Centrón et al., 2008). On a broader scale, allozyme data revealed that the North American river otters (*L. canadensis*) were sorted in eight geographic regions and that gene flow was more restricted (Serfass et al., 1998).

River otters are elusive and difficult to capture. Therefore, in this study, we used non–invasive sampling because it has demonstrated to be an efficient method to conduct genetics studies of wild animals without having to capture them, or even observe them (Dallas et al., 2003). In addition, because the same individual can be recaptured multiple times in different areas non–invasive sampling can be very useful to more directly study individual movement and dispersal. We extracted DNA from non–invasively collected fecal samples and used eight polymorphic microsatellite loci to study movement patterns and levels of genetic structure among populations of the Neotropical river otter (*L. longicaudis*) inhabiting the Lacantún River System and its tributaries in the Lacandon forest, Chiapas, Mexico. Our goal was to characterize the levels of genetic diversity from animals sampled in six major rivers to assess movement of individuals and dispersion in an environment without intense anthropogenic activities and low levels of habitat fragmentation. We hypothesized that the complex network of rivers and tributaries throughout the study area would allow neotropical river otters to move freely through the region as one large, panmictic population, especially because during the dry season, tributaries are desiccated and force animals to move back to the larger permanent rivers. To assess our hypothesis we used microsatellites specifically designed for river otters, because they have demonstrated to have sufficient power to examine fine scale genetic structure and tracking dispersion patterns of river otters among the different tributaries (Latch et al., 2008).

Material and methods

Biological sample

We sampled an area that encompassed most of the known current distributional range of the neotropical river otter (*Lontra longicaudis*) in the southern part of the Lacandon Rainforest Reserve, in the southern state of Chiapas, Mexico. The maximum linear distance between the most distant rivers is 34 km and our geographic scale is limited to the Lacandon Rainforest Reserve extension (957,240 ha). Our sampling was designed to obtain fecal samples from as many individuals as possible in an effort to accomplish an exhaustive sampling of the population and to assess individual movement patterns and dispersion in an environment with low levels of habitat fragmentation. We surveyed eight transects along the six rivers, walking a maximum distance of 10 km from the main Lacantún River, searching for otter latrines with fresh fecal material. Transects were surveyed during two field seasons in the early morning (dry and cold seasons of 2006 and 2007) to avoid the negative effect of the weather on the feces (Hájková et al., 2006). Fresh fecal samples were placed in Falcon tubes with silica gel and transported to the laboratory for analysis. In total, we collected 623 fecal samples along the Lacantún River (LcR, n = 154), Colorado River (CR, n = 65), José River (JR, n = 85), Lagartos River (LgR, n = 97), Miranda River (MR, n = 129), and Tzendales River (TR, n = 93) (fig. 1). We did not identify fecal samples using a genetic marker because visual identification of feces from neotropical river otter from latrines and by the type of food ingested (fish and crustaceans) is an established

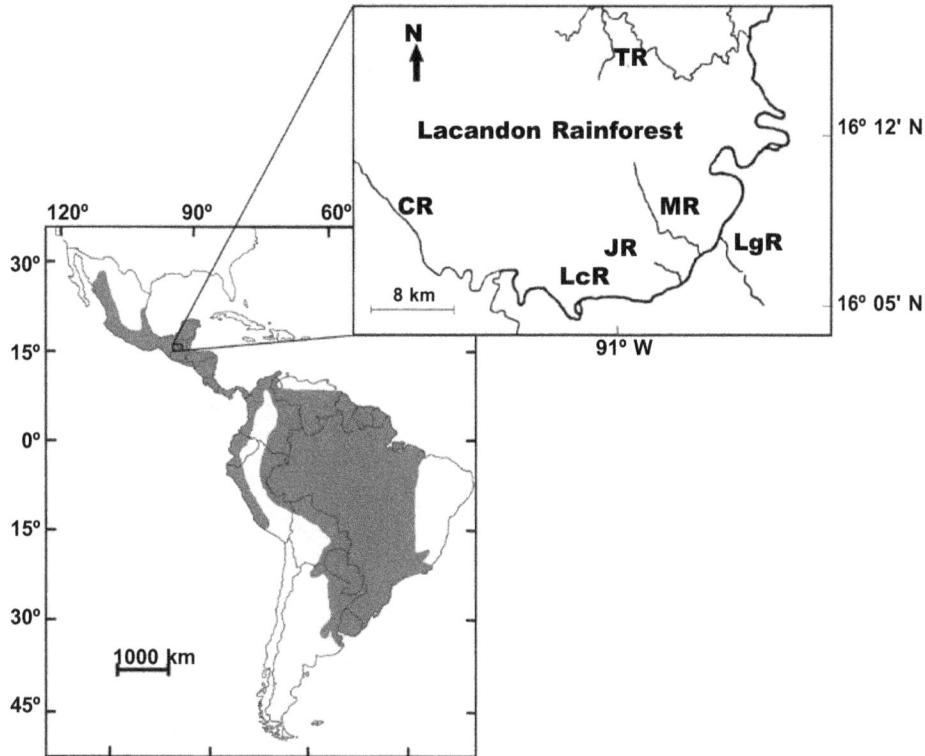

Fig. 1. Rivers sampled in the Lacandon Rainforest, Chiapas, Mexico. Location of the rivers is shown as follows: Miranda (MR), José (JR), Lacantún (LcR), Tzendales (TR), Lagartos (LgR) and Colorado (CR). We included the historical distribution map of the species, and merged the collecting site that is restricted to the extension of the Lacandon Rainforest Reserve.

Fig. 1. Ríos muestreados en la Selva Lacandona, Chiapas, México. La localización de los ríos se indica de la forma siguiente: Miranda (MR), José (JR), Lacantún (LcR), Tzendales (TR), Lagartos (LgR) y Colorado (CR). Se ha incluido el mapa de la distribución histórica de la especie y destacado el lugar de recolección que está dentro de la Reserva de la Selva Lacandona.

and very reliable method for documentation of river otters in the field (Medina–Vogel & Gonzalez–Lagos, 2008). DNA was also extracted from blood and fecal samples from five captive *L. longicaudis* housed at two different zoological parks. This DNA was used as positive controls and to validate the accuracy of our fecal DNA genotypes.

Laboratory procedures

Whole genomic DNA was extracted from fecal samples using the QIAamp DNA Stool Mini Kit and from blood using the QIAamp Tissue and Blood Kit (QIAGEN) in a separate room from PCR products (Eggert et al., 2005). Quality of DNA was assessed by electrophoresis on 1% agarose gels in combination with molecular weight standards. We optimized DNA amplification following the standard protocol suggested by Lampa et al. (2008). In total, eight polymorphic microsatellite loci designed for the sister species *L. canadensis* (Beheler et al., 2004, 2005) were optimized and amplified for this species. PCR

conditions consisted of an initial denaturation at 95°C for 2', followed by 35 cycles at 94°C for 30s, respective annealing temperature for 30s, 72°C for 30s, then 72°C for 10'. We incorporated variations for the PCR conditions (Protocol 1, 2, and 3) as suggested by Beheler et al., 2004 and Beheler et al., 2005. All reactions were performed in a Perkin Elmer 9700 Thermocycler. Amplifications of microsatellite loci were carried out in a 15 µL volume containing 30 ng of DNA, 0.3 mM of dNTP´s, 0.5 µM of each primer, 1x *Taq* buffer (2.0 µM of $MgCl_2$, 10 mM of Tris–HCl, 50 mM of KCl), 2.5x of BSA, and 1.0 U of Flexi*Taq* polymerase (PROMEGA). Each fecal sample was genotyped four independent times to validate our data and to overcome problems associated with low quality DNA obtained from non–invasive samples, and to be consistent with the multi–tube approach (Taberlet et al., 1996). We sexed samples by using the amplification of the zinc finger gene region on agarose gels, following the protocol proposed by Ortega et al. (2004). PCR products were verified by electrophoresis on 1% agarose gels stained with ethidium bromide. Fragment analysis was performed on an ABI Prism 3100 Genetic Analyzer. Computer–generated results were analysed using the GeneScan (version 2.1) software, and final allele–sizing was done using the ABI Genotyper package (version 2.1).

Genetic analysis

We estimated the proportion of polymorphic loci and the average number of alleles per locus using GDA software (Lewis & Zaykin, 2001), and the observed and the expected heterozygosity using POPGENE (Yeh & Boyle, 1997) by the algorithm proposed by Levene (1949). We analysed differences in allele frequency distribution between the different sampled sites for each locus individually and across all loci, and also tested departures from Hardy–Weinberg equilibrium and linkage disequilibrium using GENEPOP version 4.0 (Rousset, 2008). Due to the variation in sample sizes between collecting localities, inbreeding coefficients (*Fis*) were calculated using the algorithm suggested by Weir & Cockerham (1984).

Proportions of allelic dropout and false allele detection associated with the non–invasive sampling were estimated by comparing the genotypes of DNA extracts from fecal samples of captive otters with genotypes of blood DNA extracted from the same otters. The proportion of allelic dropout was estimated as the proportional number of all–PCRs in which a fragment observed in the DNA extracted from blood was undetected in PCRs derived from fecal DNA. The proportion of the false allele detection was estimated as the proportional number of all PCRs in which a fragment observed in the DNA extracted from feces was not present in PCRs derived from DNA extracted from blood (Dallas et al., 2002, 2003). Using the proportion of allele dropout and false allele, we used the program GEMINI (Valière et al., 2002) to obtain the expected proportions of incorrect genotypes estimated for the entire population. We used the same software to determine the ability of our eight microsatellites to distinguish between individuals, the probability of identity (P_{ID}) (*i.e.* the probability of different individuals sharing an identical genotype at random; Mills et al., 2000; Waits et al., 2001) and the P_{ID} between siblings using the 5 samples from captive otters caught in the same area, but under the custody of a zoological park.

We calculated population substructure and heterozygote deficits using FSTAT version 2.9.3 (Goudet, 1995). Because genotypes from all six sampling localities were in Hardy–Weinberg equilibrium, we calculated pairwise F_{st} for all sampled sites and tested the significance of gene diversity across different sites using 1,000 permutations with the same program; P–values were obtained by multiple comparisons after Bonferroni corrections. We used *Fst* as an estimator of genetic variation between populations because it is more sensitive to test the hypothesis of panmixia.

We also investigated population genetic structure using an analysis of molecular variance (AMOVA), as implemented by ARLEQUIN (version 3.01, Excoffier et al., 2005). The variance components are used to calculate differences among populations and between individuals in different rivers. To assess inter–population dispersal, we used assignment tests in the GENE-CLASS 2 software (Piry et al., 2004), by using the algorithm of Paetkau et al. (1995). These assignment tests use individual genotypes and population allele frequencies to calculate potential migrants and to allocate them to their original populations (Cornuet et al., 1999). In addition, we obtained a graph with the GENETIX v. 4.05 software (Belkhir et al., 1996–2004), and we performed a Factorial Correspondence Analysis (FCA) to visualize the distribution of sampled individuals according to their sample site.

In order to understand how genetic variation was distributed within the study area, we first used Bayesian assignment techniques to test for population structure using the program STRUCTURE (version 2.2, Pritchard et al., 2000). This method identifies clusters of genetically similar individuals from multilocus genotypes without prior knowledge of their origin and genetic relationships. We ran a series of pilot runs to estimate Pr(X|K), where X represents the data for K between 1 (the expected value if all individuals belong to the same cluster) and 6 (the maximum number of sampling localities in different rivers). In our final runs to determine the most likely K, we assumed that populations had correlated allele frequencies, inferred alpha from the data, and used a burn–in and Markov chain Monte Carlo (MCMC) of 200,000 followed by 1,000,000. Longer burn–in or MCMC did not change the results.

We also conducted estimates of population size by using the mark–recapture data of neotropical river otter genotypes presented in table 1 and the CAPWIRE software (Miller et al., 2005). This program has been recently developed to maximize the use of DNA–based mark–recapture data and performs well for smaller populations (N < 100) with substantial capture heterogeneity. The program has two models to estimate population size, both based on presence or absence of captured heterogeneity. We assumed demographic resolution on the basis of the relatively short sampling time, while geographical resolution was maximized by collecting the scats in the closest rivers. In order to remove pseudo–replicates, scats collected from the same latrine on the same day that originated from the same individual were considered as a single observation.

Results

During our scat collection surveys of the six rivers we recovered a total of 623 presumed river otter fecal samples. We attempted to isolate DNA from all 623 samples and obtained complete microsatellite and sex data for 154 fecal samples. While our amplification success was low (24.4%), these 154 samples were successfully amplified for all eight microsatellite loci designed by Beheler et al. (2004, 2005) and for a zinc finger sexing marker (Ortega et al., 2004). Furthermore, they met the scoring requirements of our strict protocol applying the multipletube approach. DNA extracted from neotropical river otter feces is

Table 1. We obtained a total of 154 genotypes for 8 microsatellite loci from the collected fecal samples. We identified 34 unique genotypes and 16 of these were recaptured many times and in many rivers (denoted with an *). This table shows the number of repeated feces collected with each of the 34 unique genotypes, the location, sex of the individual, and the field season ($^+$ 2006 and $^-$ 2007): #G. Genotype. (For abbreviations, see figure 1.)

*Tabla 1. Se obtuvo un total de 154 genotipos para los 8 loci de microsatélites a partir de las muestras fecales recogidas. Se identificaron 34 genotipos únicos, 16 de los cuales fueron recapturados en múltiples ocasiones y en varios ríos (seguidos de un *). Esta tabla muestra el número de heces recogidas repetidamente, con cada uno de los 34 genotipos únicos, la localización, el sexo del individuo y la estación del año ($^+$ 2006 and $^-$ 2007). #G. Genotipos. (Para las abreviaturas ver figura 1.)*

#G	LcR	MR	JR	CR	LgR	TR
1♀	3$^+$	0	0	0	0	0
2♀*	2$^-$	0	0	1$^-$	0	1$^+$
3♀*	4$^+$	0	0	2$^+$	0	1$^+$
4♂	5$^-$	0	0	0	0	0
5♀	0	2$^+$	0	0	0	0
6♀	0	3$^-$	0	0	0	0
7♂	0	2$^+$	0	0	0	0
8♂	0	0	0	5$^-$	0	0
9♀*	0	0	6$^+$	0	0	1$^-$
10♂	0	0	0	0	4$^-$	0
11♂	0	2$^+$	0	0	0	
12♀*	1$^-$	0	0	1$^-$	0	1$^+$
13♀*	3$^-$	0	0	2$^+$	0	0
14♂*	0	2$^-$	2$^+$	0	0	2$^-$
15♀*	2$^+$	0$^+$	2$^-$	0	3$^-$	1
16♀*	0	1$^-$	1$^-$	0	3$^-$	1$^-$
17♂	0	0	0	1$^+$	0	0
18♀*	2$^+$	0	0	2$^-$	1$^+$	0
19♀*	1$^+$	1$^-$	0	1$^+$	0	0
20♂	5	0	0	0	0	0
21♂	0	0	5$^+$	0		0
22♂*	2$^+$	3$^+$	0	0	2$^-$	0
23♂	0	0	0	2$^-$	0	0
24♀	0	0	0	0	5$^-$	0
25♀	2$^-$	0	0	0	0	0
26♀	0	0	0	5$^-$	0	0
27♂	0	0	0	0	0	3$^-$
28♂	6$^+$	0	0	0	0	0
29♀*	3$^+$	2$^-$	0	2$^-$	0	0
30♀	0	0	4$^+$	0	0	0
31♀*	3$^-$	0	2$^-$	0	0	1$^-$
32♂*	0	0	1$^+$	0	1$^+$	1$^-$
33♀*	0	2	2$^-$	0	2$^-$	0
34♂*	3$^+$	0	2$^+$	0	0	2$^-$

notoriously difficult to work with and several factors have been reported to affect PCR amplifications of microsatellite loci in otters (Hajkova et al., 2006). However, results from the comparison of our four replicates per loci suggested that we had low levels of genotyping error for these samples. In addition, PCR error values due to allelic dropout averaged 0.023 for the eight microsatellites tested on feces and blood samples of five captive individuals, and 0.065 for the field samples. For false alleles, we obtained an average value of 0.019 in the control samples, and 0.034 in the field samples. With eight microsatellites in a tissue sample set of 5 otters, we estimated that the probability of random match between unrelated individuals for all multilocus genotypes was 1.3–2.8 10^{-6} (P_{ID} unbiased), and the probability of a random match between siblings for all multilocus genotypes was 4.2–5.5 10^{-3} (P_{ID} sibs). Thus, the overall P_{ID} was low, suggesting our selected

microsatellite were adequate to differentiate between individual otters, including relatives.

Samples with complete microsatellite data came from along the LcR (n = 47), CR (n = 24), JR (n = 27), LgR (n = 21), MR (n = 20), and TR (n = 15) and represented all of the six rivers from our original surveys (fig. 1). From these 154 fecal samples, we identified 34 unique genotypes, of which 16 were found multiple times in feces from the same river and 18 were also found in multiple rivers (table 1). Among these unique genotypes, sex ratio did not differ from unit. With data for six rivers pooled, fecal samples represented 19 females and 15 males (two tailed binomial test, P = 0.72; table 1). Both sexes were found in the six distinct rivers but in a different spread pattern. Eleven males were collected only once in one river, and 4 individuals were detected at least in 2 different rivers in the area. CR was the river where we found fewest males (3), and it was the site that

Table 2. Expected (H_E) and observed (H_O) heterozygosities and number of alleles per locus (A) for every river sampled. Hardy–Weinberg Equilibrium (HWE) and inbreeding coefficient (F_{IS}) were calculated according to Weir & Cockerham (1984). Significant P values after Bonferroni corrections are denoted in bold. (For abbreviations, see figure 1.)

Tabla 2. Heterocigosidad esperada (H_E) y observada (H_O) y número de alelos por locus (A) para cada río muestreado. Se calcularon el equilibrio de Hardy–Weinberg (HWE) y el coeficiente de consanguinidad (F_{IS}) según Weir & Cockerham (1984). En negrita se consignan del valores P significativos tras las correcciones de Bonferroni. (Para las abreviaturas ver figura 1.)

Locus	A	HWE P–value	F_{IS}		LcR (n = 47)	MR (n = 20)	JR (n = 27)	CR (n = 24)	LgR (n = 21)	TR (n = 15)
RIO02	5	**0.01**	0.131	H_E	0.743	0.702	0.796	0.678	0.738	0.706
				H_O	0.643	0.71	0.437	0.5	0.667	0.667
RIO04	10	0.083	0.122	H_E	0.878	0.813	0.814	0.857	0.77	0.811
				H_O	0.857	0.684	0.75	0.75	0.6	0.833
RIO07	6	0.16	0.241	H_E	0.778	0.815	0.79	0.893	0.761	0.841
				H_O	0.571	0.632	0.812	0.25	0.467	0.667
RIO08	4	0.07	0.294	H_E	0.606	0.701	0.619	0.75	0.689	0.627
				H_O	0.071	0.5	0.5	0.75	0.533	0.583
RIO10	3	0.064	–0.047	H_E	0.593	0.412	0.458	0.679	0.515	0.598
				H_O	0.428	0.474	0.437	1	0.6	0.583
RIO11	4	**0.03**	0.139	H_E	0.518	0.685	0.704	0.536	0.687	0.746
				H_O	0.571	0.579	0.5	0.25	0.733	0.583
RIO13	3	0.16	0.049	H_E	0.5	0.547	0.627	0.571	0.653	0.648
				H_O	0.571	0.579	0.5	1	0.467	0.583
RIO16	3	0.089	0.167	H_E	0.14	0.313	0.377	0.25	0.439	0.42
				H_O	0.143	0.263	0.312	0.25	0.467	0.167
ALL	38			H_E	0.59 ± 0.2	0.62 ± 0.2	0.65 ± 0.2	0.65 ± 0.2	0.66 ± 0.1	0.67 ± 0.2
				H_O	0.48 ± 0.3	0.55 ± 0.1	0.53 ± 0.2	0.59 ± 0.3	0.57 ± 0.1	0.58 ± 0.1

did not share genotypes with other rivers. Females appeared to move more among rivers than males because we found 12 female genotypes many times in different rivers: 2 female genotypes were detected in 4 rivers, 8 were recorded in 3 distinct rivers, and 2 were present in 2 rivers. LcR had 11 out of 19 of the recorded genotypes, while Lag contained the fewest female genotypes (4). Also, 7 genotypes found in females in CR were also found in other rivers in the area, which was different from the male pattern.

We recorded 38 different alleles with an average of 4.75 ± 2.3 alleles per locus in all of the populations (table 2). The most polymorphic locus was RIO 04 with 10 alleles, and the least polymorphic were RIO 10, RIO 13, and RIO 16, each with 3 alleles. All but three loci were under HWE. The three loci that deviated from HWE showed a heterozygote deficit; none of the loci showed linkage disequilibrium. We did not observe differences in the Fis values among

the different collection sites. We used the multilocus F–statistic because it has been shown that it is not affected by sample size (Pearson coefficient $r = 0.29$, $P > 0.05$). Expected heterozygosity values ranged from a minimum 0.59 in LcR to a maximum 0.67 in TR. Observed heterozygosity ranged from 0.48 in LcR to 0.59 in CR (table 2). The results of the AMOVA analysis showed that the greatest source of genetic variation was found within rivers and not between different rivers (table 3). Results from the assignment test in GENECLASS 2 did not show genetic differences between rivers, and many of the samples were not assigned to their original sample site. The sites LgR and CR showed values equal to zero for the assignment test, while MR presented the highest assignment value (36.84). This test revealed a dispersal pattern among individuals from the different rivers, in a well–connected area, suggesting that the level of dispersion among sites is high. The FCA test showed a clear lack of genetic structure in the entire area (fig. 2).

F–statistics were not significantly different from $P < 0.05$, and two loci (RIO 07 and RIO 08) showed the largest fixation indices when compared among the six rivers (table 4). The *Fst* pairwise comparisons among rivers ranged from 0.0 to 0.018, with no genetic differentiation in any comparisons except for the LcR and JR comparison ($P < 0.05$). LcR and JR are neighbouring localities and there are no apparent barriers that can prevent dispersal, but JR is a small, seasonally dry river that is unsuitable for the neotropical river otter during the dry season. Overall, our results suggest that there are no biogeographic or ecologic barriers to dispersal because there appears to be continuous flow and *Fst* values did not differ significantly from zero between rivers.

The STRUCTURE analysis showed $\Delta K_{MAX} = 2.176$, with a high score at $K = 1$, meaning one potential genetic unit in the Lacandon area. The cluster for all individuals in the structure diagram revealed a great deal of admixture (fig. 3). The highest posterior probability was Ln Pr(X|K) = –2,107.1 for the suggested 20 runs for $K = 1$ (Pritchard et al., 2000). Our results showed similarities among individuals; these results show a high degree of admixture among the different rivers at 90% confidence intervals of the individual's posterior probabilities.

Table 3. Analysis of molecular variance (AMOVA) for all samples of *L. longicaudis*: Sv. Source of variation; Sq. Sum of squares; Vc. Variance components; Pv. Percentage variation; Ap. Among populations; Wp. Within populations.

Tabla 3. Análisis de varianza molecular (AMOVA) para todas las muestras de L. lingicaudis: *Sv. Fuente de variación; Sq. Suma de cuadrados; Vc. Componentes de la variancia; Pv. Variación de porcentages; Ap. Entre poblaciones; Wp. En una misma población.*

Sv	Sq	Vc	Pv
Ap	515.24	1.58	2.80%
Wp	10,584.34	55.13	97.20%
Total	11,099.58	56.71	

Estimates of population size using CAPWIRE yielded a total of 154 individuals in the area; we reported 34 distinct genotypes, that is, 22% of the

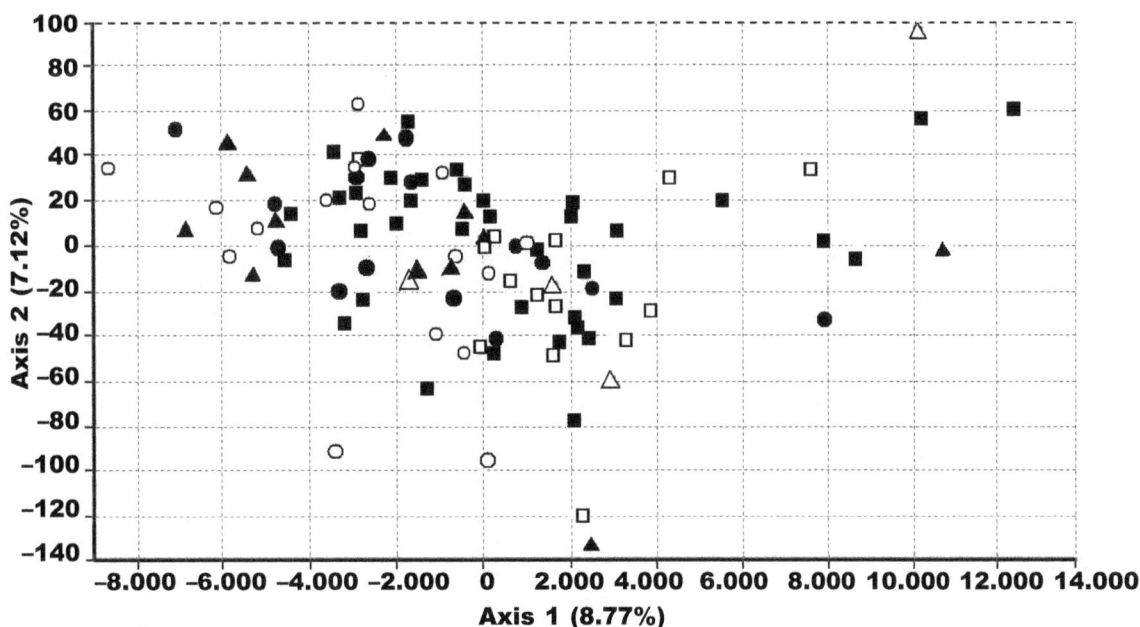

Fig. 2. Correspondence factorial analysis (CFA) for the samples of the neotropical otter in the six rivers of the area: ✱ Miranda River (MR); ▢ Lacantún River (LcR); ○ José River; ◆ Colorado River; ● Lagartos River; △ Tzendales River.

Fig. 2. Análisis de correspondencia factorial (CFA) para las muestras de la nutria neotropical en los seis ríos de la zona. Los símbolos del gráfico son: ✱ *Río Miranda (MR);* ▢ *Río Lacantún (LcR);* ○ *Río José (JR);* ◆ *Río Colorado (CR);* ● *Río Lagartos (LgR);* △ *Río Tzendales (TR).*

Table 4. Pairwise comparisons between the different rivers of the Lacandon forest. Values of F_{ST} are shown under the diagonal. In bold, the significant P–values ($P < 0.05$). (For abbreviations, see figure 1.)

Tabla 4. Comparaciones por pares entre los distintos ríos de la Selva Lacandona. Los valores de F_{ST} se encuentran por debajo de la diagonal. En negrita, los valores de P significativos (P < 0,05). (Para las abreviaturas ver figura 1.)

	MR	LcR	JR	CR	LgR	TR
MR	–	–	–	–	–	–
LcR	0.0182	–	–	–	–	–
JR	0.0172	**0.017**	–	–	–	–
CR	–0.0013	–0 .0197	0.0149	–	–	–
LgR	0.0011	0.0192	–0.0091	–0.0059	–	–
TZ	–0.0034	0.0185	–0.0182	0.0026	–0.0135	–

estimated population. The total population size of otters was subtracted from data obtained on the capture/ recapture genotypes presented in table 1. According to simulation results using CAPWIRE, conducting surveys during the two dry seasons allowed us to obtain DNA samples from 22% of the population (95% CI). With regards to the length of the sampling period needed to estimate the population size, CAPWIRE estimated a period of 45 days (CI: 27–50) which was less than our scat collecting efforts.

Discussion

Non–invasive surveys conducted in the rivers of the Lacandon rainforest revealed that otters were present in all six major rivers. We detected a total of 34 neotropical river otters over two dry seasons (19 females and 15 males) and determined that 16 of these unique genotypes were recaptured multiple times and in multiple rivers, and during both dry seasons. These results are direct evidence that individuals move freely between different rivers and suggest that there are no barriers for movement among the Lacandon River System (table 1). Furthermore, our estimates of dispersion levels between rivers suggest that they are high enough to prevent genetic differentiation. The pattern of low genetic structure is also strongly supported by the high resolution provided by microsatellite markers in these types of studies (Selkoe & Toonen 2006; Weber et al., 2009). In particular, Latch et al. (2008) found that the North American river otter (*L. canadensis*) in Louisiana is affected by a variety of landscape habitats that have structured the river otter populations and have prohibited an adequate level of gene flow, but the scale of that study is several times larger than the area surveyed in this study.

In comparison, neotropical river otters in the Lacandon rivers showed moderate levels of genetic diversity, but also showed a lack of genetic differ-

entiation among the surveyed sites. This pattern of panmixia could be explained by a recent colonization event into the area. However, given the moderate levels of allelic diversity, a recent invasion of the species seems doubtful. We did not find genetic differentiation among populations of the neotropical river otter. Our results reflected a well–interconnected network of rivers in the Lacandon forest, where individuals show high rates of dispersal. Male North American river otters tend to disperse in close populations (approx. 16 km), whereas females disperse more broadly (approx. 60 km, Blundell et al., 2002). We found that our non–invasive surveys provided direct evidence that both male and female otters move between rivers, athough females appear to move larger distances. More tracking inferences are needed, however, to confirm our results. The hydrological system in the Lacandon forest is composed of a series of permanent rivers, seasonal fishponds, channels, and marshes that provide a continuous flow of water in the area (Inda–Díaz et al., 2009). The lack of variation in *Fis* values among different localities also supports our findings of a lack of population structure. The home range for the neotropical river otter during the dry season ranges between 2 to 4 km, and can be larger during the rainy season (7 to 9 km; Gallo–Reynoso, 1989). The average distance between rivers that we surveyed was 7 km and therefore, it was not surprising that we recovered the same individuals in adjacent river systems. In well–structured populations, assignments are correctly allocated to their collecting site. Our results do not support the idea that this species is highly territorial and they are concordant with a species that has low levels of genetic structure and disperses over considerable distances (Lorenzen et al., 2008; Paetkau et al., 1995). In addition, these otters can swim a distance of 5 km every day and they may have overlapping home ranges with several individuals in the area (Gallo–Reynoso, 1989).

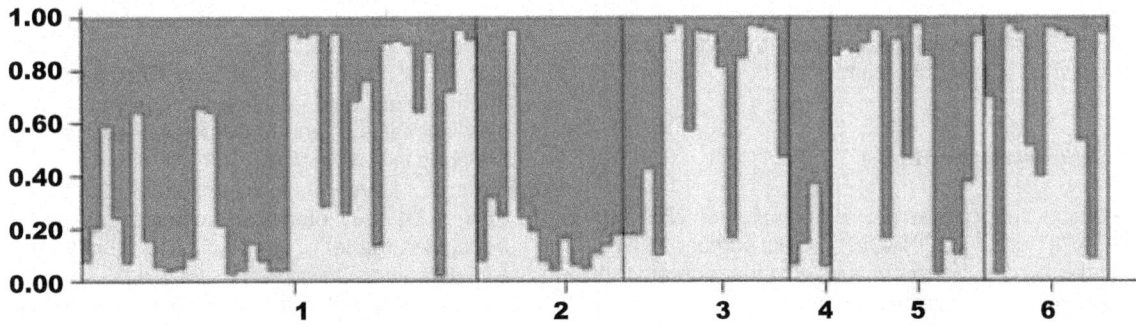

Fig. 3. Bayesian analysis of the nuclear genetic structure of river otter populations among the six sampled rivers in the Lacandon Reserve. Data are based on eight microsatellite loci with 90% probability intervals of probabilities to belong to one of each group.

Fig. 3. Análisis bayesiano de la estructura genética nuclear de las poblaciones de nutrias de los seis ríos muestreados en la Reserva de Lacandon. Los datos se basan en ocho locus microsatétile con un 90% de intérvalos de probabilidad del 90% pertenecientes a cada uno de los grupos.

Neotropical river otters are a relatively abundant species in the Lacandon forest. Because of the inhospitality of the tropical rainforest anthropogenic activities are not currently well–developed. The area has remained isolated for long periods of time, colonized only by native Lacandon tribes with low population densities. In 2000, this same area was surveyed for a dietary study based on visual observations of animals, feces and footprints. It was determined that neotropical river otters should be widely and continuously distributed along the rivers of the Lacandon forest because they are apparently connected by a complex network of tributaries without obvious subdivisions (Soler, 2004). The major geographic division in the area is found in the LcR, because it is complex and continuous and separates two major reserves: the Lacandon forest reserve in Mexico and the Peten reserve in Guatemala. For example, the distance between the CR and the TR is approximately 75 km, but both were associated with high levels of genetic diversity, and no genetic differentiation was found between them. This is consistent with current dispersion. JR is located half way between CR and TR and it showed the same levels of uniformity as the other rivers. However, JR is a seasonal river and neotropical river otters were not frequently found there. Physical barriers to the dispersal of neotropical river otters were not obvious and dispersion among the different individuals in the area appears constant.

The overall genetic diversity of the species in the area can be considered adequate to maintain a viable population, and even the smallest and most seasonal river (JR) contributed to the genetic diversity of the species. Wildlife managers should therefore focus on maintaining connectivity between the different rivers to avoid fragmentation of this large population into genetically distinct subpopulations.

To date, most studies have been conducted in the European otter, *L. lutra* (Dallas et al., 2003; Janssens et al., 2008) and the North American river otter, *L. canadensis* (Centrón et al., 2008; Latch et al., 2008), and very few studies have been conducted on populations of the neotropical river otter (*L. longicaudis*). Information concerning population dynamics of the species is scarce, and even less is known about the genetic structure of their populations. Only two studies related to the genetic structure of the neotropical river otter have been conducted previously in Brazil, but both were carried out in larger areas than our present study (Trinca et al., 2007; Weber et al., 2009). The first study found no significant mtDNA genetic differentiation between populations in the southern and south–eastern regions of Brazil and the latter study, using microsatellite markers, found significant levels of genetic differentiation in populations in the southernmost part of Brazil. This is not surprising since microsatellite markers have been widely used successfully to reveal a finer scale genetic structure in areas where mtDNA or allozymes have failed to detect it (Zhang & Hewitt, 2003). Genetic studies at different spatial scales are important because levels of variation at a large scale may help to understand how ecological and evolutionary processes act on the adaptive capacity of the species while those at a finer scale may help to elucidate patterns of kinship and social organization.

Our study shows no detectable genetic differentiation in this species among the major rivers. The geographic position of the LcR suggests that it could be managed as the main connector for this river network, and can be used as a suitable dispersal corridor for neotropical river otters in the area. As we documented that all localities have similar levels of heterozygosity, this population does not currently require translocation efforts to improve genetic diversity. However, should future efforts be required, the genetic affinities to otters from neighbouring rivers in the Peten and Calakmul reserves in Guatemala should first be investigated before they can be considered as a potential source.

Acknowledgments

Financial support was provided by project IN207607 of PAPIIT and by CONACyT (project number 90728). Lab procedures were conducted in the laboratory of Dinámica de Poblaciones y Evolución de Historias de Vida directed by Dr. Mandujano. We are also grateful to G. Rodríguez–Tapia for his technical assistance. Dr. Piñero provided valuable suggestions that greatly improved this paper. The Lombera family and Mr. Darío Meléndez helped with the collection of samples in the Lacandon rainforest. The Chapultepec Zoological Park and Aragon Zoological Park provided blood samples of neotropical river otters.

References

Beheler, A. S., Fike, J. A., Dharmarajan, D., Rhodes Jr, O. E. & Serfass, T. S., 2005. Ten new polymorphic microsatellite loci for North American river otters (*Lontra canadensis*) and their utility in related mustelids. *Molecular Ecology Notes*, 5: 602–604.

Beheler, A. S., Fike, J. A., Murfitt, L. M., Rhodes Jr, O. E. & Serfass, T. S., 2004. Development of polymorphic microsatellite loci for North American river otters (*Lontra canadensis*) and amplification in related Mustelid. *Molecular Ecology Notes*, 4: 56–58.

Belkhir, K., Borsa, P., Chikhi, L., Raufaste, N. & Bonhomme, F., 1996–2004. GENETIX 4.05, logiciel sous Windows TM pour la génétique des populations. Laboratoire Génome, Populations, Interactions, CNRS UMR 5171, Université de Montpellier II, France.

Blundell, G. M., Ben–David, M., Groves, P., Bowyer, R. T. & Geffen, E., 2002. Characteristics of sex–biased dispersal and gene flow in coastal river otters: implications for natural recolonization of extirpated populations. *Molecular Ecology*, 11: 289–303.

Centrón, D., Ramírez, B., Fasola, L., MacDonald, D. W., Chehébar, C., Schiavini, A. & Cassini, M. H., 2008. Diversity of mtDNA in southern river otter (*Lontra provocax*) from Argentinean Patagonia. *Journal of Heredity*, 99: 198–201.

Cornuet, J., Piry, S., Luikart, G., Estoup, A. & Solignac, M., 1999. New methods employing multilocus genotypes to select or exclude populations as origins of individuals. *Genetics*, 153: 1989–2000.

Dallas, J. F., Coxon, K. E., Sykes, T., Chanin, P. R. F., Marshall, F., Carss, D. N., Bacon, P. J., Piertney, S. B. & Racey, P. A., 2003. Similar estimates of population genetic composition and sex ratio derived from carcasses and feces of Eurasian otter *Lutra lutra. Molecular Ecology*, 12: 275–282.

Dallas, J. F., Marshall, F., Piertney, S. B,. Bacon, P. J. & Racey, P. A., 2002. Spatially restricted gene flow and reduced microsatellite polymorphism in the Eurasian otter *Lutra lutra* in Britain. *Conservation Genetics*, 3: 15–29.

Eggert, L. S., Maldonado, J. E. & Fleischer, R. C., 2005. Nucleic acid isolation from ecological samples–animal scat and other associated materials. *Methods in Enzymology*, 395: 73–87.

Excoffier, L., Laval, G. & Schneider, S., 2005. Arlequin v. 3.1.: An integrated software package for population genetic data analysis. *Evolutionary Bioinformatics Online*, 1: 47–50.

Gallo–Reynoso, J. P., 1989. Distribución y estado actual de la nutria o perro de agua (*Lutra longicaudis annectens* Major, 1897) en la Sierra Madre del Sur, México. MSc Thesis, Facultad de Ciencias, Universidad Nacional Autónoma de México, México.

Goudet, J., 1995. FSTAT (Version 12): a computer program to calculate F–statistics. *Journal of Heredity*, 86: 485–486.

Hájková, P., Zemanová, B., Bryja, J., Hájek, B., Roche, K., Tkadlec, E. & Zima, J., 2006. Factors affecting success of PCR amplifications of microsatellite loci for otter faeces. *Molecular Ecology Notes*, 6: 559–562.

Inda–Díaz, E., Rodiles–Hernández, R. Naranjo, E. J. & Mendoza–Carranza, M., 2009. Subsistence fishing in two communities of the Lacandon Forest, Mexico. *Fisheries Management and Ecology*, 16: 225–234.

Janssens, X., Fontaine, M. C. Michaux, J. R. Libois, R., Kermabon, J. de, Defourny, P. & Baret, P. V., 2008. Genetic pattern of the recent recovery of the European otters in southern France. *Ecography*, 31: 176–186.

Lampa, S., Gruber, B., Henle, K. & Hoehn, M., 2008. An optimisation approach to increase DNA amplification success of otter faeces. *Conservation Genetics*, 9: 201–210.

Larivière, S., 1999. *Lontra longicaudis. Mammalian Species*, 609: 1–5.

Latch, E. K., Scognamillo, D. G., Fike, J. A., Chamberlain, M. J. & Rhodes Jr, O. E., 2008. Deciphering ecological barriers to North American river otter (*Lontra canadensis*) gene flow in the Lousiana landscape. *Journal of Heredity*, 99: 265–274.

Levene, H., 1949. On a matching problem arising in genetics. *The Annals of Mathematical Statistics*, 20: 91–94.

Lewis, P. & Zaykin, D., 2001. Genetic Data Analysis: computer program for the analysis of allelic data. Version 1.0 (d16c). Free program distributed by the authors over the internet from http://lewis.eeb.uconn.edu/lewishome/software.html

Lorenzen, E. D., Arctander, P. & Siegismund, H. R., 2008. High variation and very low differentiation in the wide ranging plains zebra (*Equus quagga*): insights from mtDNA and microsatellites. *Molecular Ecology*, 17: 2812–2824.

Medina–Vogel G., & González–Lagos C., 2008. Habitat use and diet of endangered southern river otter *Lontra provocax* in a predominantly palustrine wetland in Chile. *Wildlife Biology*, 14: 211–220.

Miller, C. R., Joyce, P. & Waits, L. P., 2005. A new

method for estimating the size of small populations from genetic mark–recapture data. *Molecular Ecology*, 14: 1991–2005.

Mills, L. S., Pilgrim, K. L., Schwartz, M. K. & McKelvey, K., 2000. Identifying lynx and other North American felids based on mtDNA. *Conservation Genetics*, 1: 285–288.

Norma Oficial Mexicana (NOM–059–ECOL–1994). Norma oficial mexicana que determina las especies y subespecies de flora y fauna silvestres terrestres y acuáticas, en peligro de extinción, amenazadas, raras y las sujetas a protección especial, y que establece especificaciones para su protección. Diario Oficial de la Federación, México, Distrito Federal.

Ortega, J., Franco, M. R., Adams, B. A., Ralls, K. & Maldonado, J. E., 2004. A reliable, noninvasive method for sex determination in the endangered San Joaquin kit fox (*Vulpex macrotis mutica*) and other canids. *Conservation Genetics*, 5: 715–718.

Paetkau, D., Calvert, W., Stirling, I. & Strobeck, C., 1995. Microsatellite analysis of population structure in Canadian polar bears. *Molecular Ecology*, 4: 347–354.

Piry, S., Alapetite, A., Cornuet, J., Paetkau, D., Baudouin, L.& Estopup, A., 2004. GeneClass 2: A software for Genetic Assignment and First–Generation Migrant Detection. *Journal of Heredity*, 95: 536–539.

Pritchard, J. K., Stephens, M. & Donnelly, P., 2000. Inference of population structure using multilocus genotype data. *Genetics*, 155: 945–959.

Reid, D. G., Code, T. E., Reid, A. C. H. & Herrero, S. M., 1994. Spacing, movements, and habitat selection of the river otter in boreal Alberta. *Canadian Journal Zoology*, 72: 1314–1324.

Rousset, F., 2008. Genepop´007: a complete re–implementation of the genepop software for Windows and Linux. *Molecular Ecology Resources*, 8: 103–106.

Selkoe, K. A., & Toonen, R.J., 2006. Microsatellites for ecologists: a practical guide to using and evaluating microsatellite markers. *Ecology Letters*, 9: 615–629.

Serfass, T. L., Brooks, R. P., Novak, J. M., Johns, P. E. & Rhodes, O. E., 1998. Genetic variation among populations of river otters in North America: considerations for reintroduction projects. *Journal of Mammalogy*, 79: 736–746.

Soler, A., 2004. Cambios en la abundancia relativa y dieta de *Lontra longicaudis* en relación a la perturbación de la Selva Lacandona, Chiapas, México. Bachelor thesis, Facultad de Ciencias, Universidad Nacional Autónoma de México, México.

Taberlet, S., Griffin, S., Goossens, B., Questiau, S., Manceau, V., Escaravage, N., Waits, L. P. & Bouvet, J., 1996. Reliable genotyping of samples with very low DNA quantities using PCR. *Nucleic Acids Research*, 24: 3186–3194.Trinca, C. S., Waldemarin, H. F. & Eizirik, E., 2007. Genetic diversity of the Neotropical otter (*Lontra longicaudis* Olfers, 1818) in Southern and Southeastern Brazil. *Brazilean Journal of Biology*, 67: 813–818.

Valière N., Berthier, P., Mouchiroud, D. & Pontier, D., 2002. GEMINI: a software for testing the effects of genotyping errors and multi–tubes approach for individual identification. *Molecular Ecology Notes*, 2: 83–86.

Waits, L. P., Luikart, G. & Taberlet, P., 2001. Estimating the probability of identity among genotypes in natural populations: cautions and guidelines. *Molecular Ecology*, 19: 249–259.

Weber, L. I., Hildebrand, C. G., Ferreira, A., Pedarassi, G., Levy, J. A. & Colares, E. P., 2009. Microsatellite genotyping from faeces of *Lontra longicaudis* from southern Brazil. *Iheringia Série Zoologia Porto Alegre*, 99: 5–11.

Weir, B. & Cokerham, C., 1984. Estimating F–statistics for the analysis of population structure. *Evolution*, 38: 1358–1370.

Yeh, C. & Boyle, B., 1997. Population genetic analysis of co–dominant and dominant markers and quantitative traits. *Belgium Journal of Botany*, 129: 157.

Zhang, D. X. & Hewitt, G. M., 2003. Nuclear DNA analyses in genetic studies of populations: practice, problems and prospects. *Molecular Ecology*, 12: 563–584.

5

Effect of supplementary food on age ratios of European turtle doves (*Streptopelia turtur* L.)

G. Rocha & P. Quillfeldt

Rocha, G. & Quillfeldt, P., 2015. Effect of supplementary food on age ratios of European turtle doves (*Streptopelia turtur* L.). *Animal Biodiversity and Conservation*, 38.1: 11–21.

Abstract

*Effect of supplementary food on age ratios of European turtle doves (*Streptopelia turtur *L.).*— Many farmland birds have difficulties finding sufficient food in intensely managed agricultural ecosystems, and in more extensively worked landscapes they are often attracted to human–induced dietary sources. European turtle doves *Streptopelia turtur* feed on seeds collected on the ground, and are readily attracted to supplementary provided grain at feeding stations. Supplementary feeding is a common management practice on hunting estates around the world. This study was conducted in 40 hunting estates located in central west Spain: 20 sites where supplementary food was provided to attract turtle doves and 20 control sites without feeding stations. At sites with supplemental feeding, the field age ratio was 20% higher and the hunted age ratio was 33% higher than at control sites, indicating a positive effect of the food supplementation of the breeding success around supplemented sites. Both the amount of food provided per day and the amount of time where supplemental food was given (20–120 days) were positively correlated with the field age ratio and, less strongly, with the hunted age ratio. These data suggest that providing extra food can increase the breeding success of this species when the amount provided is sufficiently large and when supplementary food is provided early in the breeding season. However, hunting pressure was also higher at supplemented sites. Future studies should therefore closely monitor the positive and negative effects in order to ascertain which management practices will ensure the viability of these important European turtle dove populations.

Key words: Age–ratio, Hunting, Management, *Streptopelia turtur*, Food supplementation

Resumen

*Efecto de la alimentación complementaria en la razón de edad de la tórtola europea (*Streptopelia turtur *L.).*— Son numerosas las aves de los hábitats agrícolas que tienen dificultades para encontrar suficiente alimento en los ecosistemas agrícolas intensivos y que, en los hábitats explotados de forma más extensiva, suelen ser atraídas por las fuentes antrópicas de alimento. La tórtola europea, *Streptopelia turtur*, se alimenta de semillas que se hallan en el suelo y es atraída inmediatamente por los cereales que se aportan como complemento a los comederos. El aporte complementario de alimento es una práctica habitual en la gestión de los cotos de caza de todo el mundo. Este estudio se realizó en 40 cotos de caza ubicados en el centro y el oeste de España: 20 zonas en las que se aportó alimentación complementaria para atraer a las tórtolas y 20 zonas de control sin comederos. En las zonas con alimentación complementaria, las razones de edad en el campo y en las aves cazadas fueron, respectivamente, un 20% y un 33% más elevadas que en las zonas de control, lo que indica que la alimentación complementaria tiene un efecto positivo en el éxito reproductivo en torno a las zonas con aporte de alimento complementario. Tanto la cantidad de alimento suministrado por día como el período en el que se aportó (20-120 días) se correlacionaron positivamente con la razón de edad en el campo y, con menos intensidad, con la razón de edad en las aves cazadas. Estos datos sugieren que el suministro de alimento extra puede aumentar el éxito reproductivo de esta especie si la cantidad aportada es suficientemente abundante y si se empieza a proporcionar a principios de la temporada de cría. No obstante, la presión cinegética también fue mayor en las zonas con aporte de alimento complementario, por lo que sería necesario analizar minuciosamente los efectos positivos y negativos de dicho aporte con vistas a determinar qué prácticas de gestión garantizarán la viabilidad de estas importantes poblaciones de tórtola europea.

Palabras clave: Razón de edad, Caza, Gestión, *Streptopelia turtur*, Alimento complementario

Gregorio Rocha, Dept. of Agro–forestry Engineering, Univ. of Extremadura, Avda. Virgen del Puerto 2, 10600 Plasencia, Cáceres, Spain.– Petra Quillfeldt, Inst. für Tierökologie und Spezielle Zoologie, Justus Liebig Univ. Gießen, Heinrich–Buff–Ring 38, 35392 Gießen (Germany).

Corresponding author: Gregorio Rocha. E–mail: gregorio@unex.es

Introduction

The availability of food is a principal parameter determining breeding success in animals. The lack of an adequate food supply often leads to birds abandoning their brood, brood reduction, and poorer condition of offspring (Martin, 1987). Food supplementation therefore has consistent effects on parameters of breeding success in birds, such as earlier laying, larger clutch size, and accelerated population growth (reviews by Boutin, 1990; Schoech & Hahn, 2007). The use of supplemental feeding as a positive tool in species conservation has been consistently recommended (Schoech et al., 2008), although some unwanted effects may appear (Martínez–Abrain & Oro, 2013). In the case of some gallinaceous game species in Europe (pheasant *Phasianus colchicus* and quail *Coturnix coturnix*), supplementary feeding at hunting estates can influence the proportion of juveniles to adults, through an increase in reproductive success (Draycott et al., 2005; Díaz–Fernández et al., 2013). However, this practice may be negative if the hunting bag is not thoroughly controlled, because it can increase the hunting pressure excessively (Rocha & Hidalgo, 2001).

Although the distribution area of European turtle doves *Streptopelia turtur* is wide, they are restricted to warm, lowland areas, which are often agricultural areas (Cramp, 1985). In recent decades, the European turtle dove has experienced a widespread decrease both in population density and in its area of distribution (Tucker & Heath, 1994; Jarry, 1997; Browne et al., 2005). This decline has led to its inclusion as Vulnerable in the Red Book of Vertebrates of Spain (Blanco & González, 1992; Madroño et al., 2004). As a result of this, and being a hunted species in an unfavorable demographic situation, this species is the subject of a management plan by the European Commission (Boutin, 2001; Lutz & Jensen, 2007). On the Iberian peninsula, the breeding population has declined significantly, by 29.3% between 1998 and 2012 (SEO/BirdLife, 2013), and this decline has been particularly marked since 2008.

European turtle doves feed primarily on the seeds of weeds (Murton et al., 1964; Dias & Fontoura, 1996), especially at the start of the breeding season when seeds of cultivated plants are not yet available (Jiménez et al., 1992). Some studies suggested that late in the breeding season cereal seeds become available and then play a larger role (Jimenez et al., 1992; Browne & Aebischer, 2003).

On the Iberian peninsula, the European turtle dove is hunted from the second half of August to the first half of September, *i.e.* the late breeding season and the post–breeding migration. The hunting of this species is carried out using the 'fixed location' method, which takes advantage of birds passing to feeding areas such as crops and natural pastures, where they are shot at by a row of hunters. The European turtle dove plays an important role in Extremadura, where it is considered one of the main game bird species of this hunting season (Hidalgo & Rocha, 2001). However, there are as yet no studies on the economic value of this activity.

Many hunters and estate managers use supplementary food to attract and concentrate the birds (Rocha & Hidalgo, 2001). Such supplementation consists of seed mixes of various oleaginous or leguminous cereals scattered throughout the crops or natural pastures. The feeding stations usually occupy an area of between 0.2 and 5.0 hectares, although they can be larger. Currently, over 70% of the estates that hunt European turtle doves during August–September in Extremadura operate this kind of hunting management (Rocha, own data).

Food supplementation on the hunting estates can start 1 to 4 months before the hunting season (Rocha, own data). At sites with early supplementation, this covers most of the breeding season of the European turtle dove. Thus, food supplementation could influence the population dynamics of this species during the breeding season, including changes in abundance, breeding success, feeding ecology and migratory phenology.

It is known that a greater amount of food available on the estates attracts the European turtle dove since they are killed in greater numbers when extra food is provided (Rocha & Hidalgo, 2002). However, it is unknown to what extent supplemental food affects the productivity of the populations.

The main objectives of this work were to summarize data on quantity and duration of the food supplementation at hunting estates and to study how this game management practice influences reproductive success. Specifically, we tested whether the breeding success of the European turtle doves was influenced by the supplemental feeding at hunting estates by comparing the age ratio of populations of post–breeding aggregations in the second half of August. Observations of age ratios are a widely used method to estimate breeding success in birds (*e.g.,* Wagner & Stokes, 1968; Newton, 2001; Flanders–Wanner et al., 2004; Peery et al., 2007). Although observations of age ratios in the field do not provide a direct measure of productivity, they have proven to be a useful technique because they are relatively easy to apply, yet they avoid time–consuming and potentially harmful nest searches.

Methods

Study species

The European turtle dove is a migratory species that winters in the African Sahel and breeds in large parts of Europe, Asia and North Africa (Cramp, 1985).

In the Iberian peninsula, its main habitats are the areas populated by holm oaks (*Quercus ilex* L.) and cereal cultivation, where they present densities of some 2.3 birds/10 ha (Muñoz–Cobo, 2001). In the central and western part of the Iberian peninsula they nest among small to medium sized oaks, in a mix of natural pasture and cultivated habitats (Peiró, 1990; Rocha & Hidalgo, 2002). In these areas, they can reach densities of up to 10.5/10 ha (Santamaría, 2007). They also inhabit open areas with scattered trees and shrubs, riverine forests and orchards, and they are very scarce in coniferous woodlands, scrubland and all across the thermo–Mediterranean (Díaz et al.,

1996). The study area has traditionally been one of the principal nesting territories of the European turtle dove in Spain (Rocha & Hidalgo, 2002).

Study sites

The study area is located in the region of Extremadura, in the central–western Iberian peninsula (fig. 1). This region has a rich biodiversity. More than 30% of the territory is included in the *Natura 2000* ecological network (Fernández, 2004; Junta de Extremadura, 2012) and 80% of the territory is subject to hunting rights (Lázaro, 2004).

The predominant habitat on the hunting estates is the *dehesa*: open managed parkland used for livestock grazing within a savanna–like woodland of evergreen *Quercus* trees, mainly *Q. ilex* (holm oak) and *Q. suber* (cork oak). The *dehesa* is intermixed with cropland and Mediterranean woodland and scrub (Díaz et al., 1997).

The use of feeding stations for hunting had been banned in Extremadura since 2007 (included) (Junta de Extremadura, 1991, 2007, 2008), but not in the surrounding regions. Despite being banned, 62% of estate managers where hunting takes place added supplementary food during the season 2004/2005 (Hidalgo & Rocha, 2006). This hunting management is frequently used because it increases the number of birds hunted (Hidalgo & Rocha, 2001).

During the spring and summer of 2009, 40 hunting estates with *dehesa* habitats were selected throughout the region where the European turtle dove has been hunted traditionally. These estates were split in two groups according to the provision or not of supplemental food. In the first group, hunting management involved consistently adding supplementary food year after year, dispersed throughout the crops and natural pastures (group with extra food added). Supplemental food was composed of mixtures of crop seeds: wheat (*Triticum aestivum* L.), sunflower (*Helianthus annuus* L.) and vetch (*Vicia sativa* L.). The estates with feeding stations always provided the same amount of food every year, and for the same duration of time. In the control group, no supplementary food was added to the environment and the birds fed on seeds from crops and natural pastureland (group without extra food added).

The size of the estates did not differ between supplemented and control sites (mean ± sd: 661.5 ± 132.9 ha; t–test: $t_{38} = -0.84$, $P = 0.41$). The mean number of hunters per estate (± sd) was 11.8 ± 2.6, and did not differ between supplemented and control sites (t–test: $t_{38} = -0.71$, $P = 0.48$). There are no data on movements of turtle doves in Spain, but in two English populations, the mean foraging distances for radio–tagged doves were 450 m and 1,400 m (Browne & Aebischer, 2003). The sites in our study area were over 10 km distant from each other (fig. 1). We thus assume that the population recorded at each site was mainly local.

Data collection

The following data were collected at supplemented sites: the amount of food added (in kilos) and the duration (in days) of the food supplementation until the beginning of the hunt. These data were freely provided by estate staff in charge of the food supplementation. Age ratios were estimated as an indicator of the breeding success by counting the number of young birds observed after the breeding season (in the second half of August), compared to the total number of adults (young/adults), just before the beginning of the migratory season (Cramp, 1985).

Post–breeding age ratios are a common tool to estimate the reproductive productivity in monitoring programmes. Age ratios can be obtained easily and economically, while avoiding biases due to disturbance at the nest in sensitive species such as turtle doves. In the present study, age ratios were measured using two different methods based on the proportion of young to adult birds among live birds observed in the field (termed 'field age ratio') and birds killed during the hunt (termed 'hunted age ratio'), respectively.

The field age ratio was obtained by observation, from a hidden fixed position, of live birds perched, where the difference between young and adult birds could be directly ascertained as they gathered in post–reproduction groups or 'aggregates'. Field age ratios have been successfully used in work on this species in Andalucía (SEO/BirdLife, 2002). Field age ratios are important parameters when trapping methods influence the age ratio sample of the population (*e.g.,* Domènech & Senar, 1997). Hunted age ratios of turtle doves were determined in Morocco, where they were found to depend largely on reproductive productivity and hunting pressure (Hanane, 2009). To determine the 'field age ratio', we conducted observations once in each estate, from between 2 and 10 days before the hunt started. Post–reproduction groups were observed and counted in all 20 supplemented sites. However, in the control group, data were obtained from 11 of the 20 sites because post–reproduction groups were not located in the remaining nine estates. On the 11 estates, post–reproduction groups were found at feeding sites such as harvested fields, where food can be found but is not as concentrated as at supplemented sites. On observation days, one person per estate occupied an observation point (hidden) before the birds began to arrive (at dawn). This point was located near the feeding area, so that the birds were seen at a distance of 20–40 m. The observer remained there for 30 min recording the birds that came with a telescope 20–40 x, differentiating young and adult birds by identifying the presence or absence of the 'collar'. The 'collar' is a distinctive feature of the neck plumage of adult European turtle doves. It is composed of feathers that form black and white bands (Cramp 1985; Sáenz de Buruaga et al., 2001). Young European turtle doves usually have no collar, although some juveniles birds hatched from early broods may have a partial collar. Thus, birds with partial collars were always considered juveniles. Double counting was avoided by using the data of the birds recorded in the peak of simultaneous concentration. A total of 982 birds were observed (773 at supplemented sites and 209 at control sites).

The 'hunted age ratio' was determined at supplemented and control sites during the first hunting weekend in 2009 (22–23 August), in order to ensure that all of the hunted individuals belonged to the local

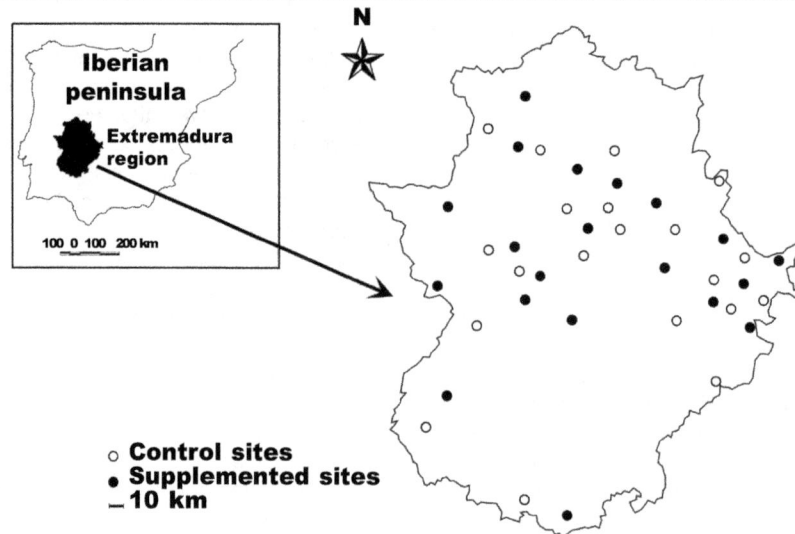

Fig. 1. Location of the region of Extremadura on the Iberian peninsula and distribution of the estates.

Fig. 1. Localización de la región de Extremadura en la península ibérica y distribución de los cotos.

breeding population and not to passing migratory populations which appear from the north from September onwards and mix with the local populations (Fernández & Camacho, 1989; Rocha & Hidalgo, 2002).

Each estate was visited on the day of the hunt in order to count the young and adult birds shot, as well as the number of hunters involved. The ratio of young to adult birds, obtained at the end of the hunt, is a simple parameter used previously on the Iberian peninsula (Gutiérrez, 2001). In total, 4,132 killed birds were observed (3,154 in the group with extra food and 987 in the group without extra food).

Data analysis

Data analyses were carried out in the R environment (R Development Core Team, 2013). Normality was tested using Kologorov–Smirnoff tests and Q–Q plots. Means were compared using unpaired Student's t–tests. General linear models were used to analyze the influence of the quantity of supplementary food and the time it was available on the field on hunted age ratios. This relationship was visualised with contour plots produced with the function vis.gam from the R package mgcv. A general lineal model was used to compare the correlations between field and hunted age ratios between the supplemented and control sites, by defining hunted age ratio as dependent, supplement (yes/no) as factorial predictor and field age ratio as covariate. The models were initially carried out with two–way interaction terms, but these were removed as they did not reach statistical significance. We used Mann Whitney U–tests to compare the number of turtle doves killed per hunter and day on the estates with extra food and those without extra food. All tests were two–tailed.

Results

Differences between supplemented sites and control sites

The mean field age ratio (fig. 2) based on the observation of post–breeding aggregates was significantly higher at supplemented sites (mean ± sd: 1.61 ± 0.19) than in control sites (1.43 ± 0.27; t–test: t_{29} = –2.13, P = 0.04). The mean hunted age ratio at supplemented sites (1.84 ± 0.22) was also significantly higher (t–test: t_{38}= –6.83, P < 0.001) than at control sites (1.38 ± 0.19).

The hunted age ratio was correlated with the field age ratio, but the regression functions differed between supplemented sites and controls (fig. 3; linear model: $F_{1,28}$= 21.8, P < 0.001; effect of field age ratio: $F_{1,28}$ = 7.9, P = 0.009).

At control sites, the hunted age ratio did not vary significantly from that obtained from the killed bird count (paired t–test: t_{10} = 0.8, P = 0.422). At supplemented sites, in contrast, the field age ratio was lower than the hunted age ratio (1.61 ± 0.19 *vs.* 1.84 ± 0.22; paired t–test: t_{19} = –3.52, P < 0.001).

The total number of birds killed per hunter varied significantly between groups (Mann–Whitney U–test: Z = –4.17, P < 0.001). At supplemented sites, an average of 12.74 ± 9.52 European turtle doves per hunter per day were killed (n = 3,154), while the average was 3.91 ± 4.54 at control sites (n = 978).

Age ratio variation within supplemented sites

The quantity of supplementary food added annually ranged from a minimum of 300 kg to a maximum of

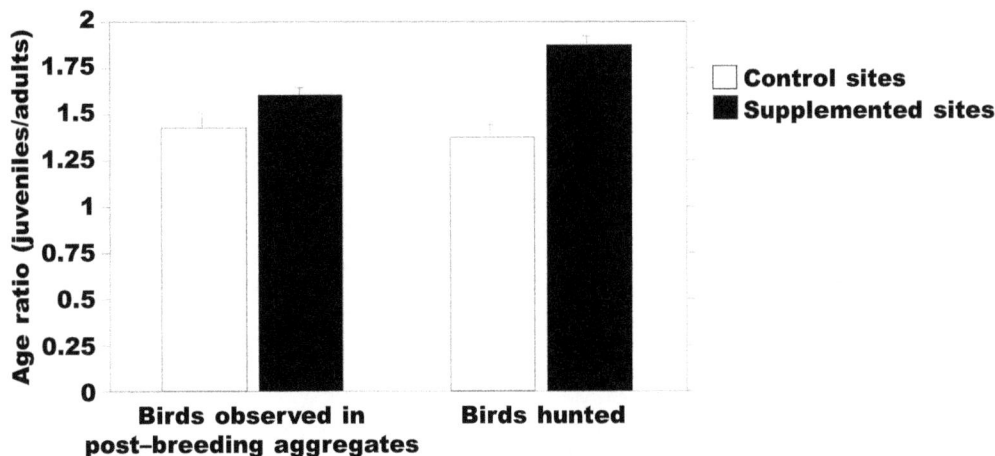

Fig. 2. Field age ratio (observed in post–reproduction aggregates) and hunted age ratio in estates with food supplementation and control sites in Extremadura, Spain, in late August 2009 (mean and standard error).

Fig. 2. Razón de edad obtenida en el campo (observada en agregados posreproductores) y razón de edad en las aves cazadas, en cotos con aporte de alimento complementario y en las zonas de control en Extremadura, en España, a finales de agosto de 2009 (media y error estándar).

1,300 kg (mean ± sd: 785 ± 275 kg). The contribution of food, from when it was added until the beginning of the hunt, had a range of 100 days, with a minimum of 20 to a maximum of 120 days (72 ± 30 days). The amount of food supplied was greater when food was supplied over a longer time ($R^2 = 0.567$, $P < 0.001$). Therefore, daily supplement rates were calculated and included as predictor variables in GLMs.

The daily amount of food provided and the duration of the supplementation correlated positively with the field and hunted age ratios (table 1, fig. 4). Together, the amount and duration of the supplementation explained 47 and 31% of the variation in field and hunted age ratios, respectively (fig. 4). The correlation coefficient was higher for field age ratios ($R^2 = 0.408$) than for hunted age ratios ($R^2 = 0.227$).

Discussion

The present data suggested a positive influence of food supplementation on the number of juveniles present at the end of the breeding season, suggesting a higher breeding success. A higher percentage of juveniles hunted, compared to field observations close–by, suggests that juveniles behave less cautiously at feeding sites or have a slower escape response than the more experienced adult birds.

Effects of food supplementation on post–breeding age ratios

The effect of artificial feeding on the turtle dove age ratio had previously been investigated over a longer period including postnuptial migration (Rocha & Hidal-

go, 2001). In the present study, in contrast, age ratio was analyzed before the onset of migration.

The hunted and field age ratios at supplemented sites in the present study were higher than in Andalusia, where they ranged from 1.15 to 1.25 (Gutiérrez, 2001). This could be explained by the positive effect of the increased availability of food, availability being a limiting factor in the dry ecosystems of these latitudes from May to September, *i.e.* during the breeding season. In this respect, Rocha & Hidalgo (2002) found that European turtle doves largely depend on the *dehesa* zones of cereal cultivation during the breeding season. This type of habitat is highly suitable because it provides abundant food and quiet and protected nest sites for the birds (Santamaria, 2007). In recent decades, the acreage used in the region for cereal cultivation has decreased notably (*e.g.* 1,500,000 ha since the 1980s across Spain; Olona, 2014), while cereal production has increased due to the agricultural intensification put in place by the Common Agricultural Policy of the European Union (Alés, 1996; Naredo, 1996; Robson, 1997). The loss has especially affected less intensively managed and marginal areas, which were a very suitable habitat for European turtle doves. Therefore, the increased age ratio of the European turtle dove in areas where supplementary food was added could be considered a response to the lack of naturally available food due to the scarcity of crops.

Several studies highlight the susceptibility of this species to agricultural changes (increased intensity, changes in crops, pesticide use, etc.), both in breeding areas and wintering quarters (Browne & Aebischer, 2004; Browne et al., 2005; Wilson & Cresswell, 2006; Eraud et al., 2009). Supplementary feeding could be used as a management tool to contribute to mitiga-

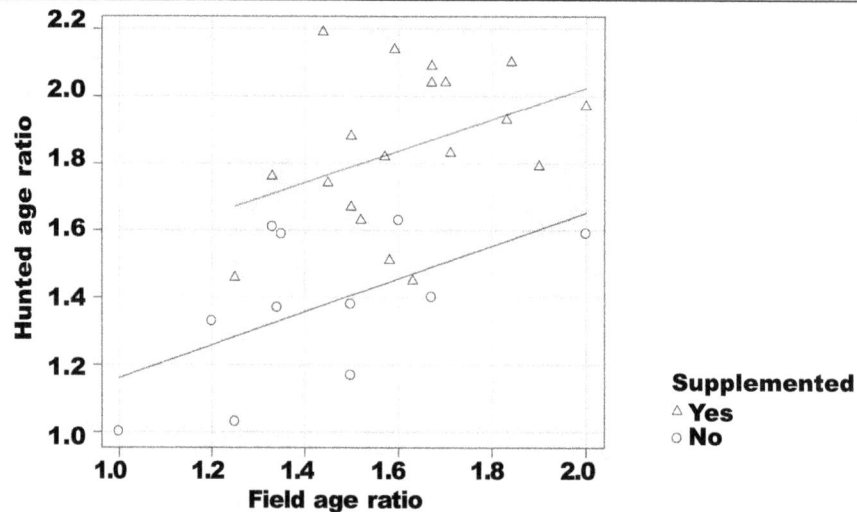

Fig. 3. Relationship between field age ratio and hunted age ratio of European turtle doves in 20 estates with food supplementation and 11 control sites in Extremadura, Spain, in late August 2009. No field age ratio could be established for the remaining nine control sites because no post–breeding aggregations were located at these sites in the days preceding the late–August hunt.

Fig. 3. Relación entre la razón de edad en el campo y en las aves cazadas de tórtola europea en 20 cotos con aporte de alimento complementario y 11 zonas de control en Extremadura, en España, a finales de agosto de 2009. No se pudieron obtener las razones de edad en el campo para las nueve zonas de control restantes porque no se localizaron agregados postreprodutores en estos sitios en los días precedentes a la caza de finales de agosto.

ting the decline of biodiversity produced by recent changes in European agricultural and livestock uses (Potts, 1997; Krebs et al., 1999; Donald et al., 2000). Likewise, planting cereal crops has been proposed as a management tool in addition to supplementary food supply, both in spring and in summer, in order to guarantee sufficient food during the breeding season and to increase productivity of the species (Rocha, 2007; Gutiérrez–Bermejo, 2009; Rocha et al., 2009). The survival rate during the first year of life of this species is very low, around 36% (Calladine et al., 1997); therefore, such measures would be effective provided they are not over–compensated by too high a hunting pressure. Mortality from hunting on breeding populations should not exceed the breeding capacity of the species. This would ensure the sustainability of the population in spite of hunting, since it would allow the annual return of birds to their breeding quarters because of the possible breeding philopatry of this species (Cramp, 1985).

Field *vs.* hunted age ratio

Regarding the methodology used to assess age ratios, the field and hunted age ratios did not differ significantly at control plots, while at supplemented sites, a greater proportion of young birds were counted when using data from the hunt as opposed to data from direct observation. Thus, supplementation resulted

in about 20% higher field age ratios, but up to 33% higher hunted age ratios. One plausible explanation for these differences may lie in the poorer escape response and lack of experience of the juveniles,

Table 1. Influence of food supplementation on the field and hunted age ratio, assessed using generalized linear models.

Tabla 1. Evaluación de la influencia del aporte de alimento complementario en la razón de edad en el campo y en las aves cazadas, utilizando modelos lineales generalizados.

GLM predictor	$F_{1,19}$	t	P
Dependent: field age ratio			
Days supplemented	15.1	5.0	0.001
Supplement/day (kg)	6.4	2.5	0.022
Dependent: hunted age ratio			
Days supplemented	4.7	2.2	0.043
Supplement/day (kg)	7.2	2.7	0.016

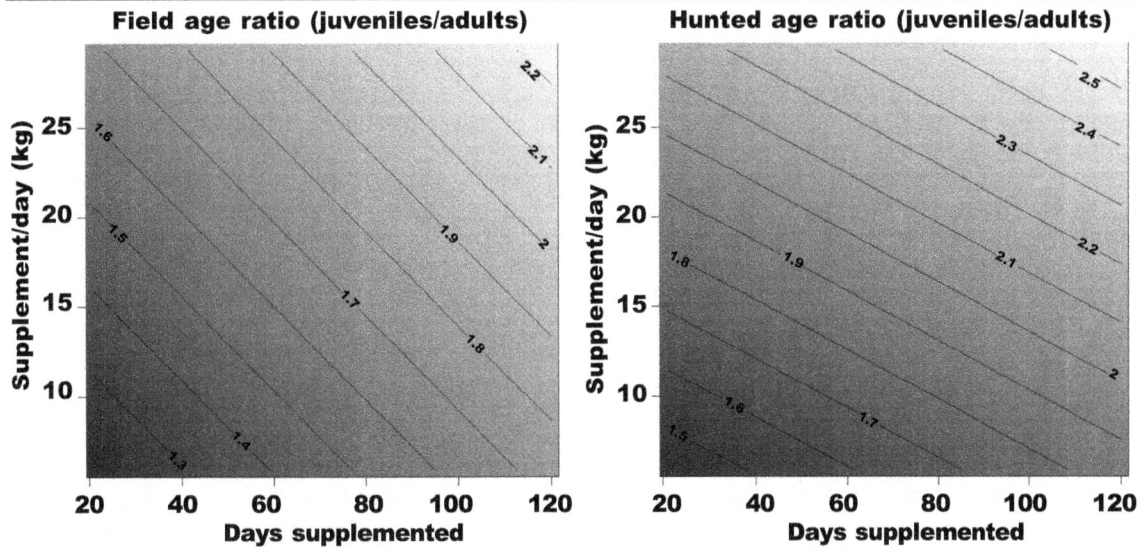

Fig. 4. Contour plot of fitted values for the age ratios.

Fig. 4. Gráfico de los valores ajustados para las razones de edad.

which are not so skilled in flight and encounter the shots of the hunters for the first time. Thus, juveniles would be killed more easily than the adults, artificially raising the ratio of young to adult birds. However, in areas with supplemental feeding, an increased hunting pressure could also partially explain the results here presented. Since the hunted age ratio was consistently larger than the field age ratio (a more reliable but more time–consuming method), an adjustment factor could be established to estimate breeding success from hunting bags. According to the present data of supplemented sites, hunted age ratios should be adjusted by a factor of 0.87 to obtain values similar to field age ratios. This correction factor would be useful for future studies.

A previous study in Extremadura reported a higher proportion of young to adult birds killed on estates with supplementary food than in control sites during the migratory period (Rocha & Hidalgo, 2001). The larger proportion of young birds shot at supplemented sites might have a negative effect on the renewal of the population, leading to an ageing population and therefore the disappearance of breeding populations in the medium to long term (Rocha & Hidalgo, 2001).

At 11 control sites, a total of 121 juveniles and 81 adults were observed (a mean of 11.0 juveniles and 8.0 adults per estate). In comparison, a total of 477 juveniles and 296 adults were observed at 20 supplemented sites (a mean of 23.9 juveniles and 14.8 adults per estate). These observation numbers are 2.2 times higher on supplemented sites for juveniles and 1.9 times higher for adults, suggesting a positive effect on breeding turtle dove populations.

However, this positive effect would be counteracted if the hunting pressure at the supplemented estates was even higher in relative terms. At the 11 control sites with observation data, a total of 410 juveniles and 281 adults were hunted (a mean of 37.3 juveniles and 25.5 adults per estate). In comparison, a total of 2021 juveniles and 1,133 adults were observed at 20 supplemented sites (a mean of 101.1 juveniles and 56.7 adults per estate). These numbers for the hunting pressure were 2.7 times higher on supplemented sites for juveniles and 2.2 times higher for adults. Based on these numbers, an increase of 2.2 (positive effect of supplementary feeding) would be counteracted by a decrease of 2.7 (negative effect of increased hunting pressure) for juveniles, thus suggesting a stronger negative effect on breeding turtle dove populations than the gain by supplemental feeding. A similar reasoning applies to adults, where an increase of 1.9 (positive effect of supplementary feeding) would be counteracted by a decrease of 2.2 (negative effect of increased hunting pressure). It has been mentioned in previous studies that increased pressure from hunting could cause serious problems for the species (Lucio & Purroy, 1992; Purroy, 1995, 1997; Rocha, 2007; Gutiérrez–Bermejo, 2009). The present data support this point of view. The average number of birds killed per hunter per day was 3 times higher on supplemented sites than on control sites. Similar figures have been recorded previously in Extremadura, where the average in other years has reached up to 4 times more (Rocha & Hidalgo, 2001). However, our methods (a single observation period per estate) were not ideal, and the relative population numbers given here are therefore tentative and should be monitored more closely to ascertain which of the two opposing effects exerts a greater influence.

Additional factors may also need to be taken into account. For example, the species is also hunted in non–breeding zones, such as the open fields of cereals and sunflowers in the southern half of Extremadura (and other areas of Iberian peninsula), where the European turtle dove is only hunted in migratory passage (Puerta, 2011). A negative effect through overhunting of young birds is expected on the migratory populations from western and central Europe (Rocha & Hidalgo, 2001).

Effects of the annual amount of food added and the duration of its availability

On the estates where a greater amount of extra food was added, a higher age ratio was observed, and this relationship was stronger for the field age ratio than for the hunted age ratio (table 1).

The field age ratio was especially strongly explained by the duration of the addition of food, suggesting that supplementation early in the breeding season had a particularly positive effect on the breeding success. European turtle doves can have up to three successive breeding attempts, and a longer supplementation would thus support early and late breeding attempts alike. In contrast, less variability was explained in the hunting age ratios by the amount of food added. This suggests that more variability in the hunting age ratios was explained by unknown factors that may also impact on the ratio of young to adult birds killed. These unknown factors could be related, among other things, to the variation in the hunting pressure applied on the estates, as occurs with other hunted species, such as the Woodcock (*Scolopax rusticola*) (Fadat, 1981). Hunting pressure may vary from estate to estate depending on variables such as the distance between hunters and the distance from the posts to the feeding zones, (the shorter the distance, the higher the pressure). These distances could facilitate or hinder the capture of young and adult birds, thereby affecting a greater variability in the proportions obtained. A further issue is the variety in the levels of marksmanship (shooting efficiency) between hunters from estate to estate.

By providing food from the beginning of June to the end of August, managers and hunters would assure the existence of food readily available for most of the feeding period of the chicks, from the eggs being hatched to their first flight and their preparation for migration.

These measures could also serve to avoid a possible reduction in the breeding season as occurred in the UK, decreasing the number of broods and causing a drop in the species' reproductive rates (Browne & Aebischer, 2004). These authors consider that agricultural intensification over the second half of the 20th century has caused a clear change in feeding habits of turtle doves by decreasing the availability of wildflower seeds (probably due to extended herbicide use, disappearance of uncultivated land, degraded field edges, etc.). This could be the ultimate reason why food supplementation could be successful in increasing productivity.

Where turtle doves are hunted, however, over–exploitation of the breeding populations of the species is possible despite, or even helped by, supplementation due to its effects as bait. This will depend on the level of extraction that the hunt exerts on the populations: we thus suggest that hunting pressure needs to be carefully controlled for this migratory species due to its vulnerability (Madroño et al., 2004). It is also possible that migratory populations suffer from the negative effects of increased hunting at supplemented sites. In any case, we do not know to what extent the positive effects on productivity can be offset or even reversed by the effect of increased hunting pressure at supplemented sites for both breeding and migratory populations.

To limit the adverse effects of this practice in Extremadura, since 2008, the addition of extra food is permitted only when the hunters are situated more than 200 meters from the edge of the area where the food is added (Junta de Extremadura, 2008). Furthermore, the distance between hunting posts has been limited to 50 meters and a quota of 15 European turtle doves/ hunter/day (reduced to 10 European turtle doves/hunter/day since 2011) has been established. In this study we did not analyse the distances between hunters and the distance between hunters and feeding stations but it would be interesting to know if these variables effectively reduce the hunting pressure.

Acknowledgments

We thank the owners of the hunting estates for kindly allowing us to work on their property. We also acknowledge the assistance of the guards and hunters who participated in recording data at the observation posts and on the hunts.

References

Alés, E. E., 1996. Cambios en el paisaje del suroeste de España: nuevos escenarios de conservación para la fauna amenazada. *Quercus,* 121: 35–39.

Blanco, J. C. & González, J. L. (Eds.), 1992. *Libro Rojo de los Vertebrados de España.* ICONA, Madrid.

Boutin, S., 1990. Food supplementation experiments with terrestrial vertebrates: patterns, problems, and the future. *Canadian Journal of Zoology,* 68(2): 203–220.

Boutin, J. M., 2001. Elements for a Turtle Dove (*Streptopelia turtur*) management plan. *Game Wildlife,* 18: 87–112.

Browne, S. J. & Aebischer, N. J., 2003. Habitat use, foraging ecology and diet of Turtle Doves *Streptopelia turtur* in Britain. *Ibis,* 145(4): 572–582.

– 2004. Temporal changes in the breeding ecology of European turtle doves *Streptopelia turtur* in Britain, and implications for conservation. *Ibis,* 146(1): 125–137.

Browne, S. J., Aebischer, N. J. & Crick, H. Q. P., 2005. Breeding ecology of Turtle Doves *Streptopelia turtur* in Britain during the period 1941–2000: an analysis of BTO nest record cards. *Bird Study*, 52: 1–9.

Calladine, J. R., Buner, F. & Aebischer, N. J., 1997. The summer ecology and habitat use of the Turtle Dove: A pilot study. *English Nature Research Reports*, 219: 87.

Cramp, S. (Ed.)., 1985. *The birds of the western Palearctic. 4.* Oxford University Press, Oxford.

Dias S., Fontoura A. P., 1996. A dieta estival da rôla-brava (*Streptopelia turtur*) no sul de Portugal. *Revista Florestal, 9:* 227–241.

Díaz, M., Asensio, B. & Tellería, J. L., 1996. *Aves Ibéricas. I No Paseriformes* (J. M. Reyero Ed.). Madrid.

Díaz, M., Campos, P. & Pulido, F. J., 1997. The Spanish dehesas: a diversity in land–use and wildlife. In: *Farming and Birds in Europe. The Common Agricultural Policy and its Implications for Bird Conservation*: 178–209 (D. J. Pain & M. W. Pienkowski, Eds.). Academic Press, London.

Díaz–Fernández, S., Arroyo, B., Casas, F., Martinez–Haro, M. & Viñuela, J., 2013. Effect of Game Management on Wild Red–Legged Partridge Abundance. *PLoS ONE*, 8(6): e66671.

Domènech, J. & Senar, J. C., 1997. Trapping methods can bias age ratio in samples of passerine populations. *Bird Study*, 44(3): 348–354.

Donald, P. F., Green, R. E. & Heath, F., 2000. Agricultural intensification and the collapse of Europe's farmland bird populations. *Proceedings of the Royal Society London*, 268: 25–29.

Draycott, R. A. H., Woodburn, M. I. A., Carroll, J. P. & Sage, R. B., 2005. Effects of spring supplementary feeding on population density and breeding success of released pheasants *Phasianus colchicus* in Britain. *Wildlife Biology*, 11: 177–182.

Eraud, C., Boutin, J. M., Riviere, M., Brun, J., Barbraud, C. & Lormee, H., 2009. Survival of Turtle Doves *Streptopelia turtur* in relation to western Africa environmental conditions. *Ibis*, 151: 186–190.

Fadat, C., 1981. Age ratio des tableaux de chasse de bécasses (*Scolopax rusticola*). Signification biologique et utilisation pour la bonne gestion des populations bécassières. *Bull. Mens. ONC. No. Sp. Scien. Techn.,* Nov. 1981: 141–172.

Flanders–Wanner, B. L., White, G. C. & Mcdaniel, L. L., 2004. Validity of prairie grouse harvest–age ratios as production indices. *Journal of Wildlife Management*, 68(4): 1088–1094.

Fernández, A., 2004. La Red Natura 2000 en Extremadura. *Foresta*, 27: 89–99.

Fernández, L. & Camacho, M., 1989. Determinación de status de la Tórtola Común *Streptopelia turtur*. ICONA, Informe inédito, Madrid.

Gutiérrez, J. E., 2001. Les populations de tourterelles des bois en Andalusie. *Faune Sauvage*, 253: 36–43.

Gutiérrez–Bermejo, L. F., 2009. Tórtolas, causas del declive y cómo gestionarlas. *Trofeo, caza y conservación*, 470: 42–47.

Hanane, S., 2009. Variabilité spatio–temporelle des âges ratios chez la Tourterelle des bois *Streptopelia turtur* dans les plaines du Souss et du Tadla (Maroc). *Go–South Bull.*, 6: 124–127.

Hidalgo, S. J. & Rocha G., 2001. Valoración de la presión cinegética sobre la Tórtola Común en Extremadura. *Naturzale*, 16: 157–171.

– 2006. *Seguimiento de la actividad cinegética sobre la tórtola común en la Media Veda de los años 2005 y 2006.* Junta de Extremadura, Informe Técnico, Mérida.

Jarry, G., 1997. Turtle Dove *Streptopelia turtur*. In: *The EBCC Atlas of European Breeding Birds: Their Distribution and Abundance*: 390–391 (W. J. M. Hagemeijer & M. J. Blair, Eds.). T and AD Poyser, London, UK.

Jiménez, R. Hodar, J. A., & Camacho, I., 1992. La alimentación estival de la Tórtola Común *Streptopelia turtur* en el sur de España. *Gibier Faune Sauvage*, 9: 119–126.

Junta de Extremadura, 1991. Ley 8/1990, de 21 de diciembre, de Caza de Extremadura. *Diario Oficial de Extremadura (14–1–1991)*, 2: 1–28.

– 2007. Orden de 14 de junio de 2007 por la que se establecen los períodos hábiles de caza durante la temporada 2007/2008 y otras reglamentaciones especiales para la conservación de la fauna silvestre de la Comunidad Autónoma de Extremadura. *Diario Oficial de Extremadura (23–6–2007)*, 72: 11162–11182.

– 2008. Orden de 17 de julio de 2008 por la que se establecen los periodos hábiles de caza durante la temporada 2008/2009 y otras reglamentaciones especiales para la conservación de la fauna silvestre de la Comunidad Autónoma de Extremadura. *Diario Oficial de Extremadura (18–7–2008)*, 139: 19700–19732.

– 2012. Superficie de las áreas protegidas de Extremadura. http://www.extremambiente.es. (Accessed on 23 June 2014).

Krebs, J. R., Wilson, J. D., Bradbury, R. B. & Siriwardena, G. M., 1999. The second silent spring? *Nature*, 400: 611–612.

Lázaro, I., 2004. La caza en Extremadura. *Foresta*, 27: 144–151.

Lucio, A. & Purroy, F. J., 1992. Caza y conservación de aves en España. *Ardeola*, 39(2): 85–89.

Lutz, M. & Jensen, F. P., 2007. Management Plan for Turtle Dove *Streptopelia turtur*, 2007–2009. European Commission. *Office for Official Publications of the European Communities*, Luxembourg. http://ec.europa.eu/environment/nature/conservation/wildbirds/hunting/docs/turtle_dove.pdf. (Accessed on 23 June 2014).

Madroño, A., González, C. & Atienza, J. C. (Eds.), 2004. *Libro Rojo de las Aves de España*. Dirección General para la Biodiversidad, SEO/BirdLife, Madrid.

Martin, T. E., 1987. Food as a limit on breeding birds: a life–history perspective. *Annual Review of Ecology and Systematics*, 18: 453–487.

Martínez–Abraín, A. & Oro, D., 2013. Preventing the development of dogmatic approaches in conservation biology: A review. *Biological Conservation*, 159: 539–547.

Muñoz–Cobo, J., 2001. Tórtola Común (*Streptopelia turtur*). In: *Libro Rojo de los Vertebrados Amenazados de Andalucía* (A. Franco & M. Rodríguez (Coord.). Consejería de Medio Ambiente, Junta de Andalucía, Sevilla.

Murton, R. K., Westwood, N. J. & Isaacson, A. J., 1964. The feeding habits of the Wood Pigeons *Columba palumbus*, Stock Dove *Columba oenas* and Turtle Dove *Streptopelia turtur*. *Ibis*, 106: 174–188.

Naredo, J. M., 1996. *La evolución de la Agricultura en España (1940–1990)*. Universidad de Granada, Granada.

Newton, I., 2001. An alternative approach to the measurement of seasonal trends in bird breeding success: a case study of the bullfinch *Pyrrhula pyrrhula*. *Journal of Animal Ecology*, 68: 698–707.

Olona, J., 2014. *Economía de la agricultura española. Evolución y tendencias*. Quasar Consultores, Zaragoza, Spain.

Potts, G. R., 1997. Cereal farming, pesticides and grey partridges. In: *Farming and Birds in Europe. The Common Agricultural Policy and its Implications for Bird Conservation*: 150–177 (D. J. Pain & M. W. Pienkowski, Eds.). Academic Press, London.

Peiró, V., 1990. Aspectos de la reproducción de la Tórtola Común *Streptopelia turtur* en Madrid. *Mediterránea serie Biológica*, 12: 89–96.

Peery, M. Z., Becker, B. H., Beissinger, S. R. & Burger, A. E., 2007. Age ratios as estimators of productivity: testing assumptions on a threatened seabird, the marbled murrelet (*Brachyramphus marmoratus*). *The Auk*, 124(1): 224–240.

Puerta, D., 2011. Tórtolas en Extremadura. Las que llegarán del norte. *Trofeo, caza y conservación*, 496: 52–64.

Purroy, F. J. (Ed.), 1995. La tórtola común. *La Garcilla*, 94: 22–23.

Purroy, F. J. (Coord.), 1997. *Atlas de las Aves de España (1975–1995)*. SEO/BirdLife, Lynx Edicions, Barcelona, Spain.

Robson, N., 1997. The evolution of the Common Agricultural Policy and the incorporation of environmental considerations. In: *Farming and Birds in Europe. The Common Agricultural Policy and its Implications for Bird Conservation*: 43–78 (D. J. Pain & M. W. Pienkowski, Eds.). Academic Press, London.

Rocha, G., 2007. Criterios de calidad en la gestión cinegética de especies migratorias. In: *Criterios para la Certificación de la Calidad Cinegética en España*: 75–80 (J. Carranza & J. M. Vargas, Eds.). Servicio de Publicaciones de la Universidad de Extremadura, Cáceres.

Rocha, G. & Hidalgo, S. J., 2001. Incidencia del uso de reclamos alimenticios sobre la Tórtola Común. *Naturzale*, 16: 147–155.

– 2002. *La Tórtola Común (*Streptopelia turtur*): Análisis de los factores que afectan a su status*. Servicio de Publicaciones de la Universidad de Extremadura, Cáceres.

Rocha, G., Merchán, T. & Hidalgo, S. J., 2009. Gestión de la tórtola común y la paloma torcaz. In: *Gestión cinegética en los ecosistemas mediterráneos*: 255–285 (J. Carranza & M. Sáenz de Buruaga, Eds.). Consejería de Medio Ambiente, Junta de Andalucía, Sevilla.

Sáenz de Buruaga, M., Lucio, A. & Purroy, F. J., 2001. *Reconocimiento de sexo y edad en especies cinegéticas*. EDILESA, León.

Santamaría, A., 2007. Análisis de la abundancia reproductora de las principales especies cinegéticas y su aprovechamiento en media veda, en varios acotados de la provincia de Cáceres. Proyecto Fin de Carrera, Universidad de Extremadura, Plasencia.

Seo/BirdLife, 2002. *Seguimiento de las poblaciones de aves fringílidas de interés canoro y de la tórtola común en Andalucía durante la temporada 2002*. Documento Técnico, Junta de Andalucía, Sevilla.

– 2013. *Resultados del programa Sacre de SEO/BirdLife*. SEO/BirdLife. Madrid. http://www.seguimientodeaves.org/ESPECIOS/docs/ESPECIES/3280_RES_SP.pdf. (Accessed on 23 June 2014).

Schoech, S. J. & Hahn, T. P., 2007. Food supplementation and timing of reproduction: does the responsiveness to supplementary information vary with latitude?. *Journal of Ornithology*, 148(2): 625–632.

Schoech, S. J., Bridge, E. S., Boughton R. K., Reynolds S. J., Atwell, J. W. & Bowman, R., 2008. Food supplementation: a tool to increase reproductive output? A case study in the threatened Florida Scrub–Jay. *Biological Conservation*, 141(1): 162–173.

Tucker, G. M. & Heath, M. F., 1994. *Birds in Europe: Their conservation status*. BirdLife Conservation Series No. 3. BirdLife International, Cambridge, UK.

Wagner, F. H. & Stokes. A. W., 1968. Indices to overwinter survival and productivity with implications for population regulation in pheasants. *Journal of Wildlife Management*, 32: 32–36.

Wilson, J. M. & Cresswell, W., 2006. How robust are Palaearctic migrants to habitat loss and degradation in the Sahel? *Ibis*, 148: 789–800.

6

European rabbit restocking: a critical review in accordance with IUCN (1998) guidelines for re–introduction

J. Guerrero–Casado, J. Letty & F. S. Tortosa

Guerrero–Casado, J., Letty, J. & Tortosa, F. S., 2013. European rabbit restocking: a critical review in accordance with IUCN (1998) guidelines for re–introduction. *Animal Biodiversity and Conservation*, 36.2: 177–185.

Abstract

European rabbit restocking: a critical review in accordance with IUCN (1998) guidelines for re–introduction.— European rabbit restocking is one of the most frequent actions in hunting estates and conservation projects in Spain, France and Portugal where rabbit is a keystone species. The aim of this work was to review current knowledge regarding rabbit restocking in accordance with the IUCN (1998) guidelines for re–introduction in order to identify gaps in knowledge and highlight the techniques that improve the overall success rate. Eight of 17 items selected from these guidelines were identified as partly studied or unknown, including important items such as the management and release of captive–reared wild rabbits, the development of transport and monitoring programs, the application of vaccine programs, and post–release long–term studies. Researchers should therefore concentrate their efforts on bridging these knowledge gaps, and wildlife managers should consider all the factors reviewed herein so as to establish accurate management guidelines for subsequent rabbit restocking programs.

Key words: Lagomorphs, Hunting management, *Oryctolagus cuniculus*, Translocation, Wildlife management.

Resumen

Repoblaciones de conejo europeo: una revisión crítica según las directrices de la IUCN (1998) para las reintroducciones.— Las repoblaciones de conejo europeo son una de las medidas más empleadas en los cotos de caza y en los proyectos de conservación en España, Francia y Portugal, donde el conejo es una especie clave. El objetivo de este trabajo consiste en revisar el conocimiento actual sobre los factores que afectan al establecimiento de las poblaciones de conejo reintroducidas según las directrices de la IUCN (1998), a fin de determinar las lagunas de conocimiento en este ámbito y destacar las técnicas que mejoran los buenos resultados reales de las reintroducciones. Ocho de los 17 puntos seleccionados de estas directrices se identificaron como desconocidos o parcialmente estudiados, incluidos importantes aspectos como el manejo y la liberación de conejos salvajes criados en cautividad, la elaboración de planes de transporte y seguimiento, la aplicación de programas de vacunación y los estudios a largo plazo posteriores a la liberación. Por lo tanto, los investigadores deben concentrar sus esfuerzos en suprimir esta falta de conocimiento y los gestores deben analizar todos los factores que aquí revisamos, con el objetivo de establecer unas directrices precisas para las futuras repoblaciones de conejo.

Palabras clave: Lagomorfos, Gestión cinegética, *Oryctolagus cuniculus*, Translocaciones, Gestión de fauna silvestre.

José Guerrero–Casado & Francisco S. Tortosa, Dept of Zoology, Univ. of Córdoba, Campus de Rabanales, E–14071, Córdoba, España (Spain).– José Guerrero–Casado, Inst. for Terrestrial and Aquatic Wildlife Research, Univ. of Veterinary Medicine Hannover, Bischofsholer Damm 15, 30173 Hannover (Germany).– Jérôme Letty, Office National de la Chasse et de la Faune Sauvage, Direction des Etudes et de la Recherche, 147 route de Lodève, F–34990, Juvignac (France).

Corresponding author: José Guerrero–Casado. E–mail: guerrero.casado@gmail.com

Introduction

Translocation of animals for conservation management is increasing worldwide due to the alarming loss of biodiversity, but success is limited (Griffith et al., 1989, Armstrong & Seddon, 2008). As many reintroduction attempts have failed, the IUCN edited guidelines for re–introductions (IUCN, 1998) to establish the knowledge needed to ensure that reintroductions meet their goal. Nevertheless, in many translocation programs, many questions remain unanswered (Armstrong & Seddon, 2008). This lack of knowledge includes data concerning the European rabbit (*Oryctolagus cuniculus*) in France, Portugal and Spain, where around half a million rabbits are translocated each year to promote the recovery of natural populations and to improve hunting stocks (Arthur, 1989; Calvete et al., 1997; Letty et al., 2008).

Rabbits are an essential element in the Mediterranean ecosystem. They play a vital role as ecosystem engineers and are key prey for more than 30 species of predator (Delibes–Mateos et al., 2008a). What is more, rabbit hunting is an economically important activity in Iberia and France (Delibes–Mateos et al., 2008a; Letty et al., 2008; Ferreira et al., 2010). The rabbit population declined drastically, however, in the 20th century, mainly as a consequence of optimal–habitat loss (Ward, 2005) and disease: recurrent outbreaks of the viral disease myxomatosis since 1952, and rabbit haemorrhagic disease (RHD) since the late 1980s (Villafuerte et al., 1995; Marchandeau et al., 1998). This sharp decline is considered a major problem for the conservation of Iberian ecosystems and hunting activity (Marchandeau, 2000; Delibes–Mateos et al., 2009), and rabbit restocking has therefore increased significantly to recover populations (Delibes–Mateos et al., 2008b). The restocking success rate, however, has often been low in traditional restocking attempts. Failure has mainly been due to a low survival rate after the simultaneous release of a large number of rabbits and to a lack of other wildlife management measures (Calvete et al., 1997). Wildlife managers have consequently started to adopt management tools to improve rabbit survival. Some of these tools, such as soft–release or habitat management, have proven to be effective, whereas other strategies, such as vaccination or quarantining, are controversial. Ferreira & Delibes–Mateos (2010) suggested that recommendations made by researchers have not been fully implemented by wildlife managers or hunters, contributing to failure. A protocol for rabbit translocation is thus clearly needed. To establish the perspectives for future research we reviewed current knowledge on rabbit restocking in accordance with the IUCN guidelines for reintroductions. We highlight the techniques and the factors that improve translocation success, and discuss the issues yet to be solved.

Data source

We used the IUCN guidelines for re–introduction as a reference guide to review current knowledge of rabbit restocking because they establish the items and the steps that restocking programs should take. These guidelines define re–introduction as an attempt to establish a species in an area that was once part of its historical range, but from which it has been extirpated or become extinct; translocation as a deliberate and mediated movement of wild individuals or populations from one part of their range to another; and restocking as an addition of individuals to increase an existing population. Whatever the case, the IUCN establishes unique guidelines for re–introductions, restocking or reinforcement and translocations, and in accordance with the literature on rabbits and Armstrong & Seddon (2008), we hereafter use the terms restocking and translocation as synonyms.

The IUCN guidelines are divided into three main sections: (1) pre–restocking activities, (2) planning, preparation and restocking phases, and (3) post–release activities. We have therefore analyzed the knowledge concerning rabbit restocking in accordance with these sections. In accordance with the suggestion of Armstrong & Seddon (2008), this paper focuses mainly on the population level, particularly on the factors and management measures that affect the critical stage of establishing a reintroduced population. Therefore, for this study, we selected only items concerning the biological, ecological and practical monitoring aspects of restocking that scientific literature should document. Once established, the subsequent persistence of the population depends on general factors of ecology requirements and classical wildlife management. A total of 17 out of 52 items were eventually selected (table 1). An issue was deemed to be partly studied if the approach was only theoretical or not fully developed, or if there was no consensus about it. We reviewed papers that addressed rabbit restocking in the scientific literature using three main web engines: Google Scholar™, ISI Web of Knowledge® and Scopus®. We searched the following words in the following combinations: 'rabbit' OR 'Oryctolagus' AND 'restocking' OR 'translocations'. To address each item involved in rabbit restocking (*e.g.* habitat management, vaccines, quarantine or stress) we performed additional searches following the same method. We also searched for data about these topics in Ph. D. Theses, books, and technical reports. Eight of the 17 items in the IUCN guidelines were identified as being poorly studied or unknown; they are summarized in table 1 and discussed in the following sections.

Pre-restocking activities: biological knowledge

The first step in a restocking program is to determine the source population from which the rabbits will be captured. This question is particularly relevant in the Iberian Peninsula, in which two rabbit subspecies coexist: *Oryctolagus cuniculus algirus*, and *O. c. cuniculus* (Branco et al., 2000). Nevertheless, Delibes–Mateos et al. (2008b) found *algirus* rabbits in localities within the *cuniculus* subspecies range, and vice–versa, as a consequence of past translocations, since in most cases rabbits are released regardless of their genetic lineage. These subspecies have differences in body size, sexual maturation and litter size (Gonçalves et al., 2002; Ferreira, 2011), difference that could affect the success of rabbit translocations and have unknown ecological and demographic consequences. Wildlife managers should therefore avoid mixing subspecies by identifying the genetic lineage of the rabbits using DNA analysis, and both the donor and the receiving populations must be located within the geographic range of the corresponding genetic lineage (Delibes–Mateos et al., 2008b). Furthermore, at the metapopulation level, although the impact of rabbit extraction on the donor population has not been empirically tested, it should also be considered because excessive captures of individuals may lead to the decline of the donor population (Cotilla & Villafuerte, 2007).

Interest in captive rearing of wild rabbit for release purposes as an alternative to capturing wild individuals has increased over the last two decades (Arenas et al., 2006). The proportion of captive–reared rabbits released in Spain likely exceeds 50% of the total number of wild rabbits released (Sánchez–García et al., 2012). Nevertheless, the success of restocking operations using captive–reared rabbit remains untested, and genetic, epidemiological and behavioural problems could be expected when hybrids between wild and domestic lineages are reared in captive intensive systems, as occurs in some farms in France and Spain (Rogers et al., 1994; Piorno, 2006). Moreover, although Arenas et al. (2006) reported management techniques that improve the reproduction of wild rabbits in captivity, training techniques to enhance rabbit restocking success in captive environments have not yet been developed. However, it seems possible to recreate wild–like environmental conditions (regarding food availability, soil type and aerial predator pressure) in breeding enclosures in situ that would enable appropriate rabbit behaviour for release purposes and greater ability to adapt to local conditions (Guerrero–Casado et al., 2013a). The sustainability of such captive–rearing populations and relevant factors to consider for translocation success —such as body condition, behaviour and age of captive–reared individuals— should be further studied.

With regard to habitat requirements, many types of habitat may be suitable for rabbits as they have a high phenotypic plasticity, however, release areas must include grazing areas, shelter to escape from predators and have soils that enable burrowing. Burrows allow a rapid increase in both population size and viability, and they provide shelter from predators. Release into an optimal habitat is expected to increase rabbit survival and to limit dispersal movements (Calvete & Estrada, 2004; Moreno et al., 2004). On the other hand, if release and capture areas have similar ecological characteristics, rabbits can be expected to adapt better to the new environment, by means of pre–adaptations to landscape, soil, flora, or parasites type (Letty et al., 2008). It is therefore advisable that donor populations should be located as close to the target area as possible (Villafuerte & Castro, 2007). However, there is a gap in knowledge concerning the possible importance of adaptations to local ecological conditions in restocking success.

Planning, preparation and restocking phases

According to scientific literature, the crux of the translocation problem is the high mortality in the first weeks after release, and the interaction between the main factors affecting rabbit survival: predation, environmental novelty and stress (Calvete et al., 1997; Moreno et al., 2004; Letty et al., 2008). Translocated animals may display high activity during the first days after release. When they are introduced into a novel habitat they are disorientated and they do not know where to feed, rest or seek refuge from predators. They may also explore the area in search of their usual landmarks or return to their previous home range (Letty et al., 2002b, 2008). Stress is an inevitable component of restocking programs, because the process of translocation involves multiple stressors: (1) capture and handling, (2) captivity or some form of prolonged restraint, (3) transport, and (4) release into an unfamiliar environment —likely the highest stressor (Letty et al., 2007; Teixeira et al., 2007; Dickens et al., 2010). This succession of events could chronically–stress translocated animals and may have a strong negative impact on their physiological condition (Cabezas et al., 2007), thus increasing their vulnerability to predation or diseases. Furthermore, eye damage, fractures, bites and wounds (Rouco, 2008) and even sporadic death (Letty et al., 2005) have been reported during the transport phase. Hence, in these phases, stress levels should be controlled by reducing handling and physical restraint (Letty et al., 2005), avoiding crowding, decreasing time between capture and release, and facilitating rapid access to high quality food (Calvete et al., 2005). Specific guidelines should specify all the points related to rabbit capture, transport and handling, with special emphasis on minimizing stress and ensuring animal welfare, since there is often little effort to reduce losses during these stages (Calvete et al., 1997). The effects of transport and handling stress, however, may be only induce temporary negative effects compared to those induced by the permanent change of area (Letty et al., 2003).

Over the last decade, various release strategies have been developed to minimize the aforementioned problems and improve rabbit survival, such as soft–release, habitat management, or predation exclusion. In the soft–release strategy, translocated rabbits are progressively acclimatised to the new environment in enclosures designed to prevent initial exploratory movements and predation mortality immediately after release, when the animals are more vulnerable. Such acclimatization highly increases the short–term rabbit survival (e.g. 82% Calvete & Estrada, 2004; 87% Rouco et al., 2010). As this gain in survival is not always clear and sometimes only temporary (Letty et al., 2000, 2008), a longer acclimatisation period is advised to increase early survival and to decrease dispersal, particularly in poor habitats (Calvete & Estrada, 2004). This approach is considered to increase initial breeding stock and overall restocking viability (Letty et al., 2008;

Table 1. Selected items of IUCN guidelines for reintroduction programs, with the references that support the information and summary and/or observation in each case.

Tabla 1. Puntos seleccionados de las directrices de la IUCN para los programas de repoblación, con las referencias que apoyan la información y el resumen y/o las observaciones en cada caso.

Pre–restocking activities

Item	Knowledge	Summary/Observations	References
Taxonomic status	Yes	Avoid the mixing of the two subspecies and perform genetic analyses	Branco et al., 2000, Delibes–Mateos et al., 2008b
Status and biology of wild population	Yes	Population crash after the appearance of viral diseases. Many populations remain at low density	Marchandeau et al., 2000; Delibes–Mateos et al., 2009; Ferreira et al., 2010
Habitat requirements	Yes	Positive effect of habitat quality. No studies on the importance of local ecological adaptations	Moreno et al., 2004; Villafuerte & Castro, 2007
Identification of previous causes of decline	Yes	Mainly viral diseases and habitat loss	Villafuerte et al., 1995; Delibes–Mateos et al., 2009
Wild population management in captivity	Partly studied	No studies on training in captivity. Possible genetic, behavioural and ecological problems with hybrid domestic rabbits	Arenas et al., 2006; Piorno, 2006; Guerrero–Casado et al., 2013a
The release of wild rabbit reared in captivity	Partly studied	No studies on restocking success. Genetic introgression of hybrid rabbits in wild populations	Piorno, 2006; Sánchez–García et al., 2012
How to minimize the infection rate	Yes	Treat for external and internal parasites before release	Cabezas & Moreno, 2007; Rouco et al., 2008

Planning, preparation and restocking phases

Item	Knowledge	Summary/Observations	References
Identification of success indicators	No	No suitable guidelines	
Design a monitoring program	Partly studied	No consensus about a standardised monitoring protocol	Fernández–de–Simón et al., 2011
Health of release stock	Yes	Positive effect of body condition and negative impact of stress	Calvete et al., 2005; Cabezas et al., 2007
Vaccination	Partly studied	Disagreements on its effectiveness. Positive effect of animals released with high natural antibody concentration	Calvete et al., 2004; Guitton et al., 2008; Ferreira et al., 2009
Quarantine	Yes	Negative effect on animals' physiological condition	Moreno et al., 2004; Calvete et al., 2005
Transport plan	Partly studied	No detailed guide. No demonstrated effect of crowding and long transports	Letty et al., 2003; Letty et al., 2005
Release strategy	Yes	Positive effect of soft–release, habitat management and predator exclusion	Calvete & Estrada, 2004; Rouco et al., 2008; Cabezas et al., 2011

Table 1. (Cont.)

Post–release activities

Item	Knowledge	Summary/Observations	References
Demographic, ecological and behavioral studies	Partly studied	Restocked rabbit's show the same behaviour in the long–term as wild individuals	Rouco et al., 2011b; Ruiz–Aizpurua et al., 2013
Long–term adaptations	Partly studied	Low breeding contributions. No studies of the interactions with the resident congeners	Letty et al., 2002a
Investigation of mortalities	Yes	Mainly predation, environmental novelty and stress. Survival and dispersal after release in short–term are well documented	Calvete et al., 1997; Letty et al., 2008; Cabezas et al., 2011

Rouco et al., 2010). More recently, large *in situ* breeding enclosures have been widely used in predator conservation projects to enhance rabbit availability (Ward, 2005; Ferreira & Delibes–Mateos, 2010) and may be a highly effective way of establishing a new population. The role of fences is not only to reduce mortality due to terrestrial predators and dispersal movements but also to establish a captive *in situ* breeding stock so that young individuals will naturally disperse and settle in the surrounding areas (Letty et al., 2006; Rouco et al., 2008; Guerrero–Casado et al., 2013b). Furthermore, as rabbits translocated to a new environment are highly vulnerable to predation in the short–term, soft–release or long–lasting acclimatization in predator–free enclosures should effectively minimise the impact of predation without concentrating efforts on removal of predators. The impact of predation can also be reduced by selecting areas with a high portion of natural shelter (Calvete & Estrada, 2004) or by increasing the shelter availability through habitat management (Cabezas et al., 2011).

Habitat management is a highly effective and widespread practice in rabbit restocking (Catalán et al., 2008; Ferreira & Alves, 2009; Ferreira et al., 2013). If rabbits are released in a sub–optimal habitat, habitat management should occur prior to release so as to create feeding habitats and provide shelter through scrubland management and/or the construction of artificial warrens where refuge is scarce (Ferreira et al., 2013). Rabbit abundance and survival rate is significantly higher when the translocation is carried out in areas improved by the creation of pastureland and provision of artificial warrens (Cabezas & Moreno, 2007; Cabezas et al., 2011). Releasing rabbits into artificially constructed warrens is a common practice that also enhances the availability of shelter and breeding sites. Put simply, it is preferable to build many small warrens rather than a few large warrens (Rouco et al., 2011a). These smaller structures, preferentially built with tub-

ing, should be close enough to each other (Barrio et al., 2009) to allow a small population to settle, and they should be located in areas with adequate food and shelter (Fernández–Olalla et al., 2010). Detailed guidelines on how to conduct habitat management can be found in several technical documents (Anomynous, 2003; Ferreira & Alves, 2006; San–Miguel, 2006; Guil, 2009).

The risk of disease is another threat that may jeopardize wild rabbit translocations. Many translocation therefore include the vaccination of rabbits against myxomatosis and RHD virus (Delibes–Mateos et al., 2008b) even though its effectiveness in the field is controversial. Despite some possible short–term negative effects, an overall positive effect of vaccination has been recorded in free–ranging rabbit populations (Cabezas et al., 2006; Calvete, 2006; Guitton et al., 2008; Ferreira et al., 2009). Vaccination in translocation may have some drawbacks: its short–term negative effect may negatively affect the physiological condition of rabbits and increase early mortality risks (Calvete et al., 2004); the immune response depends on body condition and may be decreased by the stress induced by translocation (Cabezas et al., 2006); the vaccine may cause an immunosuppressive effect in individuals with a poor physiological condition (Calvete et al., 2004); and its effectiveness may be reduced in immunized individuals or, for RHD, in case of a significant evolution of the virus (Le Gall–Reculé et al., 2011). On the other hand, translocation is a rare case in which vaccination may be relevant, and indeed, it could be crucial for population establishment in case of subsequent disease outbreak. The effectiveness of vaccination campaigns should be high since it is possible to vaccinate the whole 'population' (released individuals). However, as a subsequent booster is not feasible, a long–lasting positive effect of a single vaccination and of the related immunity in wild rabbit seems questionable. The exact impact of vaccination on the fitness

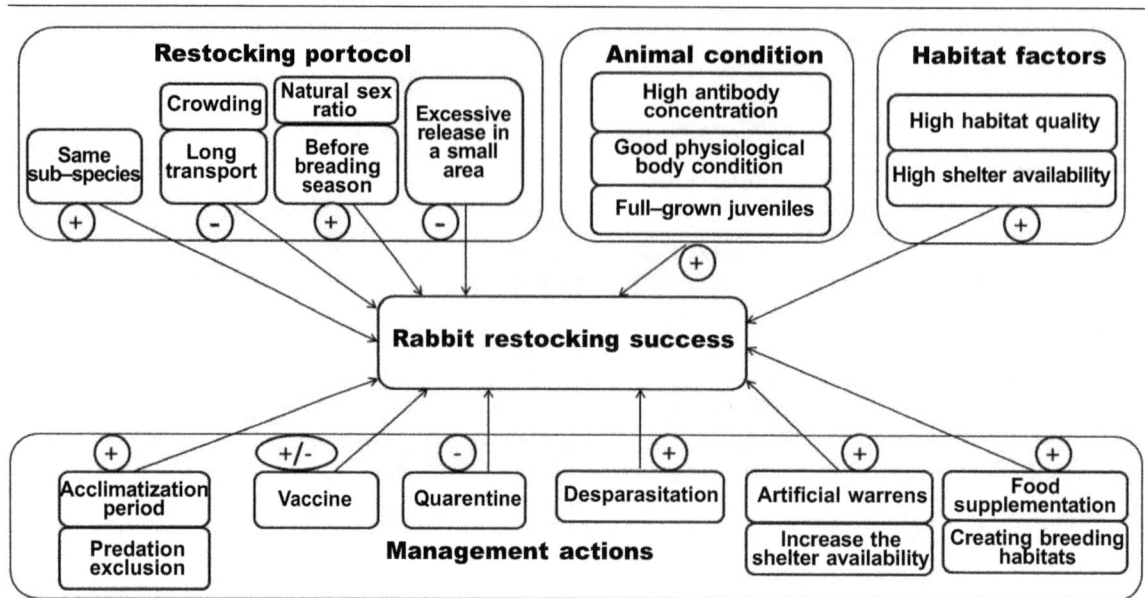

Fig. 1. Summary of the factors that affect wild rabbit restocking success. The symbols + and − indicate positive or negative relationships with the restocking success.

Fig. 1. Resumen de los factores que afectan al éxito de las repoblaciones de conejo. Los símbolos + y − indican una relación positiva o negativa con los buenos resultados de las repoblaciones.

of translocated rabbits, and the related cost–benefit ratio for restocking success has therefore yet to be adequately and experimentally addressed.

Given that long–term survival is positively correlated with antibody concentrations before release (Cabezas et al., 2011), Rouco (2008) proposed releasing rabbits with naturally high antibody concentrations as an alternative to vaccine. This might be feasible in the wild, since just after the annual outbreak of diseases, most individuals have natural antibodies in high density rabbit populations (Cotilla et al., 2010). Thus, vaccination is unnecessary when restocking with such individuals (Calvete, 2006), and vaccination protocols would only be necessary if donor populations have low antibody prevalence (Cabezas et al., 2006). However, the precise monitoring of antibody prevalence in wild populations may not be easy to carry out in the field.

In turn, in some restocking programs, rabbits are kept in quarantine for several days to ensure the effect of the vaccine and to make sure that animals do not incubate the viral diseases. Such captivity periods, nevertheless, induce stress, loss of body mass, abortion in pregnant females, and other possible physiological disorders (Calvete et al., 2005). This management tool therefore controls diseased or injured animals (Calvete et al., 1997) but does not generally increase restocking success (Calvete & Estrada, 2004). Restocking operations have also been shown to be a potential means of introducing pathogens into resident populations (Haz et

al., 2001; Reglero et al., 2007; Navarro–González et al., 2010). Hence, all rabbits should be treated for external and internal parasites before release in order to minimise the possibility of disease and parasite transmission. As regards the health of the stock released, selecting animals with a good body condition (those with a good index of fat, and free of traumatic injuries, cachexia, or high parasite levels) enhances the probability of survival (Calvete et al., 2005; Cabezas et al., 2006).

Other factors suggested to affect restocking success are release timing, sex ratio, age, and the number of rabbits released. For demographical reasons, the release of rabbits before the breeding season could lead to higher population growth (Cotilla & Villafuerte, 2007), whereas releasing rabbits during the breeding season (when social stress is high) might increase agonistic behaviour and direct competition among rabbits (Moreno et al., 2004). The timing of the release could also affect the translocation success if, for instance, the impact of predation depends on the season (the availability of food and cover differs between seasons). Regarding age, the model of Cotilla & Villafuerte (2007) indicated that success would be maximized by releasing only adult rabbits (at least 4 months old). However, the fitness of individuals to translocation may play a role; full–grown juveniles might be less affected by translocation than adults and better able to adapt to the new situation (Mauvy et al., 1991; Letty et al., 2008). It is therefore necessary to clarify the optimal age of release animals. As a general rule, it is advisable to release rabbits in optimal numbers and in a natural sex ratio so at to attain viable population dynamics after release (*e.g.* 1:1 Moreno et al., 2004; Cabezas & Moreno, 2007).

Post–release activities

Little is known with regard to demographic, ecological and behavioral long–term adaptations in released populations. Some works, however, have suggested that restocked rabbits exhibit the same behavior in the long–term as wild individuals (Rouco et al., 2011b; Ruiz–Aizpurua et al., 2013). Indeed, social behavior can also affect restocking success (Ruiz–Aizpurua, 2013), although Letty et al. (2006, 2008) did not record a clear difference in the survival rate between individuals released in familiar groups (captured in the same warren) and unfamiliar groups, suggesting that the translocation process destabilizes previous social relationships. Earlier studies have also suggested a low breeding contribution of introduced individuals during the first months after release (Letty et al., 2002a). These points should thus be clarified in further research to understand the behavior of the rabbits released and their interactions with the resident congeners.

Finally, it is necessary to identify short– and long–term indicators to assess the outcome of the translocation in agreement with aims and objectives. Rabbit translocation often lacks careful monitoring (Cabezas & Moreno, 2007). A standardized monitoring protocol is needed to acquire reliable data (rabbit abundance, special distribution, time scale,...) on restocking success. To correctly assess restocking effectiveness, wildlife managers could monitor rabbit abundance before and after release using indices based on transect counts or pellet counts (Fernández–de–Simón et al., 2011).

Conclusions

Despite the relevance of rabbit restocking activities, eight of the items considered important in the IUCN re–introduction guidelines are only partly answered. Conservationists, hunters, wildlife managers and researchers should thus concentrate their efforts on bridging these knowledge gaps and implementing scientific recommendations to establishing accurate management guidelines for subsequent rabbit restocking. We suggest that the overall success rate would be improved by: (1) establishing a long period of acclimatization; (2) selecting a high quality habitat or enhancing its carrying capacity with artificial warrens, food supplementation or scrub management; (3) avoiding the mixing of two subspecies; (4) selecting animals with a good body condition and antibody concentration; (5) reducing predation risk and stress; (6) releasing full–grown rabbits in a natural sex ratio before the breeding season; and (7) avoiding the simultaneous release of an excessive number of animals in a small area (fig. 1).

Acknowledgements

We would like to thank I. C. Barrio, A. J. Carpio and L. Ruiz–Aizpurua for their useful comments and S. Crespo for her support. The Andalusia Environmental Government funded this work through a project for the conservation of the Black Vulture in Cordoba province.

References

Anomynous, 2003. *Le Lapin de garenne. Éléments techniques pour un repeuplement efficace.* Fédération Régionale des Chasseurs de Bretagne.

Arenas, A. J., Astorga, R. J., García, I., Varo, A., Huerta, B., Carbonero, A., Cadenas, R. & Perea, A., 2006. Captive breeeding of wild rabbits: techniques and population dynamics. *Journal of Wildlife Management,* 76: 1801–1804.

Armstrong, D. P. & Seddon, P. J., 2008. Directions in reintroduction biology. *Trends in Ecology & Evolution,* 23: 20–25.

Arthur, C., 1989. Les opérations de repeuplement hivernal en lapins de garenne. *Bulletin Mensuel de l'Office National de la Chasse,* 139: 15–28.

Barrio, I., Bueno, C. & Tortosa, F., 2009. Improving predictions of the location and use of warrens in sensitive rabbit populations. *Animal Conservation,* 12: 426–433.

Branco, M., Ferrand, N. & Monnerot, M., 2000. Phylogeography of the European rabbit (*Oryctolagus cuniculus*) in the Iberian Peninsula inferred from RFLP analysis of the cytochrome b gene. *Heredity,* 85: 307–317.

Cabezas, S., Blas, J., Marchant, T. A. & Moreno, S., 2007. Physiological stress levels predict survival probabilities in wild rabbits. *Hormones and Behavior,* 51: 313–320.

Cabezas, S., Calvete, C. & Moreno, S., 2006. Vaccination success and body condition in the European wild rabbit: applications for conservation strategies. *Journal of Wildlife Management,* 70: 1125–1131.

– 2011. Survival of translocated wild rabbits: importance of habitat, physiological and immune condition. *Animal Conservation,* 14: 665–675.

Cabezas, S. & Moreno, S., 2007. An experimental study of translocation success and habitat improvement in wild rabbits. *Animal Conservation,* 10: 340–348.

Calvete, C., 2006. The use of immunization programs in wild populations: modelling effectiveness of vaccination campaigns against rabbit haemorrhagic disease. *Biological Conservation,* 130: 290–300.

Calvete, C., Angulo, E., Estrada, R., Moreno, S. & Villafuerte, R., 2005. Quarantine length and survival of translocated European wild rabbits. *Journal of Wildlife Management,* 69: 1063–1072.

Calvete, C. & Estrada, R., 2004. Short–term survival and dispersal of translocated European wild rabbits. Improving the release protocol. *Biological Conservation,* 120: 507–516.

Calvete, C., Estrada, R., Osácar, J. J., Lucientes, J. & Villafuerte, R., 2004. Short–term negative effects of vaccination campaigns against myxomatosis and viral hemorrhagic disease (VHD) on the survival of European wild rabbits. *Journal of Wildlife Management,* 68: 198–205.

Calvete, C., Villafuerte, R., Lucientes, J. & Osácar, J. J., 1997. Effectiveness of traditional wild rabbit restocking in Spain. *Journal of Zoology,* 241: 271–277.

Catalán, I., Rodríguez–Hidalgo, P. & Tortosa, F. S., 2008. Is habitat management an effective tool for wild rabbit (*Oryctolagus cuniculus*) population reinforcement? *European Journal of Wildlife Research,* 54: 449–453.

Cotilla, I., Delibes–Mateos, D., Ramírez, E., Castro, F., Cooke, B. D. & Villafuerte, R., 2010. Establishing a serological surveillance protocol for rabbit hemorrhagic disease by combining mathematical models and field data: implication for rabbit conservation. *European Journal of Wildlife Research,* 56: 725–733.

Cotilla, I. & Villafuerte, R., 2007. Rabbit conservation: models to evaluate the effects of timing of restocking on recipient and donor populations. *Wildlife Research,* 34: 247–252.

Delibes–Mateos, M., Delibes, M., Ferreras, P. & Villafuerte, R., 2008a. Key role of European rabbits in the conservation of the Western Mediterranean Basin hotspot. *Conservation Biology,* 22: 1106–1117.

Delibes–Mateos, M., Ferreras, P. & Villafuerte, R., 2009. European rabbit population trends and associated factors: a review of the situation in the Iberian Peninsula. *Mammal Review,* 39: 124–140.

Delibes–Mateos, M., Ramírez, E., Ferreras, P. & Villafuerte, R., 2008b. Translocations as a risk for the conservation of European wild rabbit *Oryctolagus cuniculus* lineages. *Oryx,* 42: 259–264.

Dickens, M. J., Delehanty, D. J. & Michael–Romero, L., 2010. Stress: an inevitable component of animal translocation. *Biological Conservation,* 143: 1329–1341.

Fernández–De–Simón, J., Díaz–Ruiz, F., Cirilli, F., Tortosa, F. S., Villafuerte, R., Delibes–Mateos, D. & Ferreras, P., 2011. Towards a standardized index of European rabbit abundance in Iberian Mediterranean habitats. *European Journal of Wildlife Research,* 57: 1091–1100.

Fernández–Olalla, M., Martínez–Jauregui, M., Guil, F. & San Miguel–Ayanz, A., 2010. Provision of artificial warrens as a a means to enhance native wild rabbit populations: what type f warren and where should they be sited? *European Journal of Wildlife Research,* 56: 829–837.

Ferreira, C., 2011. Relationships between predation risk, disease and fitness in the wild rabbit: management implications. Ph. D. Thesis, Univ. de Castilla–La Mancha, Spain.

Ferreira, C. & Alves, P. C., 2006. *Gestão de populações de coelho–bravo (Oryctolagus cuniculus algirus).* Federação Alentejana de Caçadores
– 2009. Influence of habitat management on the abundance and diet of wild rabbit (*Oryctolagus cuniculus algirus*) populations in Mediterranean ecosystems. *European Journal of Wildlife Research,* 55: 487–496.

Ferreira, C. & Delibes–Mateos, D., 2010. Wild Rabbit Management in the Iberian Peninsula: state of the art and future perspectives for iberian lynx conservation. *Wildlife Biology in Practice,* 3: 48–66.

Ferreira, C., Paupério, J. & Alves, P. C., 2010. The usefulness of field data and hunting statistics in the assessment of wild rabbit (*Oryctolagus cuniculus*) conservation status in Portugal. *Wildlife Research,* 37: 223–229.

Ferreira, C., Ramírez, E., Castro, F., Ferreras, F., Alves, P. C., Redpath, S. M. & Villafuerte, R., 2009. Field experimental vaccination campaings against myxomatosis and their effectiviness in the wild. *Vaccine,* 27: 6998–7002.

Ferreira, C., Touza, J., Rouco, C., Díaz–Ruiz, F., Fernandez–De–Simón, J., Ríos–Saldaña, C. A., Ferreras, P., Villafuerte, R. & Delibes–Mateos, M., 2013. Habitat management as a generalized tool to boost European rabbit Oryctolagus cuniculus populations in the Iberian Peninsula: a cost–effectiveness analysis. *Mammal Review.* DOI:10.1111/mam.12006

Gonçalves, H., Alves, P. C. & Rocha, A., 2002. Seasonal variation in the reproductive activity of the wild rabbit (*Oryctolagus cuniculus algirus*) in a Mediterranean ecosystem. *Wildlife Research,* 29: 165–173.

Griffith, B., Scott, J. M., Carpenter, J. W. & Reed, C., 1989. Translocation as a species conservation tool: status and strategy. *Science,* 245: 477–480.

Guerrero–Casado, J., Ruiz Aizpurua, L., Carpio, A. J. & Tortosa, F. S., 2013a. Factors affecting wild rabbit production in extensive breeding enclosures: how we can optimixe efforts? *World Rabbit Science,* 21: 193–199.

Guerrero–Casado, J., Ruiz–Aizpurua, L. & Tortosa, F. S., 2013b. The short–term effect of total predation exclusion on wild rabbit abundance in restocking plots. *Acta Theriologica,* 58: 415–418.

Guil, F., 2009. *Actuaciones de fomento del conejo de monte.* Madrid: Real Federación Española de Caza–Fundación CBD–Hábitat, Madrid.

Guitton, J.–S., Devillard, S., Guénézan, M., Fouchet, D., Pontier, D. & Marchandeau, S., 2008. Vaccination of free–living juvenile wild rabbits (*Oryctolagus cuniculus*) against myxomatosis improved their survival. *Preventive Veterinary Medicine,* 84: 1–10.

Haz, P., Alvarez, F., Freire, M., Barcena, F. & Sanmartin, M., 2001. Effects of restocking rabbits on the helminth fauna of wild rabbit population in the Northwest Iberian Peninsula. *Acta Parasitologica,* 46: 306–312.

IUCN, 1998. *Guidelines for Re–introductions.* Prepared by the IUCN/SSC Reintroductions Specialist Group – IUCN, Gland, Switzerland and Cambridge, UK.

Le Gall–Reculé, G., Zwingelstein, F., Boucher, S., Le Normand, B., Plassiart, G., Portejoie, Y., Decors, A., Bertagnoli, S., Guérin, J. L. & Marchandeau, S., 2011. Detection of a new variant of rabbit haemorrhagic disease virus in France. *Veterinary Record,* 168: 137–138.

Letty, J., Aubineau, J., Berger, F. & Marchandeau, S., 2006. Repeuplements de lapins de garenne : enseignements des suivis par radio–pistage. *Faune Sauvage,* 274: 76–88.

Letty, J., Aubineau, J. & Marchandeau, S., 2005. Effect of storage conditions on dispersal and short–term survival of translocated wild rabbits *Oryctolagus cuniculus*. *Wildlife Biology,* 11: 249–255.

– 2008. Improving rabbit restocking success: a review of field experiments in France. In: *Lagomorph biology: evolution, ecology and conservation:* 327–348 (P. C. Alves, N. Ferrand & K. Hacklánder, Eds.). Springer–Verlag, Berlin.

Letty, J., Aubineau, J., Marchandeau, S. & Clobert, J., 2003. Effect of translocation on survival in wild rabbit (*Oryctolagus cuniculus*). *Mammalian Biology. Zeitschrift für Säugetierkunde,* 68: 250–255.

Letty, J., Hirvert, J., Queney, G., Aubineau, J., Monnrerot, M. & Marchandeau, S., 2002a. Assessment of genetic introgression due to a wild rabbit restocking. *Zeitschrift fur Jagdwissenschaft,* 48: 33–41.

Letty, J., Marchandeau, S. & Aubineau, J., 2007. Problems encountered by individuals in animal translocations: Lessons from field studies. *Ecoscience,* 14: 420–431.

Letty, J., Marchandeau, S., Clobert, J. & Aubineau, J., 2000. Improving translocation success: an experimental study of anti–stress treatment and release method for wild rabbits. *Animal Conservation,* 3: 211–219.

Letty, J., Marchandeau, S., Reitz, F., Clobert, J. & Sarrazin, F., 2002b. Survival and movements of translocated wild rabbits (*Oryctolagus cuniculus*). *Game and Wildlife Science,* 19: 1–23.

Marchandeau, S., 2000. Le lapin de garenne : enquête nationale sur les tableaux de chasse à tir – Saison 1998–1999. 2000, 251 – Enquête nationale sur les tableaux de chasse à tir – Saison 1998–1999):. *Faune Sauvage (Cahiers techniques),* 251: 18–25.

Marchandeau, S., Chantal, J., Portejoie, Y., Barraud, S. & Chaval, Y., 1998. Impact of viral hemorrhagic disease on a wild population of European rabbits in France. *Journal of Wildlife Diseases,* 34: 429–435.

Marchandeau, S., Chaval, Y. & Le Goff, E., 2000. Prolongued decline in the abundance of wild European rabbits *Oryctolagus cuniculus* and high immunity level over three years following the arrival of rabbit haemorrhagic disease. *Wildlife Biology,* 6: 141–147.

Mauvy, B., Peroux, R., Lartiges, A. & Sidaine, M., 1991. Repeuplements en lapins de garenne: résultats des essais effectués dans le nord du Massif Central. Première partie: la survie et la dispersion des animaux lâchés. *Bulletin Mensuel Office National de la Chasse,* 157: 9–20.

Moreno, S., Villafuerte, R., Cabezas, S. & Lombardi, L., 2004. Wild rabbit restocking for predator conservation in Spain. *Biological Conservation,* 118: 183–193.

Navarro–González, N., Serrano, E., Casas–Díaz, E., Velarde, R., Marco, L., Rossi, L. & Lavín, S., 2010. Game restocking and the introduction of sarcoptic mange in wild rabbit in north–eastern Spain. *Animal Conservation,* 13.

Piorno, V., 2006. Gestión cinegética y conservación del conejo de monte. Ph. D. Thesis, Universidad de Vigo, Spain.

Reglero, M., Vicente, J., Rouco, C., Villafuerte, R. &

Gortazar, C., 2007. *Trypanosoma* spp. infection in wild rabbits (*Oryctolagus cuniculus*) during a restocking program in Southern Spain. *Veterinary Parasitology,* 149: 178–184.

Rogers, P. M., Arthur, C. P. & Soriguer, R. C., 1994. The rabbit in continental Europe. In: *The European rabbit: the history and biology of a successful colonizer:* 22–63 (C. M. King & H. V. Thompson Eds.). Oxford Univ. Press.

Rouco, C., 2008. Restauración de las poblaciones de conejo de monte y mejora de la gestión para su conservación. Ph. D. Thesis, Univ. de Castilla–La Mancha, Spain.

Rouco, C., Ferreras, P., Castro, F. & Villafuerte, R., 2008. Effect of terrestrial predator exclusion on short term survival of translocated European wild rabbits. *Wildlife Research,* 35: 625–632.

– 2010. A longer confinement period favors European sild rabbit (*Oryctolagus cuniculus*) survival during soft releases in low –cover habitats. *European Journal of Wilidlife Research,* 56: 215–219.

Rouco, C., Villafuerte, R., Castro, F. & Ferreras, P., 2011a. Effect of artificial warren size on a restocked European wild rabbit population. *Animal Conservation,* 14: 117–123.

– 2011b. Response of naïve and experienced European rabbits to predator odour. *European Journal of Wildlife Research,* 57: 395–398.

Ruiz–Aizpurua, L., 2013. Study of territorial behaviour and stress for the improvement of European wild rabbit restocking programs. Ph. D. Thesis, Univ. de Córdoba, Spain.

Ruiz–Aizpurua, L., Planillo, A., Carpio, A. J., Guerrero–Casado, J. & Tortosa, F. S., 2013. The use of faecal markers for the delimitation of the European rabbit's social territories (*Oryctolagus cuniculus* L.). *Acta Ethologica,* 16: 157–162.

San–Miguel, A., 2006. *Manual para la gestión del hábitat el lince ibérico (Lynx pardinus Temminck) y de su presa principal, el conejo de monte (Oryctolagus cuniculus L.).* Fundación CBD–Habitat, Madrid.

Sánchez–García, C., Alonso, M. E., Díez, C., Pablos, M. & Gaudioso, V. R., 2012. An approach to the statistics of wild lagomorph captive rearing for releasing purposes in Spain. *World Rabbit Science,* 20: 49–56.

Teixeira, C. P., De Azevedo, C. S., Mendl, M., Cipreste, C. F. & Young, R. J., 2007. Revisiting translocation and reintroduction programmes: the importance of considering stress. *Animal Behaviour,* 73: 1–13.

Villafuerte, R., Calvete, C., Blanco, J. C. & Lucientes, J., 1995. Incidence of viral haemorrhagic disease in wild rabbit populations in Spain. *Mammalia,* 59: 651–659.

Villafuerte, R. & Castro, F., 2007. Repoblaciones. In: *Claves para o éxito na mellora das poboacións de coello en Galicia:* 46-53. Federación Galega de Caza, España.

Ward, D., 2005. *Reversing rabbit decline. One of the biggest challenges for nature conservation in Spain and Portugal.* Technical report. World Conservation Union, Gland, Switzerland.

7

Spatial distribution patterns of terrestrial bird assemblages on islands of the Sabana–Camagüey Archipelago, Cuba: evaluating nestedness and co–occurrence patterns

C. A. Mancina, D. Rodríguez Batista & E. Ruiz Rojas

Mancina, C. A., Rodríguez Batista, D. & Ruiz Rojas, E., 2013. Spatial distribution patterns of terrestrial bird assemblages on islands of the Sabana–Camagüey Archipelago, Cuba: evaluating nestedness and co–occurrence patterns. *Animal Biodiversity and Conservation*, 36.2: 195–207.

Abstract

Spatial distribution patterns of terrestrial bird assemblages on islands of the Sabana–Camagüey Archipelago, Cuba: evaluating nestedness and co–occurrence patterns.— Using distribution data of 131 terrestrial bird species on 17 islands of the Archipelago Sabana–Camagüey, Cuba, we tested for non–randomness in presence–absence matrices with respect to co–occurrence and nestedness. We conducted separate analyses for the whole assemblage and sub–matrices according to trophic levels and residence status (breeding and migratory). We also explored the influence of weighting factors such as island area and isolation. The C–occurrence analyses were susceptible to the species subsets and the weighting factors. Unweighted analyses revealed a significant negative co–occurrence pattern for the entire assemblage and for most sub–matrices. The area weighted analyses always indicated strong non–random structure. However, an analysis with intra–guild species pairs showed that most pairs were randomly assembled; very few pairs had a significant segregated pattern. Bird assemblages followed a nested subset structure across islands. Nestedness was strongly correlated with area and unrelated with island isolation. Overall, this study suggests that terrestrial bird assemblages were shaped by extinction processes mediated through area effects rather than interspecific trophic guild competition. Data suggest that conservation of largest islands will guarantee high terrestrial bird richness on the archipelago.

Key words: Archipelago, Birds, Community ecology, Cuba, Macroecology, Null models.

Resumen

Patrones de distribución espacial de las agrupaciones de aves terrestres en las islas del archipiélago Sabana–Camagüey, Cuba: evaluación de los patrones de anidamiento y de coexistencia.— Se emplearon datos de distribución de 131 especies de aves terrestres en 17 islas del archipiélago Sabana–Camagüey para analizar la no aleatoriedad en las matrices de presencia y ausencia con respecto a la coexistencia y el anidamiento. Los análisis se realizaron para todo el conjunto y para submatrices por grupos tróficos y estados de residencia (especies migratorias y reproductivas). Además, se analizó la influencia de factores de ponderación, como el área y el aislamiento de las islas. El patrón de coexistencia fue sensible a los grupos de especies y los factores de ponderación. Los análisis no ponderados revelaron un patrón de coexistencia significativamente negativo para todo el conjunto y la mayoría de los grupos. Cuando se usó el área de las islas como factor siempre se observó una estructura no aleatoria de las agrupaciones. Sin embargo, dentro de los gremios tróficos la mayoría de los pares de especies mostraron un patrón aleatorio y muy pocos pares tuvieron un patrón significativamente segregado. La distribución de las aves terrestres sigue una estructura anidada. El anidamiento estuvo fuertemente correlacionado con el área y no presentó relación con el aislamiento de las islas. De manera general este estudio sugiere que las agrupaciones de aves terrestres en este archipiélago están más estructuradas por procesos de extinción selectiva relacionados con el área de las islas, que por la competición interespecífica dentro de gremios tróficos. Los datos sugieren que la conservación de las islas de mayor área podría garantizar una elevada riqueza de especies en el archipiélago.

Palabras claves: Archipiélago, Aves, Ecología de comunidades, Macroecología, Modelos nulos.

Carlos A. Mancina & Daysi Rodríguez Batista, Inst. de Ecología y Sistemática, carretera de Varona km 3 ½ Capdevila, Boyeros, A. P. 8029, C. P. 10800, La Habana, Cuba; Edwin Ruiz Rojas, Centro de Estudios y Servicios Ambientales, Villa Clara (Cuba).

Corresponding author: Carlos A. Mancina. E–mail: mancina@ecologia.cu

Introduction

Recognition of patterns in ecological communities and understanding the mechanisms that produce these patterns are fundamental goals of ecology and conservation biology. An essential question is whether communities are composed of random species assemblages or whether deterministic processes such as competition influence the composition of species within communities. Diamond (1975) expanded this approach with his analyses of the distribution of terrestrial bird species on islands of the Bismarck Archipelago. He found that interspecific interactions determine non–random co–occurrence patterns and proposed rules known as assembly rules, including the checkerboard distribution, forbidden species combinations, and so on. Diamond`s rules and other more recent community assembly rules (such as favored states, food–web structure, guild proportionality, and nested subset) are frequently examined in studies of metacommunities and community ecology (Fortuna et al., 2010; Beaudrot et al., 2013; Henriques–Silva et al., 2013).

Many studies have relied on null models to test the community structure. Null models are randomization methods that exclude a target mechanism to determine whether a specific no–random pattern can be generated (Connor & Simberloff, 1979; Gotelli & Graves, 1996).Two of the most widely applied models are species co–occurrence (Gotelli, 2000) and nestedness (Patterson & Atmar, 1986). The co–occurrence patterns are attributed to competitive inter–specific interactions or environmental factors. Several co–occurrence indices are used to quantify patterns in presence–absence matrices, in many instances relating the observed patterns to Diamond`s assembly rules (Gotelli & McCabe, 2002; Collins et al., 2011; Wang et al., 2011). Nested species subsets are a common pattern of community assembly characteristic of many types of fragmented landscapes and insular systems. Nestedness is a condition in which species distributions occur hierarchically so that the fauna of species–poor islands comprise a perfect subset of the fauna on increasingly species–rich islands (Ulrich et al., 2009). In contract with co–occurrence models, nestedness is not directly related to competition events such as the structuring mechanism of communities. Rather, nested patterns could be related with differential colonization or extinction of species, passive sampling, and carrying capacities, distance or area effects (Patterson & Atmar, 1986; Wright et al., 1998; Ulrich et al., 2009).

Many studies of avian communities on archipelagos or isolated habitats have shown more segregated patterns of co–occurrence than expected by chance, suggesting that interespecific interactions are an underlying mechanism in structuring of bird communities (Stone & Roberts, 1992; Gotelli & McCabe, 2002; Feeley, 2003). Besides, nested patterns of insular bird assemblages are common and have been related to extinction and colonization processes (Lomolino, 1996), habitat nestedness (Calmé & Desrochers, 1999; Wang et al., 2011) and passive sampling (Wright et al., 1998).

The Sabana–Camagüey Archipelago (hereafter SCA) constitutes the largest system of islands or cays in the Caribbean region (Alcolado et al., 2007). Several studies have contributed to the knowledge of avian richness of some islands (*e.g.* Garrido, 1973; Sánchez et al., 1994; Wallace et al., 1996, 1999; Rodríguez, 2000; Sánchez & Rodríguez, 2001; González et al., 2008), and 241 bird species have been reported from this archipelago, representing 65% of the whole Cuban ornithofauna (Rodríguez et al., 2007). However, the distribution patterns and factors that determine the species richness on these islands remain unexplored. The high species richness of birds, their geographic position, and the high number of islands that differ in area and landscape characteristics make this archipelago an appropriate scenario to test hypotheses on assembly and structure of bird communities. In this study, we used null model analysis to test for patterns of species co–occurrence and nestedness with data on presence–absence of terrestrial birds from a set of 17 islands from SCA. We explored the potential role of extinction and colonization events as underlying mechanism in the structure of bird assemblages by analyzing correlations of nestedness and island traits, such as area and isolation.

Material and methods

Study area and avifauna data

The Sabana–Camagüey Archipelago (SCA) is a chain of 2,515 islands or cays along 465 km of the north coast of Cuba; total area of the SCA is c. 3,414 km^2. The islands range in area from < 0.1 km^2 to 680 km^2 Cayo Romano, the largest island of the SCA. The landscape heterogeneity and flora diversity tend to be higher on larger islands such as Sabinal, Coco, Romano and Guajaba (Priego–Santander et al., 2004). The vegetation is diverse and several plant communities have been described for the SCA. The mangrove forest is widespread along coasts and constitutes the main coverage on the smallest islands. The most extensive plant formations are the semi–deciduous and dry evergreen forests, xerophytic scrubs, and sandy coastal vegetation (Alcolado et al., 2007).

Data of bird communities across SCA were gathered from an extensive review of literature and our field data. Although information is available about the bird fauna of 86 islands, we selected only 17 islands because these have more complete information about their avian communities (largest number of surveys across several years and seasons). The selected islands range in area from 0.27 to 680 km^2 and are separated between 0.5 and 33 km from the main island of Cuba (table 1, fig. 1). These variables were obtained from digital maps using the software DIVA–GIS v 7.5 (Hijman et al., 2005).

We selected only terrestrial species because their assemblages should depend on the islands as breeding or feeding sites. The data were organized as a presence–absence matrix in which each row represents a species and each column an island. To ensure that the results were not biased by the inclusion of species with very different strategies in the habitat use, we

Table 1. Characteristics of the study islands in the Sabana–Camagüey Archipelago and number of terrestrial birds on each island: A. Area (in km²); I. Isolation (in km); N. Number of species. (Isolation is given as the nearest distance to the main island of Cuba.)

Tabla 1. Características de las islas estudiadas del archipiélago Sabana–Camagüey y número de especies de aves por islas: A. Área (en km²); I. Aislamiento (en km); N. Número de especies. (El valor de aislamiento es la distancia más cercana a la isla de Cuba.)

Island	A (km²)	I (km)	N
Aguada	2.29	5.47	36
Coco	334.52	21.43	117
Cruz	26.14	29.95	49
Ensenachos	1.45	27.75	39
Fábrica	0.79	4.07	40
Francés	6.22	26.66	40
Guajaba	105.2	10.47	89
Guillermo	15.65	24.42	63
Las Brujas	7.23	24.79	61
Lucas	3.16	5.74	45
Mégano Grande	7.55	31.8	24
Palma	0.27	0.49	46
Paredón Grande	10.71	32.99	84
Romano	680	14.16	88
Sabinal	338.3	2.16	90
Salinas	1.08	4.72	41
Santa María	21.9	28.69	85

generated presence–absence submatrices for two species subsets: 1) breeding *vs.* migratory species (including winter residents), and 2) four trophic guilds of breeding birds (omnivores, predators, insectivores and phytophagous) based on our field observations and published data (*e.g.* Kirkconnell et al., 1992). Vagrants, transients or very rare migrants in the Cuban archipelago were excluded from data analyses (Llanes et al., 2002; Garrido & Kirkconnell, 2010).

Co–occurrence and nestedness analysis

To estimate whether bird species co–occurred more or less than expected by chance, we used the checkerboard score (C–score) index (Stone & Roberts, 1990). C–score measures the average number of 'checkerboard units' among all possible pairs of species. This index measures the extent to which species

are segregated across islands but does not require perfect checkerboard distributions; the C–score should be significantly larger than expected by chance in communities structured by interspecific interactions (Gotelli, 2000).

The C–score index was compared to those of 5,000 randomly assembled communities using the software EcoSim 7.0 (Gotelli & Entsminger, 2001). We used the sequential–swap algorithm to generate random null matrices (Manly, 1995). Simulated matrices were generated under two null models that differ in the way row and column totals are treated: 1) a fixed–fixed (FF) algorithm, where both the row and the column totals of the original matrix are fixed (the biological justification for this model is that it preserves in the null matrices the observed differences between sites in species richness —column totals— and observed differences among species in their frequency of occurrence or row totals, Gotelli, 2000); and 2) a fixed–weighted (FW) algorithm (Gotelli & Entsminger, 2001; Jenkins, 2006), where columns are weighted by factors that during randomization contribute to inter–island differences in community composition. We separately used two weighting factors: the island area (FW_{area}) and the distance ($FW_{isolation}$) from the main island of Cuba (used as an isolation criterion).

We calculated a standardized effect size (SES) as ([observed score–mean simulated score]/standard deviation of simulated score); SES indicates the number of standard deviations that the observed index is above or below the mean index of simulated matrices. Non–random matrices generally have SES for the C–score > |2| (Gotelli & McCabe, 2002). In addition, we used Bayes methods implemented by Gotelli & Ulrich (2010) to identify particular species pairs for each trophic guild that contributes to observed patterns, and to determine those random, segregated or aggregated species pairs. We used the criteria mean–based (Bayes M criterion) and confidence interval (Bayes CL criterion) (Gotelli & Ulrich, 2010) calculated with the software 'Pairs' (Ulrich, 2008).

For nestedness analysis, we used a metric–based on overlap and decreasing fill, NODF (Almeida–Neto et al., 2008). NODF calculates nestedness independently among rows and columns, evaluating nestedness only among islands (*i.e.* species richness) or only among species (*i.e.* species occupancy). NODF varies from 0 to 100 and higher values indicate more nested assemblage. The nestedness significance was estimated on 1,000 random matrices. We used a null model with a fixed–equiprobable algorithm, where the number species in an island is allowed to vary during randomization; this random model represents a scenario where the probability of colonization of all species is equal for all islands (Gotelli, 2000; Ulrich et al., 2009). Nestedness analyses and randomizations were conducted using the software 'NODF' (Almeida–Neto & Ulrich, 2011).

To explore the role of extinction and colonization events upon nestedness we used Spearman rank correlations between island rank order in the maximally packed matrix, and island area and isolation, respectively (Patterson & Atmar, 2000). This method has

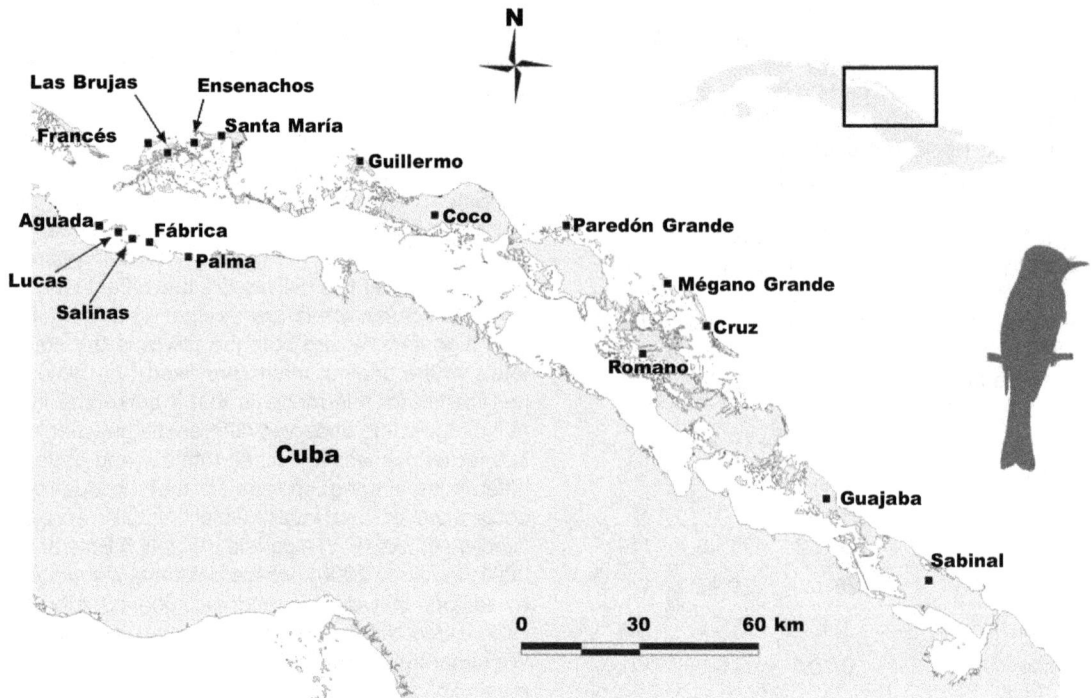

Fig. 1. Map of the Sabana–Camagüey Archipelago, Cuba; islands included in the study are named.

Fig. 1. Mapa del archipiélago Sabana–Camagüey, Cuba; se indican las islas incluidas en el estudio.

proven useful for indicating the possible mechanisms involved in a nested pattern; for example, a significant correlation between isolation and maximal nestedness will be related with immigration or colonization events. However, correlation with the island area suggests that extinction processes should determine the nested pattern (Lomolino, 1996; Patterson & Atmar, 2000; Fernández–Juricic, 2002).

Results

A total of 131 terrestrial bird species were found to inhabit the islands considered in this study (appendix 1). There are similar numbers of breeding (67 species, 51.2%) and migrant (64 species, 48.8%) species; the species number ranges from 24 to 117 species across islands (table 1). Species richness on the islands is significantly correlated with island area ($p < 0.001$; both variables in logarithm) which explained 56% of the variance. Species richness is not correlated with the island isolation ($p = 0.7$).

Co–occurrence patterns

Our results were influenced by the type of null model algorithm used (table 2). The observed C–scores for most subsets, under the F–F model, were significantly higher than expected by chance, suggesting segrega-

ted patterns of species co–occurrence. The C–score did not differ from null model figures only for phytophagous and predators, indicating random species co–occurrence. When the island area was used as a weighting factor (FW_{area}), all subsets were significant (segregated patterns), being stronger (Z value > 10) for the whole assemblage, and for breeding and omnivorous species. On the other hand, when using isolation as the weighting factor ($FW_{isolation}$) the null hypothesis was not rejected, suggesting random co–occurrence patterns. The C–score was found to be marginally significant only for the omnivorous species (observed score = 5.27, expected score = 3.92, $p = 0.04$), suggesting a weak pattern of interspecific segregation.

Analysis of species pairs showed that most of them were randomly assembled. For each trophic guild, very few pairs had a significant segregated pattern. Neither species pairs showed an aggregated pattern. The highest percentage of species pairs occurred for omnivorous and insectivorous species; the Bayes confidence interval criterion identified only 4.5% and 2.2% of segregated pairs, respectively. Table 3 shows the significantly segregated species pairs with highest values of C–Score; other species–pairs such as Gray Kingbird (*Tyrannus dominicensis*) – Oriente Warbler (*Teretistris fornsi*), Mangrove Cuckoo (*Coccyzus minor*) – Cuban Tody (*Todus multicolor*), and Smooth Billed Ani (*Crotophaga ani*) – Bahama Mockingbird (*Mimus gundlachii*), had values significantly

Table 2. Results from the analysis of species co–occurrence of terrestrial bird assemblages inhabiting 17 islands of the Sabana–Camagüey Archipelago. The observed C–Score, the values expected by chance, and standardized effect size (in parentheses) are shown for each species subset. (Significant results in bold.)

Tabla 2. Resultados de los análisis de coexistencia de las especies en las agrupaciones de aves terrestres que habitan en las islas del archipiélago Sabana–Camagüey. Para cada subgrupo de especies se muestran el valor del conteo C observado, los valores esperados por efecto del azar y el valor del tamaño del efecto estandarizado (entre paréntesis). (Los resultados significativos se indican con negritas.)

| Subset (# species) | Observed C–score | Simulated C–scores | | |
		F–F	FW_{area}	$FW_{isolation}$
All species (131)	3.41	3.18 (**6.61**)	1.02 (**19.98**)	3.98 (–2.65)
Migratory (64)	2.71	2.55 (**2.65**)	1.19 (**7.62**)	5.11 (–6.73)
Breeding (67)	2.83	2.61 (**5.17**)	0.85 (**13.21**)	3.04 (–0.74)
Phytophagous (16)	0.82	0.71 (1.27)	0.37 (**2.36**)	1.43 (–1.68)
Omnivorous (15)	5.27	4.86 (**2.54**)	1.02 (**11.22**)	3.92 (**1.76**)
Predators (10)	2.22	2.08 (0.72)	0.92 (**2.63**)	3.82 (–1.61)
Insectivorous (14)	2.84	2.60 (2.13)	0.93 (**5.22**)	2.33 (0.81)

segregated but with low C–scores, suggesting weakly segregated patterns between these species pairs.

Nestedness

The entire community of terrestrial birds showed a significantly nested pattern (NODF = 78.41, $p < 0.0001$). The breeding subset of bird species showed higher degrees of nestedness than the migratory assemblage. The degree of nestedness of species richness among islands (columns) was higher than the degree of nestedness in species occupancy (rows) for whole assemblage and for migratory and breeding subset separately (table 4). Spearman rank correlations between species order in the maximally nested matrix with island area and isolation indicate that area is the most important factor in nestedness (table 5). The analysis suggests that the distance to main island of Cuba has no influence on the degree of nestedness of avian assemblages from the Sabana–Camagüey Archipelago.

Table 3. Species pairs with the highest and most significant figures of C–Score (Obs.) denoting segregated distribution patterns. The number of occurrences and the number of islands with joint occurrences (U) are shown in brackets. For each species pairs, the values expected by chance (Sim.) and standardized effect size (SES) are shown.

Tabla 3. Parejas de especies con los índices de conteo C (Obs.) más elevados y significativos, indicando patrones significativamente segregados. Se muestran entre paréntesis el número de observaciones y la cantidad de islas donde coexisten (U). Para cada pareja de especies se muestra los valores esperados por efecto del azar (Sim.) y el valor del tamaño del efecto estandarizado (SES).

Species 1	Species 2	U	Obs.	Sim. (SES)
Mimus gundlachii (8)	*Dives atroviolaceus* (5)	0	1.0	0.212 (**5.42**)
Mimus gundlachii (8)	*Priotelus temnurus* (4)	0	1.0	0.215 (**4.88**)
Icterus melanopsis (7)	*Dives atroviolaceus* (5)	1	0.68	0.227 (**3.07**)
Glaucidium siju (9)	*Accipiter striatus* (3)	1	0.59	0.082 (**3.79**)
Geotrygon chrysia (7)	*Tiaris bicolor* (3)	1	0.57	0.072 (**4.31**)

Table 4. Results of nestedness analyses for the terrestrial bird assemblages on islands of the Sabana–Camagüe Archipelago. The table shows observed (NODF$_{obs}$) and expected by chance (NODF$_{sim}$) values, and also the degree of nestedness independently for columns and rows. (The standardized effect size is shown in brackets; all combinations were significantly nested, in bold.)

Tabla 4. Resultados de los análisis de anidamiento de las agrupaciones de aves terrestres en las islas del archipiélago Sabana–Camagüey. Se muestran los valores del índice observado (NODF$_{obs}$) y los valores esperados por efecto del azar (NODF$_{sim}$), así como el grado de anidamiento para filas y columnas de forma independiente. (Entre paréntesis se muestra el valor del tamaño del efecto estandarizado; todas las combinaciones fueron significativamente anidadas, en negrita.)

	Total		Columns		Rows	
	NODF$_{obs}$	NODF$_{sim}$	NODF$_{obs}$	NODF$_{sim}$	NODF$_{obs}$	NODF$_{sim}$
All species	78.41	61.26 (**72.75**)	83.58	62.79 (**11.62**)	78.32	61.23 (**73.2**)
Migratory	78.15	51.36 (**49.43**)	80.47	53.12 (**13.29**)	78.00	51.24 (**55.17**)
Breeding	84.46	69.13 (**31.37**)	86.15	68.75 (**6.88**)	84.36	69.15 (**37.64**)

Discussion

Our analyses show that terrestrial bird species co–occurred less frequently than expected by chance on islands from the Sabana–Camagüey Archipelago, suggesting that these avian communities are probably structured by negative interspecific interactions. However, similar to other studies (*e.g.* Meyer & Kalko, 2008), the results were susceptible to the species subsets and the weighting factors. When the fixed–fixed model was used we found random co–occurrence patterns for predators and phytophagous species. However, using area as weighting, all subsets showed significant segregated co–occurrence patterns. Contrarily, weighting analyses by island isolation showed random patterns for most species subsets. This result suggests that because of the short distance between the archipelago and the main island of Cuba, the differential dispersal abilities of the bird species would not be an important factor in the structure of the avian assemblages. On the other hand, island area and other attributes associated with of area, such as landscape diversity or the number of plant formations (see Priego–Santander et al., 2004), have a more important role structuring the bird communities of the Sabana–Camagüey Archipelago.

Although the avifauna assemblages showed a wide segregated pattern, we found that species pairs, within each trophic guild, showed random patterns. This result suggests competitive exclusion could be rare in these bird assemblages. A similar result was obtained for other avifauna on archipelagos (Gotelli & Ulrich, 2010; Collins et al., 2011), and might reflect widespread, but weak species interactions or mechanisms of species segregation that are not related to direct species interactions but to historical events or resource abundance (Gotelli & McCabe, 2002; Gotelli & Ulrich, 2010).

The analyses indicate a strong nested structure in the entire assemblage and for breeding and migratory birds. Common and widespread species (*e.g.* Greater Antillean Grackle, Cuban Emerald, Yellow Warbler, etc.) tended to comprise the avifauna of islands with lesser species richness, while richer islands included these species in addition to other rare species or with restricted ranges. We found that island nested rank order was significantly correlated with the rank order of island area but not with island isolation. This result, together with the significant species–area relationship, suggests that the terrestrial bird assemblages at SCA are structured through

Table 5. Results of Spearman Rank correlations of island order in the maximally nested matrix with the values of area and isolation; *p* values were generated by 1,000 Monte Carlo simulations: A. Area; I. Isolation.

Tabla 5. Resultados de la correlación por rangos de Spearman entre los valores ordinales que le corresponde a cada isla en la matriz de máximo anidamiento y sus valores de área y aislamiento; los valores p fueron generados por 1.000 simulaciones de Monte Carlo: A. Área; I. Aislamiento.

Subset	A r_s	*p*	I r_s	*p*
All species	–0.76	0.0003	0.078	0.76
Migratory	–0.74	0.0005	–0.24	0.34
Breeding	–0.73	0.00004	0.16	0.53

local extinction rather than through colonization or immigration processes from the main island of Cuba (Lomolino, 1996; Wright et al., 1998).

The lower nested pattern observed in the migratory assemblages would be related to habitat generalists with high dispersal abilities (*e.g.* some wintering migrant passerines such as Black and White Warbler, Palm Warbler, American Redstart; Rappole, 1995; Wallace et al., 1996; Latta et al., 2003). The highest nested patterns of breeding birds would be related to the low habitat diversity or limited resource abundance on the small islands, although these would be limiting factors mainly for those breeding species with large area requirements or habitat specialists (*e.g.* Gundlach`s Hawk, Zapata Sparrow, Fernandina`s Flicker, Cuban Grassquit, etc.). Among the islands smaller than 15 km^2, Cayo Paredón Grande had the highest species richness, with 84 bird species. This island has unusually high landscape heterogeneity and floristic diversity (Priego–Santander et al., 2004), supporting the idea that habitat diversity is an important factor in explaining the distribution and species richness on the archipelago.

Ours results are consistent with several studies that show that nested avian assemblages on islands or fragmented habitat are apparently shaped by selective extinction processes through island or patch area and the habitat diversity effects rather than interspecific guild competition. (Fernández–Juricic, 2000; Feeley, 2003; Wang et al., 2011). The strong nested patterns and significant species–area relationships of the avian assemblages suggest, from a conservation perspective, that the protection of the largest islands with the most species rich assemblages (*e.g.* Romano, Sabinal, Coco, Guajaba and Santa María) will warrant high terrestrial bird richness. However, an adequate conservation strategy will be to conserve small and large islands with the purpose of maintaining a high heterogeneity in the environmental conditions on the Sabana–Camagüey Archipelago (Fischer & Lindenmayer, 2005).

Acknowledgements

We are grateful to Luis M. Carrascal, Eduardo E. Iñigo–Elias and Cayetano Casado for their reviews and constructive comments on drafts.

References

A.O.U. (American Ornithologist Union), 2011. Check–list of North American Birds. http://www.aou.org/checklist

Alcolado, P., García, E. E. & Arellano–Acosta, M., 2007. *Ecosistema Sabana–Camagüey. Estado actual, avances y desafíos en la protección y uso sostenible de la biodiversidad*. Editorial Academia. La Habana.

Almeida–Neto, M., Guimaraes, P., Guimaraes, P. R., Loyola, R. D. & Ulrich, W., 2008. A consistent metric for nestedness analysis in ecological systems: reconciling concept and measurement. *Oikos*, 117: 1227–1239.

Almeida–Neto, M. & Ulrich, W., 2011. A straightforward computational approach for measuring nestedness using quantitative matrices. *Environmental Modelling and Software*, 26: 173–178.

Beaudrot, L., Struebig, M. J., Balen, S., Meijaard, E., Husson, S. & Marshall, A. J., 2013. Co–occurrence patterns of Bornean vertebrates suggest competitive exclusion is strongest among distantly related species. *Oecologia*. DOI: 10.1007/s00442–013–2679–7

Calmé, S. & Desrochers, A., 1999. Nested bird and micro–habitat assemblages in a peatland archipelago. *Oecologia*, 118: 361–370.

Collins, M. D., Simberloff, D. & Connor, E. F., 2011. Binary matrices and checkerboard distributions of birds in the Bismarck Archipelago. *Journal of Biogeography*, 38: 2373–2383.

Connor, E. F. & Simberloff, D., 1979. The assembly of species communities: chance or competition. *Ecology,* 60: 1132–1140.

Diamond, J. M., 1975. Assembly of species communities. In: *Ecology and Evolution of Communities*: 342–444 (M. L. Cody & J. M. Diamond, Eds.). Harvard Press, Cambridge.

Feeley, K., 2003. Analysis of avian communities in Lake Guri, Venezuela, using multiple assembly rule models. *Oecologia*, 137: 104–113.

Fernández–Juricic, E., 2000. Bird community composition patterns in urban parks of Madrid: the role of age, size and isolation. *Ecological Research*, 15: 373–383.

– 2002. Can human disturbance promote nestedness? A case study with breeding birds in urban habitat fragments. *Oecologia*, 131: 269–278.

Fischer, J. & Lindenmayer, D. B., 2005. Perfectly nested or significantly nested–an important difference for conservation management. *Oikos*, 109: 485–494.

Fortuna, M. A., Stouffer, D. B., Olesen, J. M., Jordano, P., Mouillot, D., Krasnov, B. R., Poulin, R. & Bascompte, J., 2010. Nestedness versus modularity in ecological networks: two sides of the same coin? *Journal of Animal Ecology,* 79: 811–817.

Garrido, O. H., 1973. Anfibios, reptiles y aves del Archipiélago de Sabana–Camagüey, Cuba. *Torreia*, 27: 1–72.

Garrido, O. H. & Kirkconnell, A., 2010. *Aves de Cuba*. Comstock Publishing Associates. Cornell Univ.

González, H., Pérez, E., Rodríguez, P. & Barrio, O., 2008. Composición y abundancia de las comunidades de aves terrestres residentes y migratorias en cayo Sabinal, Cuba. *Poeyana*, 496: 23–32.

Gotelli, N. J., 2000. Null model analysis of species co–occurrence patterns. *Ecology*, 81: 2606–2621.

Gotelli, N. J. & Entsminger, G. L., 2001. EcoSim: Null models software for ecology. Version 7.0. Acquired Intelligence Inc. &Kesey–Bear. http://homepages.together.net/~gentsmin/ecosim.htm

Gotelli, N. J. & Graves, G. R., 1996. *Null Models in Ecology*. Smithsonian Institution. Press, Washington D.C.

Gotelli, N. J. & McCabe, D. J., 2002. Species co–occurrence: a meta–analysis of J. M. Diamond's assembly rules model. *Ecology*, 83: 2091–2096.

Gotelli, N. J. & Ulrich, W., 2010. The empirical Bayes approach as a tool to identify non–random species associations. *Oecologia*, 162: 463–477.

Henriques–Silva, R., Lindo, Z. & Peres–Neto, P., 2013. A community of metacommunities: exploring patterns in species distributions across large geographical areas. *Ecology*, 94: 627–639.

Hijmans, R. J., Guarino, L., Bussink, C., Mathur, P., Cruz, M., Barrentes, I. & Rojas, E., 2005. DIVA–GIS. Vsn. 7.5. A geographic information system for the analysis of species distribution data. (manual available at: http://www.diva–gis.org).

Jenkins, D. G., 2006. In search of quorum effects in metacommunity structure: species co–occurrence analyses. *Ecology*, 87:1523–1531.

Kirkconnell, A., Garrido, O. H., Posada, R. M. & Cubillas, S., 1992. Los grupos tróficos en la avifauna cubana. *Poeyana*, 415: 1–21.

Latta, S. C., Rimmer, C. C. & Mcfarland, Y. K. P., 2003. Winter bird communities in four habitats along an elevational gradient on Hispaniola. *TheCondor*, 105: 179–197.

Llanes, A., González, H., Sánchez, B. & Pérez, E., 2002. Lista de las aves registradas para Cuba. In: *Aves de Cuba*: 147–155 (H. González, Ed.). UPC Print, Vaasa, Finland.

Lomolino, M. V., 1996. Investigating causality of nestedness of insular communities: selective immigrations or extinctions? *Journal of Biogeography*, 23: 699–703.

Manly, B. F. J., 1995. A note on the analysis of species co–occurrences. *Ecology*, 76: 1109–1115.

Meyer, C. F. J. & Kalko, E. K. V., 2008. Bat assemblages on Neotropical land–bridge islands: nested subsets and null model analyses of species co–occurrence patterns. *Diversity and Distributions*, 14: 644–654.

Patterson, B. D. & Atmar, W., 1986. Nested subsets and the structure of insular mammalian faunas and archipelagos. *Biological Journal of the Linnean Society*, 28: 65–82.

– 2000. Analyzing species composition in fragments. In: *Isolated vertebrate communities in the tropics*: 1–16 (G. Rheinwald, Ed.). *Proceedings 4th International Symposium. Bonn Zool. Monogr.*, 46.

Priego–Santander, A. G., Palacio–Prieto, J. L., Moreno–Casasola, P., López–Portillo, J. & Geissert, D., 2004. Heterogeneidad del paisaje y riqueza de flora: su relación en el Archipiélago de Camagüey, Cuba. *Interciencia*, 29: 138–144.

Rappole, J. H., 1995. *Ecology of migratory birds: a Neotropical perspective*. Smithsonian Institution Press, Washington DC.

Rodríguez, D., 2000. Composición y estructura de las comunidades de aves en tres formaciones vegetales de Cayo Coco, Archipiélago de Sabana–Camagüey, Cuba. Ph D. Thesis, Instituto de Ecología y Sistemática, La Habana, Cuba.

Rodríguez, D., Martínez, M., Arias, A., Ruiz E., Llanes, A., Pérez, E., Rodríguez, P., Socarrás, E., González, H., Parada, A., Barrios, O., Vilma, E., Chamizo, A. & Mancina, C. A., 2007. Vertebrados terrestres. In: *Ecosistema Sabana–Camagüey: Estado actual, avances y desafíos en la protección y uso sostenible de la biodiversidad*: 31–37 (P. M. Alcolado, E. E. García & M. Arellano–Acosta, Eds.). Editorial Academia, La Habana.

Sánchez, B. & Rodríguez, D., 2001. Avifauna asociated with the aquatic and coastal ecosystems of Cayo Coco, Cuba. *Pitirre*, 13: 68–75.

Sánchez, B., Rodríguez, D. & Kirkconnell, A., 1994. Avifauna de los cayos Paredón Grande y Coco durante la migración otoñal de 1990 y 1991. *Avicennia*, 1:31–38.

Simaiakis, S. M., Dretakis, M., Barboutis, C., Katritis, T., Portolou, D. & Xirouchakis, S., 2012. Breeding land birds across the Greek islands: a biogeographic study with emphasis on faunal similarity, species–area relationships and nestedness. *Journal of Ornithology*. DOI: 10.1007/s10336–011–0803–1

Stone, L. & Roberts, A., 1990. The checkerboard score and species distributions. *Oecologia*, 85: 74–79.

– 1992. Competitive exclusion or species aggregation? An aid in deciding. *Oecologia*, 91: 419–424.

Ulrich, W., 2008. Pairs – a FORTRAN program for studying pair–wise species associations in ecological matrices. www.uni.torun.pl/~ulrichw

Ulrich, W., Almeida–Neto, M. & Gotelli, N. J., 2009. A consumer's guide to nestedness analysis. *Oikos*, 118: 3–17.

Wallace G. E., González, H., McNicholl, M. K., Rodríguez, D., Oviedo, R., Llanes, A. & Sánchez, B., 1996. Winter surveys of forest–dwelling neotropicalmigrant birds and resident birds in three regions of Cuba. *The Condor*, 98: 745–768.

Wallace, G. E., Wallace, E. A. H., Froehlich, D. R., Ealker, B., Kirkconnell, A. & Socarrás, E., 1999. Hermit Thrush and Black–throated Gray Warbler, new for Cuba, and other significant bird records from Cayo Coco and vicinity, Ciego de Avila province, Cuba, 1995–1997. *Florida Field Naturalist*, 27: 37–51.

Wang, Y., Chen, S. & Ding, P., 2011. Testing multiple assembly rule models in avian communities on islands of an inundated lake, Zhejiang Province, China. *Journal of Biogeography*, 38: 1330–1344.

Wright, D. H., Patterson, B. D., Mikkelson, G., Cutler, A. & Atmar, W., 1998. A comparative analysis of nested subset patterns in species composition. *Oecologia*, 113:1–20.

Appendix 1. List of terrestrial bird species included in this study, ordering species according to the maximally nested matrix. Status (B. Breeding, M. Migratory). Trophic group (TG: O. Omnivores, P. Predators, I. Insectivores, Ph. Phytophagous). The last column shows the number of islands where each species was recorded. Nomenclature follows A.O.U. (2011).

Apéndice 1. Listado de las especies de aves terrestres incluidas en este estudio, ordenadas acorde a su posición en la matriz de máximo anidamiento. Status (B. Reproductora, M. Migratoria). Grupo trófico (TG: O. Omnívora, P. Depredadora, I. Insectívora; Ph. Fitófaga). En la última columna se indica el numero de islas donde la especie han sido registrada. Taxonomía según la A.O.U. (2011).

Common name	Scientific name	Status	TG	Islands
Greater Antillean Grackle	*Quiscalus niger*	B	O	17
Turkey Vulture	*Cathartes aura*	B	P	17
White Crowned Pigeon	*Patagioenas leucocephala*	B	Ph	17
Cuban Pewee	*Contopus caribaeus*	B	I	17
Cuban Emerald	*Chlorostilbon ricordii*	B	Ph	17
Commonm Ground Dove	*Columbina passerina*	B	Ph	17
Yellow Warbler	*Setophaga petechia*	B	I	17
Yellow–faced Grassquit	*Tiaris olivaceus*	B	Ph	16
American Kestrel	*Falco sparverius*	B	P	16
Black and White Warbler	*Mniotilta varia*	M	I	16
Northern Mockingbird	*Mimus polyglottos*	B	O	16
Palm Warbler	*Setophaga palmarum*	M	I	16
American Redstart	*Setophaga ruticilla*	M	I	16
Western Spindalis	*Spindalis zena*	B	Ph	16
La Sagra's Flycatcher	*Myiarchus sagrae*	B	I	16
Cuban Bullfinch	*Melopyrrha nigra*	B	Ph	16
Ovenbird	*Seiurus aurocapilla*	M	I	15
Cuban Green Woodpecker	*Xiphidiopicus percussus*	B	O	15
Gray Kingbird	*Tyrannus dominicensis*	B	I	15
White–winged Pigeon	*Zenaida asiatica*	B	Ph	15
Loggerhead Kingbird	*Tyrannus caudifasciatus*	B	I	15
Black–throated Blue Warbler	*Setophaga caerulescens*	M	I	15
Common Yellowthroat	*Geothlypis trichas*	M	I	15
Mourning Dove	*Zenaida macroura*	B	Ph	15
Killdeer	*Charadrius vociferus*	B	O	14
Great Lizard–Cuckoo	*Coccyzus merlini*	B	P	14
Gray Catbird	*Dumetella carolinensis*	M	O	14
Black–whiskered Vireo	*Vireo altiloquus*	B	I	14
Red–legge Thrush	*Turdus plumbeus*	B	O	14
Smooth Billed Ani	*Crotophaga ani*	B	O	13
Cuban Vireo	*Vireo gundlachii*	B	I	13
Yellow–bellied Sapsucker	*Sphyrapicus varius*	M	I	13
Zenaida Dove	*Zenaida aurita*	B	Ph	12
Northern Parula	*Setophaga americana*	M	I	12
Praire Warbler	*Setophaga discolor*	M	I	12

Appendix 1. (Cont.)

Common name	Scientific name	Status	TG	Islands
Northern Waterthrush	*Parkesia noveboracensis*	M	I	12
Barn Owl	*Tyto alba*	B	P	12
Indigo Bunting	*Passerina cyanea*	M	Ph	11
West Indian Woodpecker	*Melanerpes superciliaris*	B	O	11
Yellow–throated Vireo	*Vireo flavifrons*	M	I	11
Yellow–throated Warbler	*Setophaga dominica*	M	I	11
Crested Caracara	*Caracara cheriway*	B	P	10
Merlin	*Falco columbarius*	M	P	10
Magnolia Warbler	*Setophaga magnolia*	M	I	10
Cape May Warbler	*Setophaga tigrina*	M	I	10
Red–tailed Hawk	*Buteo jamaicensis*	B	P	10
Mangrove Cuckoo	*Coccyzus minor*	B	I	10
Blue–gray Gnatcatcher	*Polioptila caerulea*	M	I	10
Peregrine Falcon	*Falco peregrinus*	M	P	9
Cuban Pygmy–Owl	*Glaucidium siju*	B	P	9
Cuban Tody	*Todus multicolor*	B	I	9
Blue Grosbeak	*Passerina caerulea*	M	Ph	9
Cave Swallow	*Petrochelidon fulva*	B	I	9
Painted Bunting	*Passerina ciris*	M	Ph	9
Antillean Nighthawk	*Chordeiles gundlachii*	B	I	8
Greater Antillean Nighthawk	*Caprimulgus cubanensis*	B	I	8
Bahama Mockingbird	*Mimus gundlachii*	B	O	8
Cuban Gnatcatcher	*Polioptila lembeyei*	B	I	8
Worm–eating Warbler	*Helmitheros vermivorum*	M	I	7
Northern Flicker	*Colaptes auratus*	B	O	7
White–eyed Vireo	*Vireo griseus*	M	I	7
Key West Quail–Dove	*Geotrygon chrysia*	B	Ph	7
Bobolink	*Dolichonyx oryzivorus*	M	O	7
Blackpoll Warbler	*Setophaga striata*	M	I	7
Yellow–billedCuckoo	*Coccyzus americanus*	B	I	7
Cuban Oriole	*Icterus melanopsis*	B	O	7
Oriente Warbler	*Teretistris fornsi*	B	I	7
Baltimore Oriole	*Icterus galbula*	M	O	6
Yellow–rumped Warbler	*Setophaga coronata*	M	I	6
Scarlet Tanager	*Piranga olivacea*	M	Ph	6
Chuck–will'swidow	*Caprimulgus carolinensis*	M	I	6
Rose–breasted Grosbeak	*Pheucticus ludovicianus*	M	Ph	6
Ruddy Quail–Dove	*Geotrygon montana*	B	Ph	6
Tawny–shouldered Blackbird	*Agelaius humeralis*	B	O	6
Cuban Blackbird	*Dives atroviolaceus*	B	O	5
Cuban Martin	*Progne cryptoleuca*	B	I	5

Appendix 1. (Cont.)

Common name	Scientific name	Status	TG	Islands
Barn Swallow	Hirundo rustica	M	I	5
Tree Swallow	Tachycineta bicolor	M	I	5
Bay–breasted Warbler	Setophaga castanea	M	I	5
Bananaquit	Coereba flaveola	M	I	5
Prothonotary Warbler	Protonotaria citrea	M	I	5
Black–throated Green Warbler	Setophaga virens	M	I	5
Hooded Warbler	Setophaga citrina	M	I	5
Burrowing Owl	Athene cunicularia	B	P	5
Northern Harrier	Circus cyaneus	M	P	5
Cuban Crow	Corvus nasicus	B	O	4
Grasshopper Sparrow	Ammodramus savannarum	M	Ph	4
Eastern Meadowlark	Sturnella magna	B	O	4
Orchard Oriole	Icterus spurius	M	O	4
Blackburnian Warbler	Setophaga fusca	M	I	4
Swainson's Warbler	Limnothlypis swainsonii	M	I	4
Golden–winged Warbler	Vermivora chrysoptera	M	I	4
Bare–legged Owl	Gymnoglaux lawrencii	B	P	4
Summer Tanager	Piranga rubra	M	Ph	4
Cuban Trogon	Priotelus temnurus	B	O	4
Swainson`sThrush	Catharus ustulatus	M	O	4
Eastern Wood–Pewee	Contopus virens	M	I	4
Red–eyed Vireo	Vireo olivaceus	M	I	4
Scaly–naped Pigeon	Patagioenas squamosa	B	Ph	4
Gundlach's Hawk	Accipiter gundlachi	B	P	3
Broad–winged Hawk	Buteo platypterus	B	P	3
Black–faced Grassquit	Tiaris bicolor	B	Ph	3
Sharp–shinned Hawk	Accipiter striatus	B	P	3
Plain Pigeon	Patagioenas inornata	B	Ph	3
Savannah Sparrow	Passerculus sandwichensis	M	Ph	3
Chestnut–sided Warbler	Setophaga pensylvanica	M	I	3
Louisiana Waterthrush	Parkesia motacilla	M	I	3
Short–eared Owl	Asio flameus	B	P	3
Gray–cheeked Thrush	Catharus minimus	M	O	3
Kentucky Warbler	Geothlypis formosa	M	I	2
Orange–crowned Warbler	Oreothlypis celata	M	I	2
Tennessee Warbler	Oreothlypis peregrina	M	I	2
Nashville Warbler	Oreothlypis ruficapilla	M	I	2
Wilson's Warbler	Cardenilla pusilla	M	I	2
Stygian Owl	Asio stygius	B	P	2
Veery	Catharus fuscescens	M	O	2
Wood Thrush	Hylocichla mustelina	M	O	2

Appendix 1. (Cont.)

Common name	Scientific name	Status	TG	Islands
Thick–billed Vireo	*Vireo crassirostris*	B	I	2
Philadelphia Vireo	*Vireo philadelphicus*	M	I	2
Cedar Waxwing	*Bombycilla cedrorum*	M	Ph	2
Cuban Grassquit	*Tiaris canorus*	B	Ph	2
Zapata Sparrow	*Torreornis inexpectata*	B	O	2
Clay–colored Sparrow	*Spizella pallida*	M	Ph	2
Northern Rough–winged Swallow	*Stelgidopteryx serripennis*	M	I	2
Shiny Cowbird	*Molothrus bonariensis*	B	O	2
Yellow–breasted Chat	*Icteria virens*	M	I	2
Northern Bobwhite	*Colinus virginianus*	B	Ph	1
Fernandina's Flicker	*Colaptes fernandinae*	B	I	1
Red–legged Honeycreeper	*Cyanerpes cyaneus*	B	Ph	1
Ruby–throated Hummingbird	*Archilochus colubris*	M	Ph	1
Sandhill Crane	*Grus canadensis*	B	O	1

8

Dalechampii oak (*Quercus dalechampii* Ten.), an important host plant for folivorous lepidoptera larvae

M. Kulfan, M. Holecová & P. Beracko

Kulfan, M., Holecová, M. & Beracko, P., 2013. Dalechampii oak (*Quercus dalechampii* Ten.), an important host plant for folivorous lepidoptera larvae. *Animal Biodiversity and Conservation*, 36.1: 13–31.

Abstract

*Dalechampii oak (*Quercus dalechampii *Ten.), an important host plant for folivorous lepidoptera larvae.*— We conducted a structured analysis of lepidoptera larvae taxocenoses living in leaf bearing crowns of Dalechampii oak (*Quercus dalechampii* Ten.) in nine study plots in the Malé Karpaty Mountains (Central Europe). The differences between lepidoptera taxocenoses in individual oak stands were analyzed. A total of 96 species and 2,140 individuals were found. Species abundance peaked in May, while number of species and species diversity reached the highest values from April to May and from April to June, respectively. Abundance showed two notable peaks in flush feeders and in late summer feeders. Lepidoptera taxocenosis in the study plot Horný háj (isolated forest, high density of ants) differed significantly from all other taxocenoses according to Sörensen's index of species similarity, species diversity, analysis of similarity on the basis of permutation and pairwise tests (ANOSIM), seasonal variability of species composition, and NMDS ordination.

Key words: Moths, Caterpillars, *Q. dalechampii*, Malé Karpaty Mountains, SW Slovakia.

Resumen

*El roble de dalechampii (*Quercus dalechampii *Ten.), una importante planta hospedadora de las larvas de lepidópteros filófagos.*— Llevamos a cabo un análisis estructurado de las taxocenosis de larvas de lepidópteros que viven en las copas del roble de dalechampii (*Quercus dalechampii* Ten.) en nueve parcelas del estudio en los Pequeños Cárpatos (Europa central). Se analizaron las diferencias entre las taxocenosis de lepidópteros de cada roble. Se hallaron 96 especies y 2.140 individuos. La abundancia de especies alcanzó su valor más elevado en mayo, mientras que el número y la diversidad de especies fueron máximos desde abril hasta mayo y desde abril hasta junio, respectivamente. La abundancia mostró dos máximos notables en las larvas que se alimentan durante la brotación y las que se alimentan al final del verano. La taxocenosis de los lepidópteros en la parcela del estudio Horný háj (un bosque aislado con una elevada densidad de hormigas) difirió significativamente de las demás taxocenosis según el índice de Sörensen para la similitud de las especies, la diversidad de las especies, el análisis de la similitud sobre la base de las pruebas de permutación y las pruebas de pares (ANOSIM), la variabilidad estacional de la composición de especies y el escalamiento multidimensional no métrico (NMDS por sus siglas en inglés).

Palabras clave: Polillas, Orugas, *Q. dalechampii*, Pequeños Cárpatos, Eslovaquia sudoccidental.

Miroslav Kulfan & Pavel Beracko, Dept. of Ecology, Fac. of Natural Sciences, Comenius Univ., Mlynská dolina B–1, SK–84215 Bratislava, Slovakia.– Milada Holecová, Dept. of Zoology, Fac. of Natural Sciences, Comenius Univ., Mlynská dolina B–1, SK–84215 Bratislava, Slovakia.

Corresponding author: M. Kulfan. E–mail: kulfan@fns.uniba.sk

Introduction

Oaks belong to the woody plants that host the richest insect assemblages in Central Europe (Patočka et al., 1999). Lepidoptera larvae have been shown to be the most important group of oak defoliators (Patočka et al., 1962, 1999). About 250 lepidoptera species are known to damage the assimilation tissue of oaks in Central Europe (Patočka et al., 1999; Reiprich, 2001).

Lepidoptera fauna on some oak species in Central Europe have been relatively well studied (Patočka et al., 1962, 1999; Csóka, 1990–1991, 1998a, 1998b; Kulfan, 1990, 1997; Kulfan, 1992; Kulfan et al., 1997, 2006; Kulfan & Degma, 1999; Turčáni et al., 2009, 2010; Parák et al., 2012, etc.). Taxocenoses of lepidoptera caterpillars on three oak species from Slovakia and the Czech Republic (*Quercus robur, Q. petraea* and *Q. cerris*) have been used to explain why there are so many species of herbivorous insects in tropical rainforests (Novotny et al., 2006).

However, the lepidoptera fauna related to *Q. dalechampii* growths has been poorly explored in Europe. A total of nine lepidoptera miner species from families Nepticulidae, Tischeriidae and Gracillariidae have been recorded on *Q. dalechampii* in southern Slovakia (Arborétum Čífáre) (Skuhravý et al., 1998). Kollár (2007) mentions the species *Phyllonorycter roboris* (lepidoptera miner) as a pest of *Q. dalechampii* in Slovakia. Stolnicu (2007) studied lepidoptera leaf-miners on *Q. dalechampii* in Romania. Kulfan (2012) partially studied economically most important pest species on *Q. dalechampii* in Central Europe.

Dalechampii oak (*Quercus dalechampii* Ten.) is one of the most common oaks in Europe and is naturally distributed in Western Italy, Sicily, Greece, Albania, Montenegro, Macedonia, Bosnia & Herzegovenia, Serbia, Slovenia, Austria, Hungary, Slovakia, Romania, and Bulgaria.

The main aims of the present study were: (i) to analyze the structure taxocenoses, alpha diversity and representation of trophic groups and seasonal guilds of lepidoptera en bloc on Dalechampii oak; (ii) to complete data concerning biodiversity of lepidoptera species feeding on oaks in Central Europe; and (iii) to highlight the differences among the individual study plots representing various types of oak forests, with emphasis on fragmentation, forest age and crown canopy.

Material and methods

Material was collected by the beating method into a tray of 1 m diameter (one quantitative sample = beating from 25 branches) on nine selected plots at regular 2–weekly intervals from April to October 2000–2002. Samples were taken from branches at a height of about 1–2.5 m above ground with varying exposure to cardinal points. Larvae were identified using the keys by Gerasimov (1952), Patočka (1954, 1980) and Patočka et al. (1999). Seasonal guilds of lepidoptera caterpillars were established according to Turčáni et al. (2009).

The complete linkage clustering in combination with Sörensen's index and Wishart's similarity ratio was used to classify the taxocenoses. Visualization of dendrograms was done by computer program Syn–tax, Version 5.0 (Podani, 1993). Diversity of taxocenoses was characterised using Pielou's index of equitability, Shannon–Wiener's index of total species diversity, and Simpson's index of dominance (Poole, 1974; Ludwig & Reynolds, 1988). Shannon–Wiener diversity indices were compared using the *t*–test (Poole, 1974). Ordination was carried out with non–metric multidimensional scaling (NMDS) using the Bray–Curtis dissimilarity coefficient. One–way analysis of similarities (ANOSIM) was used to identify difference in species variability of the lepidoptera taxocenosis in the study plots during the year. Hierarchical (nested) ANOVA was used to examine spatial (locality) and temporal (sampling months) variation in the distribution of the total abundance, number of species, taxa and species diversity of lepidoptera. The model contained factors (terms) representing the effects of locality and sampling date nested in locality. Multiple sample comparisons were used to identify significant differences in the number of individuals, number of species and species diversity between localities and sampling months. The hypothesis that occurrences of three types of feeding specialization are randomly distributed throughout the vegetation season was tested according to Poole & Rathcke (1979). Differences of means and dispersion of species numbers in feedings groups were analyzed by Tukey's pairwise comparison and Levene's test in ANOVA, respectively. Analyses of variance and Tukey's pairwise comparison were used to identify differences between the number of species and the number of individuals in seasonal gilds. The nomenclature and systematic classification of the lepidoptera species were used according to Laštůvka & Liška (2011). The trophic groups of lepidoptera larvae were established according to Brown & Hyman (1986). The map (fig. 1) and pedological and phytocoenological characteristics of the investigated area are given in detail by Zlinská et al. (2005).

Voucher specimens (in ethanol) are deposited at the Faculty of Natural Sciences, Comenius University, Bratislava, Slovakia.

Study area

The lepidoptera larval stages on *Quercus dalechampii* were studied in the territories of the Protected Landscape Area of Malé Karpaty and Trnavská pahorkatina hills situated in the centre of Europe in the western part of Slovakia. The vast majority of the plots are located in the southern to northern part of the Malé Karpaty Mountains (Mts.) at altitudes of 240–350 m a.s.l. and an average annual temperature of 8–9°C. Study plots in Trnavská pahorkatina hilly land are situated near the Malé Karpaty Mts. at an altitude of 240 m. The annual precipitation in both territories is about 650–800 mm.

Study plots (abbreviation of study plot in parentheses used in the text):

Vinosady (VI), 48° 19' N, 17° 17' E, 280 m a.s.l.: a 60–80–year–old forest at the foot of the Kamenica

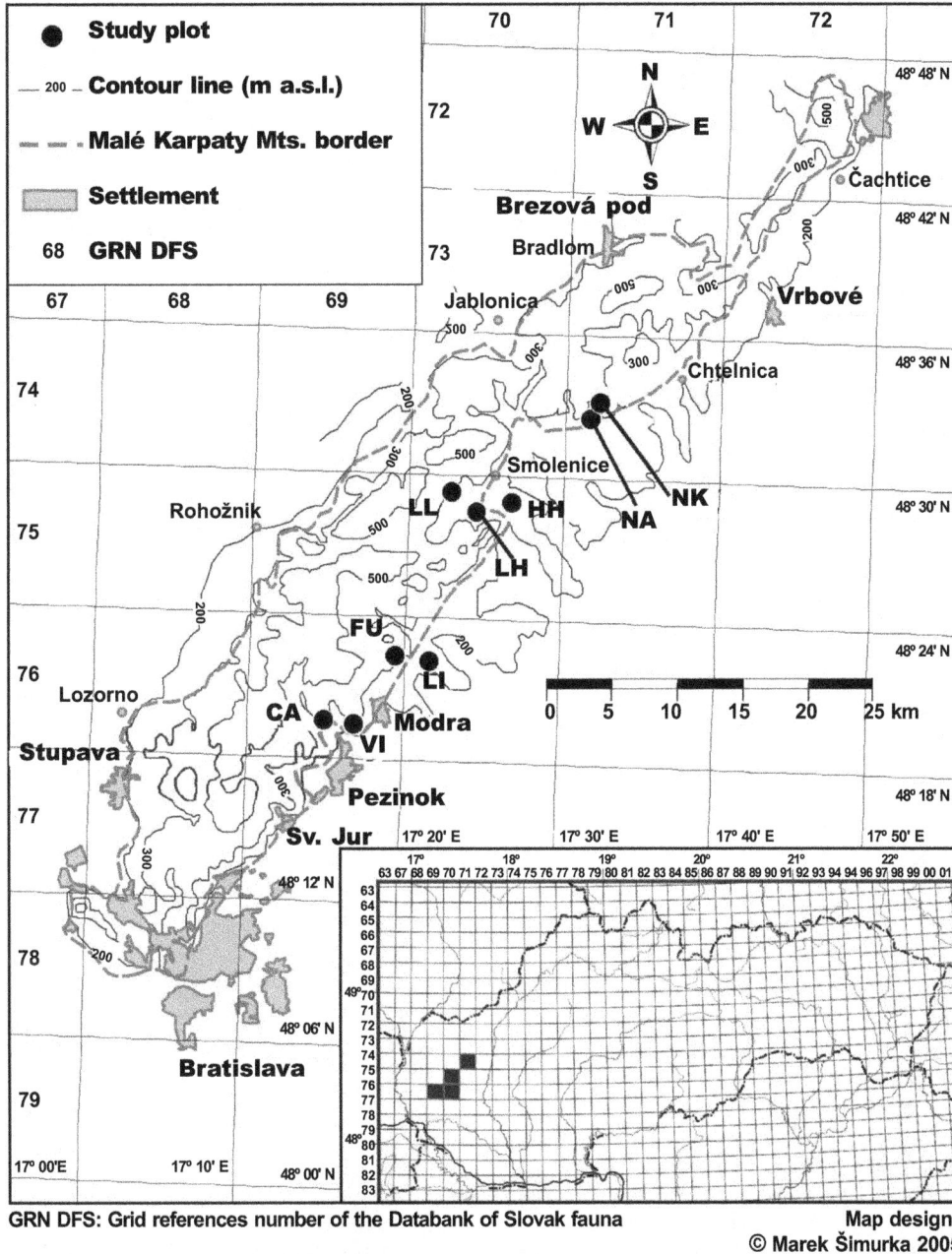

Fig. 1. Study area with location of the study plots.

Fig. 1. Área del estudio con la ubicación de las parcelas del estudio.

hill, NW and W oriented, with drier subxerophilous meadows and shrub complexes. Besides *Quercus dalechampii*, the tree stratum consists of *Q. cerris* and *Acer campestre*.

Cajla (CA), 48° 20' N, 17° 16' E, 260–280 m a.s.l.: an 80–100–year–old forest at the foot of the Malá cajlanská homola hill, S oriented and neighbouring meadows and vineyards on S and E, from N and W closed forest complexes. *Quercus dalechampii* and *Carpinus betulus* predominate in the tree layer.

Fúgelka (FU), 48° 22' N, 17° 19' E, 350 m a.s.l.: an 80–100–year–old forest near the Dubová village, S oriented. Besides *Quercus dalechampii*, the tree stratum consists of *Acer pseudoplatanus*.

Lindava (LI) (Nature Reserve), 48° 22' N, 17° 22' E, 240 m a.s.l.: an 80–100 (120)–year–old forest near the village of Píla. *Quercus dalechampii* and *Q. cerris* predominate in the tree layer.

Horný háj (HH), 48° 29' N, 17° 27' E, 240 m a.s.l.: a larger complex of an island forest 60–80–years old near the village of Horné Orešany, surrounded by fields and vineyards, W and SW oriented. *Quercus cerris, Q. dalechampii, Carpinus betulus* and *Fraxinus excelsior* predominate in the tree layer.

Lošonec–lom quarry (LL), 48° 29' N, 17° 23' E, 340 m a. s. l.: an 80–100–year–old forest SW oriented, neighbouring with mesophilous meadows and pastures. The tree layer consists of *Quercus dalechampii, Q. cerris* and *Carpinus betulus.* The leaf litter, herbage undergrowth and trees are strongly covered with calcareous dust from a nearby quarry.

Lošonský háj (LH) (Nature Reserve), 48° 28' N, 17° 24' E, 260 m a.s.l.: an 80–100–year–old oak–hornbeam forest NE oriented, surrounded by closed forest complexes. *Quercus dalechampii, Q. cerris* and *Carpinus betulus* predominate in the tree stratum.

Naháč–Kukovačník (NA), 48° 32' N, 17° 31' E, 300 m a.s.l.: a small forest island, approximately 40–60–year–old surrounded by fields and pastures, NE oriented. *Quercus dalechampii, Q. cerris* and *Carpinus betulus* predominate in the tree layer.

Naháč–Katarínka (NK) (Nature Reserve), 48° 33' N, 17° 33' E, 340 m a.s.l.: a 40–60–year–old forest NW oriented, surrounded by closed forest ecosystems. *Quercus dalechampii* and *Carpinus betulus* predominate in the canopy.

Only abbreviations of the study plots are used in the following text.

The study plots LI and HH are situated in Trnavská pahorkatina hills and the others are in the Malé Karpaty Mts.

Results

From 2000–2002, a total of 2,140 Lepidoptera larvae were collected in nine study plots with *Quercus dalechampii.* They represented 96 species from 17 families (appendix 1). The families Geometridae, Noctuidae and Tortricidae encompassed the highest number of species found (27, 23, and 13, respectively) (appendix 1). The lowest number of species (18 species) were found in HH (appendix 1). Six species (*Coleophora siccifolia, Lomographa temerata, Peribatodes rhomboidaria, Acronicta auricoma, Orthosia opima* and *Amata phegea*) were found on oaks for the first time in Slovakia (cf. Hrubý, 1964; Patočka et al., 1999). *A. phegea* is one of six species presenting first records of lepidoptera larvae feeding on oaks. This species probably entered the oak crown from the surrounding low vegetation because it has not been found previously on trees according to the literature (Reiprich, 2001).

The most abundant families were Geometridae and Noctuidae (appendix 1, table 1). The families Tortricidae and Erebidae achieved relatively high dominance, mainly due to the species *Aleimma loeflingiana* (Tortricidae) and *Lymantria dispar* (Erebidae) (appendix 1, table 1). Species with dominance higher than 10% were *Lymantria dispar* in HH, *Operophtera brumata* in CA (calamitous oak pests), *Cosmia trapezina* in LI, *Aleimma loeflingiana* in FU (an important pest of oaks) and *Cyclophora linearia* in HH (cf. Patočka et al., 1999; appendix 1).

Table 1. Family dominance (%) of lepidoptera larvae on *Quercus dalechampii* in the Malé Karpaty Mountains in 2000–2002 (based on total number of individuals).

Tabla 1. Dominancia por familia (%) de las larvas de lepidópteros que se encontraron en Quercus dalechampii *en los Pequeños Cárpatos entre los años 2000 y 2002 (con respecto al número total de individuos).*

Family / year	2000	2001	2002	Total
Psychidae	0.00	0.17	0.00	0.05
Bucculatricidae	0.00	0.00	0.43	0.19
Gracillariidae	0.17	0.00	0.00	0.05
Ypsolophidae	2.48	2.15	1.29	1.87
Chimabachidae	2.74	1.49	1.12	1.73
Peleopodidae	8.85	1.65	1.25	3.46
Coleophoridae	8.77	3.63	4.09	5.28
Gelechiidae	0.17	0.83	0.22	0.37
Tortricidae	6.79	19.64	9.68	11.68
Lycaenidae	0.33	0.33	0.43	0.37
Pyralidae	1.82	0.50	1.29	1.21
Drepanidae	0.33	0.33	0.11	0.23
Geometridae	32.62	39.11	42.58	38.79
Notodontidae	5.46	0.17	0.22	1.68
Erebidae	7.95	5.94	9.04	7.85
Nolidae	3.48	1.98	1.72	2.29
Noctuidae	18.05	22.11	26.56	22.90
No individuals	604	606	930	2,140

The species *Lymantria dispar, Cyclophora linearia, Pseudoips prasinana* and *Carcina quercana* reached the highest dominance on the species poorest study plot HH when compared with other plots (appendix 1).

Characteristic species of the plot LL covered with calcareous dust are as follows: *Tortrix viridana, Conobathra tumidana, Aleimma loeflingiana, Agriopis leucophaearia* and *Alsophila aceraria.* Three lepidoptera species, *Archips podana, Eudemis profundana* and *Apocheima hispidaria* (appendix 1), were found only in this plot but abundance was low.

Lepidoptera species *Agriopis marginaria, Cosmia trapezina, Orthosia cruda* and *Lymantria dispar* (apendix 1) were typical of the lighter, sparser and younger oak stands (study plots NK, LI, CA, VI).

The vast majority of Lepidoptera belonged to the monovoltine species with main occurrence in spring. Further oligophagous species (*Cyclophora linearia* and *Ennomos erosaria*) and polyphagous species (*Parectropis similaria* and *Colocasia coryli*) belonged to the bivoltine species. *Watsonalla binaria* proved to be trivoltine species (appendix 1).

Most species found belonged to the trophic group of generalists (64 species). Narrow oligophages (18 species) feeding on oaks are considered to be typical oak species. Only six species belonged to wider oligophages.

The value of Shannon–Wiener´s diversity index of the richest lepidoptera taxocenosis (NK, H' = 3.428) and the poorest taxocenosis (HH, H' = 2.505) was statistically significantly different from other taxocenoses (T–test, P < 0.05) (table 2). A detailed algorithm is given by Poole (1974). The richest taxocenosis NK includes 462 individuals representing 52 species; of these, seven species dominate at least 5%. The poorest taxocenosis HH includes only 44 individuals belonging to 18 species; 4 of these species dominate over 5% (appendix 1).

Poor qualitative–quantitative taxocenosis of lepidoptera larvae on island forest HH is also expressed by Simpson's index of dominance (c = 0.126) where dominance is concentrated in a small number of species (appendix 1). In other taxocenoses, dominance is spread to more co–dominant species (Simpson's index of dominance values from 0.044 to 0.086). The value of equitability was highest at FU, NK and NA (table 2).

A dendrogram based on the qualitative representation (Sörensen's index, complete linkage) separated the lepidoptera taxocenosis on the study plot HH (isolated forest, high density of ants, the lowest diversity of species) (fig. 2). Based on a qualitative–quantitative similarity (Wishart's similarity ratio, complete linkage), the hierarchical classification divided the lepidoptera taxocenoses into two clusters connected on the relatively low level of similarity (fig. 3). The first cluster consisted of the taxocenoses HH and NA (island forests) with the lowest figures for abundance and individuals (44 and 133, respectively). The second cluster had two subclusters and included other taxocenoses. The first subcluster contained the taxocenoses from the denser and older plots (LL. Study plot affected by calcium dust deposition and with higher canopy cover of shrub story; LH. Lot with higher canopy cover of wood species crowns; and FU. Plot with higher canopy cover of both shrub story and wood species crowns). The second subcluster may be formed from the taxocenoses on lighter and younger plots (NK, LI, CA and VI)

The NMDS showed plot HH was set apart from all the other study plots (fig. 4). The study plot NA was also separated (although less marked so) as confirmed by Wishart's index.

Table 2. Species diversity test and basic characteristics of caterpillar taxocenoses at study plots in 2000–2002: H'. Shannon's index of species diversity; e. Pielou's index of equitability; c. Simpson's index of dominance. (T–test values of H' are under the diagonal and degrees of freedom are above it; the testing process is detailed in Materials and methods; significance levels: *** P < 0.001; ** 0.001 < P < 0.01; * = 0.01 < P < 0.05; ns = 0.05 < P (non–significant); for abbreviations of the study plots see Material and methods).

*Tabla 2. Prueba de la diversidad de especies y características básicas de las taxocenosis de orugas en las parcelas del estudio entre los años 2000 y 2002. H'. Índice de Shannon para la diversidad de especies; e. Índice de Pielou para la equidad; c. Índice de Simpson para la dominancia. (Los valores de H' de la prueba t se encuentran debajo de la diagonal y los grados de libertad, encima; el proceso de la prueba se detalla en el apartado Material and methods; niveles de significación: *** P < 0,001; ** 0,001 < P < 0,01; * = 0,01 < P < 0,05; ns = 0,05 < P (no significativo); para consultar las abreviaturas de las parcelas del estudio, ver Material and methods).*

		VI	CA	FU	LI	HH	LL	LH	NA	NK
	e	0.851	0.801	0.872	0.809	0.867	0.838	0.849	0.867	0.868
	c	0.066	0.086	0.063	0.063	0.126	0.067	0.063	0.066	0.044
	H'	3.097	3.065	3.101	3.212	2.505	3.091	3.194	3.197	3.428
VI	3.1	0	541.083	349.481	636.538	55.534	445.853	356.73	228.02	670.81
CA	3.07	0.343ns	0	421.247	582.174	65.467	495.143	431.18	294.16	477.82
FU	3.1	0.048ns	0.346ns	0	396.331	64.992	372.919	350.76	265.84	289.7
LI	3.21	1.356ns	1.495ns	1.127ns	0	59.702	491.438	402.8	259.76	603.69
HH	2.51	3.549***	3.218**	3.429**	4.162***	0	63.662	68.671	78.673	51.015
LL	3.09	0.071ns	0.246ns	0.106ns	1.263ns	3.387**	0	386.06	272.78	381.14
LH	3.19	0.99ns	1.174ns	0.842ns	0.182ns	3.901***	0.957ns	0	284.73	299.12
NA	3.2	0.908ns	1.092ns	0.792ns	0.131ns	3.764***	0.894ns	0.029ns	0	192.36
NK	3.43	4.681***	4.198***	3.784***	2.773**	5.653**	4.013***	2.567**	2.201*	0

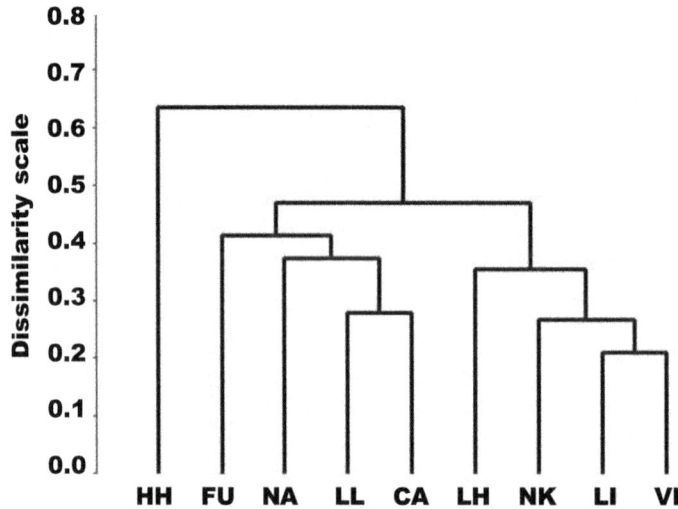

Fig. 2. Classification of lepidoptera taxocenoses on individual study plots according to species presence/absence (Sörensen's index).

Fig. 2. Clasificación de las taxocenosis de lepidópteros en cada una de las parcelas del estudio en función de la presencia o ausencia de las especies (índice de Sörensen).

Table 3 shows the overall result of the permutation test and pairwise ANOSIMs between all pairs of groups (provided as post–hoc test). Significant comparisons (at $P < 0.05$) are shown in bold.

Analysis of similarity based on seasonal variability of species composition distinguished two significant different lepidoptera taxocenoses. The lepidoptera taxocenosis of the HH had significantly lower abundance and number of species than the taxocenoses of the other eight study plots (table 4).

Generally, lepidoptera larvae were weakly represented in HH because of the occurrence of numerous

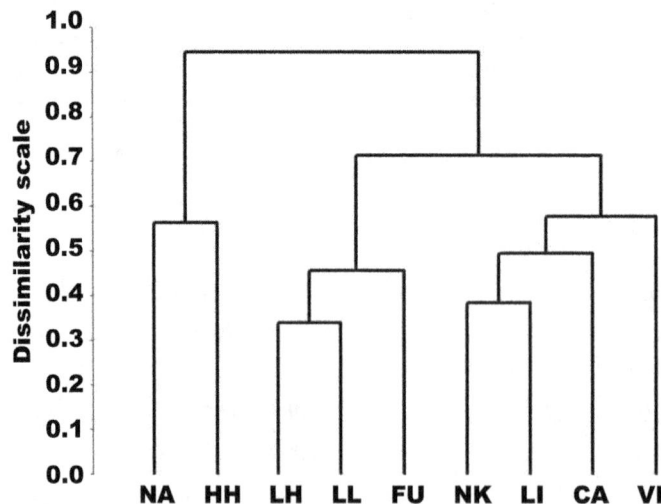

Fig. 3. Classification of lepidoptera taxocenoses on individual plots according to qualitative–quantitative similarity (Wishart's index).

Fig. 3. Clasificación de las taxocenosis de lepidópteros en cada una de las parcelas del estudio en función de la similitud cualitativa y cuantitativa (índice de Wishart).

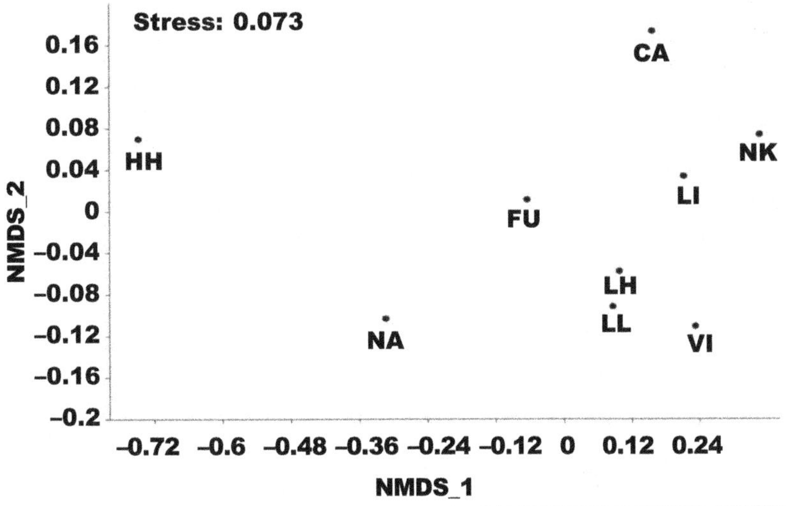

Fig. 4. Nonmetric multidimensional (NMDS) scaling plot based on Bray–Curtis similarities for species abundance data from nine study plots. (For abbreviations of the study plots see Material and methods.)

Fig. 4. Gráfico del escalamiento multidimensional no métrico (NMDS) de las similitudes de Bray–Curtis a partir de los datos sobre la abundancia de las especies obtenidos en nueve parcelas del estudio (para consultar las abreviaturas de las parcelas del estudio, véase el apartado Material and methods).

colonies of ants as predators of lepidoptera larvae in this plot (appendix 1).

The seasonal effect was reflected in all three examined parameters of the taxocenosis (table 4). The species abundance peaked in May, while number of species and species diversity reached the highest values from April to May and from April to June, respectively (fig. 5).

Figure 6 shows the number of species and the number of individuals in seasonal guilds. The number of species and the abundance showed two clear peaks in flush feeders and in late summer feeders. The number of flush feeder species was significantly higher than the number of species in other seasonal guilds (table 5).

Table 3. Results of analysis of similarity (ANOSIM): permutation number: 1,000; mean rank within: 5,047; mean rank between: 5,241; $R = 0.037$, $P < 0.01$. (For abbreviations of the study plots see Material and methods.)

Tabla 3. Resultados del análisis de similitud (ANOSIM): número de permutaciones: 1.000; rango medio dentro: 5.047; rango medio entre: 5.241; R = 0,037; P < 0,01. (Para las abreviaturas de las parcelas del estudio, véase el apartado Material and methods).

	VI	CA	LI	LL	NA	NK	HH	FU	LH
VI	–	0.16	0.39	0.73	0.15	0.23	**0.02**	0.06	0.18
CA		–	0.58	0.67	0.77	0.46	**0.01**	0.68	0.21
LI			–	0.68	0.15	0.32	**0.01**	0.73	0.17
LL				–	0.38	0.62	**0.02**	0.83	0.33
NA					–	0.48	**0.01**	0.18	0.22
NK						–	**0**	0.3	0.3
HH							–	0.09	0.11
FU								–	0.3
LH									–

Table 4. Comparison of abundance, species number and species diversity in spatial and temporal scaling of studied lepidoptera taxocenoses in the hierarchical (nested) analysis of variance (ANOVA): * P < 0.05, ** P < 0.01; SSq. Sum of squares; MSq. Mean squares; SS. Sampling site; SD. Sampling date.

*Tabla 4. Comparación de la abundancia, el número de especies y la diversidad de especies en las escalas espacial y temporal de las taxocenosis estudiadas de lepidópteros en el análisis jerárquico (anidado) de la varianza (ANOVA): * P < 0,05, ** P < 0,01; SSq. Suma de los cuadrados; MSq. Media de los cuadrados; SS. Lugar de muestreo; SD. Fecha de muestreo.*

	SSq	df	MSq	F–statistic	P–value	Mann–Whitney pairwise comparison
Abundance						
SS	13336.73	8	1667.09	4.29614	P < 0.01	HH < CA, VI, FU, LI, NK, NA, LI*
SD	61095.30	54	1131.39	3.51589	P < 0.01	June–October < April–May*
Number of species taxa						
SS	268.798	8	33.600	3.0216	P < 0.01	HH < CA, VI, FU, LI, NK, NA, LI*
SD	4419.608	54	81.845	7.3602	P < 0.01	June–October < April–May*
Species diversity						
SS	6.50334	8	0.81292	1.2961	P = 0.2571	– –
SD	86.24383	54	1.59711	4.9800	P < 0.01	August–October < April–June**

The occurrence in time of lepidoptera species with two types of feeding specialization (generalists, narrow oligophagous) was non–randomly distributed throughout the season (table 6). The number of species in these feeding groups peaked in May. On the other hand, species in the wider oligophagous feeding groups exploited time in a random way.

Discussion

Taxocenoses of lepidoptera larvae observed on *Quercus dalechampii* in Malé Karpaty Mts. can be compared with taxocenoses on *Q. cerris* that were studied under similar conditions. A comparison shows similarities and differences (cf. Kulfan et al., 2006). Species richness was higher on *Q. dalechampii* (96 species on *Q. dalechampii* compared to 58 species on *Q. cerris*). The lowest number of species in both types of taxocenoses was found in the study plot HH. *Lymantria dispar* and *Operophtera brumata* belonged to the most abundant species both on *Q. dalechampii* and on *Q. cerris*. In the study plot HH, *L. dispar* reached a higher dominance on *Q. cerris* than on *Q. dalechampii* (cf. Kulfan et al., 2006). Regarding cumulative dominance, the families Ypsolophidae, Pyralidae and Drepanidae predominated on *Q. cerris*. On the contrary, the families Peleopodidae and Chimabachidae were noticeably more common on *Q. dalechampii* (cf. table 1, Kulfan et al., 2006).

In general, when compared with other areas of Slovakia, the abundance of lepidoptera larvae found on *Q. dalechampii* corresponds to the latent phase of the gradation cycle on oaks. No marked outbreaks of folivorous lepidoptera larvae have been observed on oaks in Slovakia since 1990 (cf. Kulfan, 1990, 1998, 2002; Kulfan, 1992; Kulfan et al., 1997, 2006).

The values of species diversity of lepidoptera taxocenoses on *Q. dalechampii* are characterized by a greater variance than the values of diversity of taxocenoses on *Q. petraea* in the Malé Karpaty Mts. Diversity of lepidoptera taxocenoses on *Q. petraea* reached annual values ranging from 3.042 to 3.296 (Kulfan, 1990). The smallest diversity of lepidoptera taxocenoses on oaks in the Malé Karpaty Mts.was found on *Q. cerris* and it achieved a value of 2.230 (Kulfan et al., 2006) for the three–year period.

As a rule, there is notable spring peak in abundance of lepidoptera caterpillars on oaks in Central Europe (Kulfan, 1992; Kulfan, 1983, 1990; Parák et al., 2012, etc.). This was also confirmed by the research on *Q. dalechampii*. Other peaks in caterpillar abundance on *Q. dalechampii* were not found throughout the growing season. Two peaks in the number of lepidoptera caterpillars during the season with prevalence in spring time were found on some oak species in Central Europe (Kulfan, 1992; Kulfan, 1983, 1990). The abundance of lepidoptera taxocenoses on *Q. petraea* throughout the growing season in the Malé Karpaty Mts in Slovakia has been found to have a very marked peak in spring (May–June) and a less noticeable peak in autumn (September). However, it is interesting that the autumn peak of total abundance of lepidoptera caterpillars on *Q. petraea* in the old oak stand in 1978 was more noticeable when compared to the spring peak (cf. Kulfan, 1983).

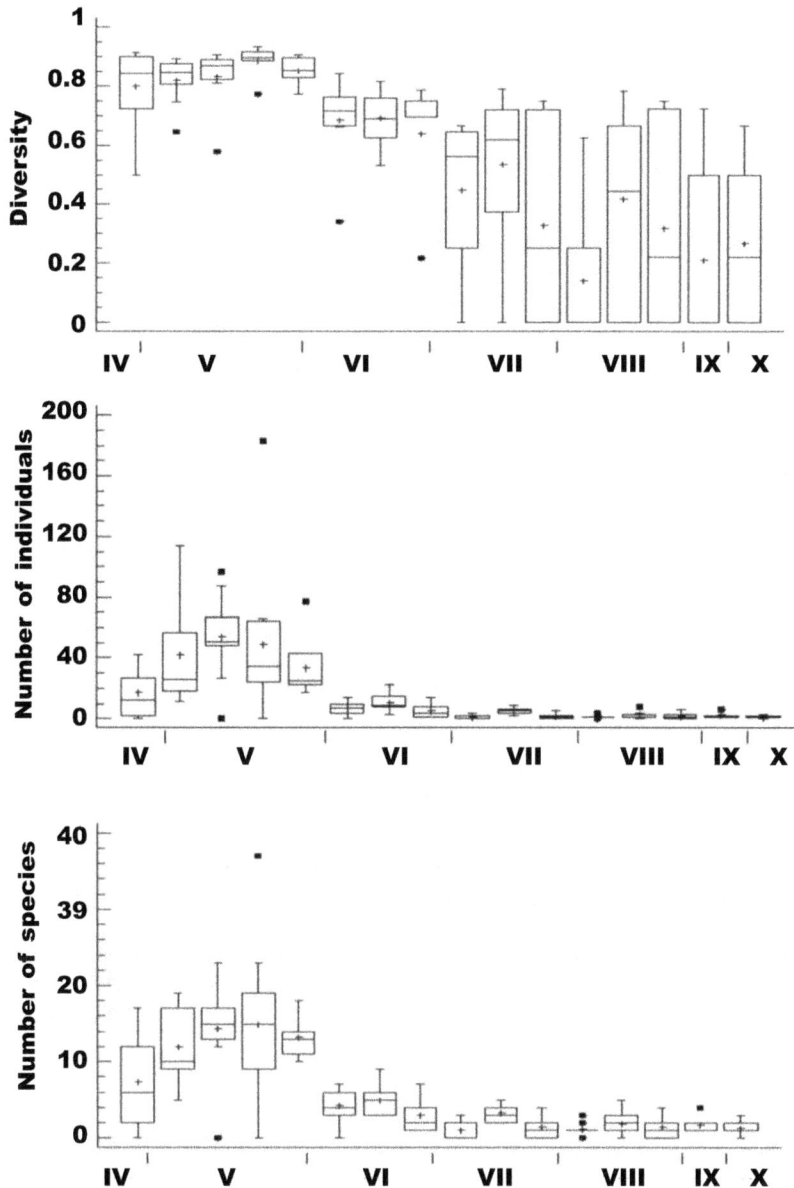

Fig. 5. Box plots showing the monthly variation of diversity, abundance and number of species of lepidoptera taxocenosis: IV–X. Months of presence of lepidopteran caterpillars during the season.

Fig. 5. Diagramas de caja en los que se muestra la variación mensual de la diversidad, la abundancia y el número de especies de la taxocenosis de los lepidópteros: IV–X. Meses de presencia de las orugas de lepidópteros durante la estación.

This was probably caused by unfavorable weather conditions in spring.

Yoshida (1985) in northern Japan presented the highest abundance of lepidoptera caterpillars on oaks in summer. This difference compared to our results may be caused by different climate, because the frosts in May in northern Japan have a large impact on leaf phenology, which is associated with the development of the spring taxocenoses of caterpillars.

In oak forest on Mont Holomontas (Mediterranean area, Greece) even three peaks in insect abundance (consisting mainly of lepidoptera larvae) on six oak species were found (Kalapanida & Petrakis, 2012).

Not only the species abundance but also the species richness and diversity of lepidoptera species on *Q. dalechampii* culminated in the vernal aspect. The marked increase of species diversity of lepidoptera taxocenoses on *Q. dalechampii* was in spring. A similar

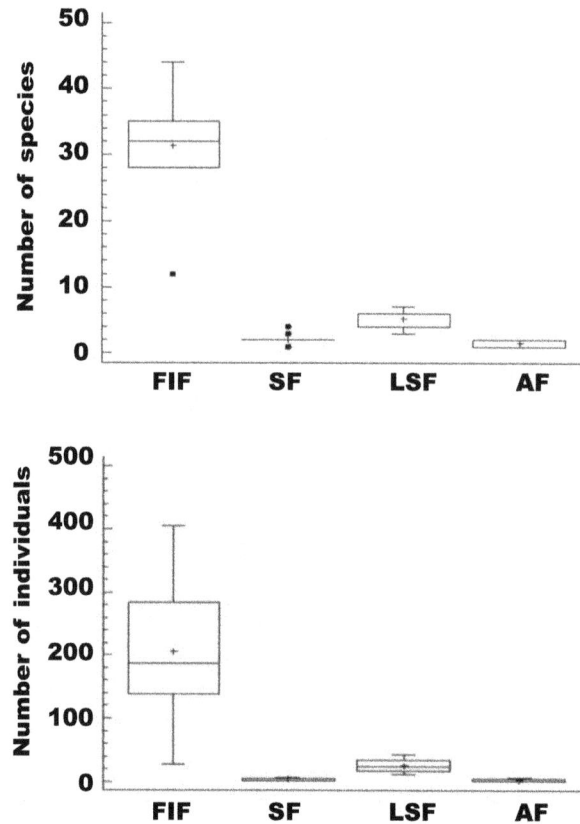

Fig. 6. Box plots showing the effects of seasonality on species number and abundance of lepidoptera taxocenosis: FIF. Flush feeders; SF. Summer feeders; LSF. Late summer feeders; AF. Autumn feeders.

Fig. 6. Diagramas de caja en los que se muestra los efectos de la estacionalidad en el número de especies y en la abundancia de la taxocenosis de los lepidópteros: FIF. Se alimentan durante la brotación; S. Se alimentan en verano; LSF. Se alimentan al final del verano; AF. Se alimentan en otoño.

Table 5. Results of one–way ANOVA on differences between seasonal guilds in the number of species taxa and number of individuals. The post hoc multiple sample comparison test (Tukey's pairwise comparison) for differences in mean number of species taxa and number of individuals between seasonal guilds: * $P < 0.05$, ** $P < 0.01$; FIF. Flush feeders; LSF. Late spring feeders; SF. Summer feeders; AF. Autumn feeders.

*Tabla 5. Resultados de la ANOVA simple de las diferencias existentes entre los gremios estacionales en cuanto el número de taxones y el número de individuos. La prueba múltiple de comparación a posteriori de Tukey (comparación por pares de Tukey) de las diferencias en el número medio de taxones y el número de individuos entre gremios estacionales: * P < 0,05; ** P < 0,01; FIF. Se alimentan durante la brotación; LSF. Se alimentan al final de la primavera; SF. Se alimentan en verano; AF. Se alimentan en otoño.*

	F–statistic	df	P–value	Tukey's pairwise comparison
Number of species in seasonal guilds	85.49	35	$P < 0.01$	SF, LSF, AF < FIF**
Number of individuals in seasonal guilds	25.44	35	$P < 0.01$	SF, LSF, AF < FIF**
				SF, AF < LSF**

Table 6. Results obtained from the Poole–Rathcke method used to segregate the moths in time. The null hypothesis (H1) states that the dispersion is not significantly different from random and the second null hypothesis (H2) that the two means and dispersions are not significantly different: N. Number of species; OV. Observed variance; EV. Expected variance; DR. Dispersion ratio; RD. Random dispersion (significance of H1); HM. Homogeneity of means; HD. Homogeneity of dispersion (significance of H2). (* P < 0.05, ** P < 0.01, ns. Non–significant)

Tabla 6. Resultados obtenidos con el método de Poole–Rathcke empleado para segregar las polillas en el tiempo. La hipótesis nula (H1) afirma que la dispersión no es significativamente distinta de la aleatoria y la segunda hipótesis nula (H2), que las dos medias y las dos dispersiones no son significativamente diferentes: N. Número de especies; OV. Varianza observada; EV. Varianza esperada; DR. Razón de la dispersión; RD. Dispersión aleatoria (significación de H1); HM. Homogeneidad de las medias; HD. Homogeneidad de la dispersión (significación de H2). (P < 0,05; ** P < 0,01; ns. No significativa).*

	N	OV	EV	DR	RD
Feeding specialization					
Generalists	64	9.87	0.19	51.94736842	**
Narrow oligophagous	18	18.7	0.27	69.25925926	**
Wider oligophagous	6	15.33	0.82	18.69512195	ns

	HM	HD
Feeding specialization compared		
Generalist/narrow oligophagous	Q = 8.54 P < 0.01	W = 2.14 P = 0.15
Generalist/wider oligophagous	Q = 8.74 P < 0.01	W = 13.55 P < 0.01
Narrow oligophagous/wider oligophagous	Q = 0.19 P = 0.98	W = 14.92 P < 0.01

course of diversity was observed on four oak species in the Borská nížina Lowland in Slovakia. (Kulfan & Degma, 1999).

Southwood et al. (2005) found distinct seasonal patterns in species richness of the arthropod fauna on four oak species in the U.K. In terms of species richness, the values showed a general trend peaking in summer and early autumn, but biomass peaked in May on the native oak species, mainly due to lepidoptera larvae.

A relatively steady decrease in the individuals from early spring to autumn is well known from the 'Quercus type' of host tree (Niemelä & Haukioja, 1982). These authors suggested that this effect was due to a decline in available resources. Feeny (1970) and Kamata & Igarashi (1996) stated that tougher leaves with a higher tannin concentration contributed to the lower richness of Lepidoptera later during the growing season. A negative correlation between some specialist oak feeders and condensed tannins in the canopy of Quercus alba and understorey of Q. velutina was found (Forkner et al., 2004). Their results generally indicated a negative response from both specialists and generalists to condensed tannins.

A higher number of flush feeders in spring compared to the number of species in other seasonal guilds on Q. dalechampii was also found on three oak

species in Borská nížina lowland (western Slovakia, Central Europe), where the greatest proportion of flush feeders was found on Q. robur (cf. Turčáni et al., 2009).

Acknowledgements

This work was supported by the Slovak Grant Agency for Science (Grant Nos 1/0124/09, 1/0137/11, 2/0035/13 and 1/0066/13). We are grateful to Jaroslav Fajčík for technical assistance.

References

Brown, V. K. & Hyman, P. S., 1986. Successional communities of plants and phytophagous Coleoptera. *Journal of Ecology,* 7: 963–975.

Csóka, G., 1990–1991. Oak hostplant records of Hungarian Macrolepidoptera. *Erdészeti Kutatások,* 82–83: 89–93.

– 1998a. Oak Defoliating Insects in Hungary. In: *Proceedings: Population Dynamics, Impacts, and Integrated Management of Forest Defoliating Insects:* 334–335 (M. L. McManus & A. M. Liebhold, Eds.). USDA Forest Service General Technical

Report NE–247, Hungary.

– 1998b. A Magyarországon honos tölgyek herbivir rovaregyüttese. *Erdészeti Kutatások*, 88: 311–318.

Feeny, P., 1970. Seasonal changes in oak leaf tannins and nutrient as a cause of spring feeding by winter moth caterpillars. *Ecology*, 51: 565–581.

Forkner, R. E., Marquis, R. J. & Lill, J. T., 2004. Feeny revisited: condensed tannins as anti–herbivore defences in leaf–chewing herbivore communities of *Quercus. Ecological Entomology*, 29: 174–187.

Gerasimov, A. M., 1952. *Insects. Lepidoptera 1(2). Caterpillars 1. Fauna of Soviet Union 56*. Zoological Institute, Academy of Sciences USSR, Moscow, Leningrad.

Hrubý, K., 1964. *Prodromus lepidopter Slovenska.* Vydavateľstvo SAV, Bratislava.

Kalapanida, M. & Petrakis, V., 2012. Temporal partitioning in an assemblage of insect defoliators feeding on oak on a Mediterranean mountain. *European Journal of Entomology*, 109: 55–69.

Kamata, N. & Igarashi, I., 1996. Seasonal and annual change of a folivorous insect guild in the Siebolds beech forests associated with the outbreaks of the beech caterpillar, *Quadricalcarifera punctatella* (Motschulsky) (Lep., Notodontidae). *Journal of Applied Entomology*, 120: 213–220.

Kollár, J., 2007. The harmful entomofauna of woody plants in Slovakia. *Acta entomologica serbica*, 12: 67–79.

Kulfan, J., 1992. Zur Struktur und Saisondynamik von Raupenzönosen (Lepidoptera) an Eichen. *Biológia, Bratislava*, 47: 653–661.

Kulfan, M., 1983. The seasonal dynamics of lepidoptera caterpillar communities in the Malé Karpaty mountains. *Biologia, Bratislava*, 38: 1003–1009.

– 1990. *Communities of Lepidoptera caterpillars (Lepidoptera) on broadleaf tree species of Malé Karpaty*. Veda, Vydavateľstvo SAV, Bratislava.

– 1997. Lepidoptera living on oaks in Southwestern Slovakia lowlands. *Folia faunistica Slovaca*, 2: 85–92.

– 1998. Lepidoptera as pests of oak crowns in Southwestern Slovakia lowlands. *Folia faunistica Slovaca*, 3: 119–124.

– 2002. Lepidoptera larvae communities on *Quercus robur* and *Cerasus mahaleb* in NNR Devínska Kobyla (SW Slovakia). *Folia Faunistica Slovaca*, 7: 55–60.

– 2012. Structure of lepidopterocenoses on oaks *Quercus dalechampii* and *Q. cerris* in cenral Europe and estimation of the most important species. *Munis Entomology & Zoology*, 7: 732–741.

Kulfan, M. & Degma, P., 1999. Seasonal dynamics of lepiropteran larvae communities diversity and equitability on oaks in the Borská nížina lowland. *Ekologia*, 18: 100–105.

Kulfan, M., Holecová, M. & Fajčík, J., 2006. Caterpillar (Lepidoptera) communities on European Turkey oak (*Quercus cerris*) in Malé Karpaty Mts (SW Slovakia). *Biologia, Bratislava*, 61: 573–578.

Kulfan, M., Šepták, L. & Degma, P., 1997. Lepidoptera larvae communities on oaks in SW Slovakia.

Biologia, Bratislava, 52: 247–252.

Laštůvka, Z. & Liška, J., 2011. *Annotated checklist of moths and butterflies of the Czech republic (Insecta: Lepidoptera)*. Biocont Laboratory, Brno.

Ludwig, J. A. & Reynolds, J., 1988. *Statistical ecology: a primer of methods and computing*. Whiley – Interscience Public, New York.

Niemelä, P. & Haukioja, E., 1982. Seasonal patterns in species richness of herbivores: Macrolepidoptera larvae on Finnish deciduous trees. *Ecological Entomology*, 7: 169–175.

Novotny, V., Drozd, P., Miller, S. E., Kulfan, M., Janda, M., Basset, Y. & Weiblen, G. D., 2006. Why are there so many species of herbivorous insects in tropical rainforests? *Science*, 313: 1115–1118.

Parák, M., Kulfan, J. & Svitok, M., 2012. Spoločenstvá húseníc (Lepidoptera) na troch druhoch dubov (Quercus spp.) z oblasti Čachtických Karpát (západné Slovensko). *Folia faunistica Slovaca*, 17: 247–256.

Patočka, J., 1954. *Caterpillars on oak trees in Czechoslovakia*. Štátne pôdohospodárske nakladateľstvo, Bratislava.

– 1980. *Die Raupen und Puppen der Eichenschmetterlinge Mitteleuropas*. Paul Parey Verlag, Germany.

Patočka, J., Čapek, M. & Charvát, K., 1962. *Contribution to the knowledge of the crown arthropod fauna on oaks of Slovakia, in particular with regard to the order Lepidoptera*. Veda, Vydavateľstvo SAV, Bratislava.

Patočka, J., Krištín, A., Kulfan, J. & Zach, P., 1999. *Die Eichenschädlinge und ihre Feinde*. Ústav ekológie lesa SAV, Zvolen.

Podani, J., 1993. *Syn–tax. Version 5.0. Computer programs for Multivariate Data Analysis in Ecology and Systematics. User's guide*. Scientia Publishing, Budapest.

Poole, R. W., 1974. *An Introduction to Quantitative Ecology*. McGraw–Hill, New York.

Poole, R. W. & Rathcke, B.–J., 1979. Regularity, randomness and aggregation in flowering phonologies. *Science*, 203: 470–471.

Reiprich, A., 2001. *Die Klassifikation der Schmetterlinge der Slowakei laut den Wirten (Nährpflanzen) ihrer Raupen*. Správa Národného parku Slovenský raj vo vydavateľstve SZOPK, Spišská Nová Ves.

Skuhravý, V., Hrubík, P., Skuhravá, M. & Požgaj, J., 1998. Occurrence of insect associated with nine *Quercus* species (Fagaceae) in cultured plantations in southern Slovakia during 1987–1992. *Journal of Applied Entomology*, 122: 149–155.

Southwood, T. R. E., Wint, G. R. W., Kennedy, C. E. J. & Greenwood, S. R., 2005. The composition of the arthropod fauna of the canopies of some species of oak (*Quercus*). *European Journal of Entomology*, 102: 65–72.

Stolnicu, A. M., 2007. Leaf–mining insects encountered in the forest reserve of Hârboanca, Vaslui county. *Analele Ştiinţifice ale Universităţii, Al. I. Cuza" Iaşi, s. Biologie animală*, 53: 109–114.

Turčáni, M., Patočka, J. & Kulfan, M., 2009. How

do lepidopteran seasonal guilds differ on some oaks? – A case study. *Journal of Forest Science*, 55: 578–590.

– 2010. Which factors explain lepidopteran larvae variance in seasonal guilds on some oaks? *Journal of Forest Science*, 56: 68–76.

Yoshida, K., 1985. Seasonal population trends of macrolepidopterous on oak trees in Hokaido, northern Japan. *The Entomological Society of Japan*, 53: 125–133.

Zlinská, J., Šomšák, L. & Holecová, M., 2005. Ecological characteristic of studied forest communities of an oak–hornbeam tier in SW Slovakia. *Ekológia (Bratislava)*, 24 (Suppl. 2): 3–19.

Appendix 1. The list of the lepidoptera species recorded in the nine study plots in the Malé Karpaty Mountains on *Quercus dalechampii* with dominance (%), months of occurrence of larvae (MO), trophic group (TG: S2. Narrow oligophagous; S3. Wider oligophagous species; G. Generalists; U. Unknown), and larval trophic specialization and seasonal guilds (SG: FIF. Flush feeders; LSF. Late spring feeders; SF. Summer feeders; AF. Autumn feeders). (For abbreviations of study plots see Material and methods.)

Families and species	VI	CA	FU	LI
Psychidae				
Sterrhopterix fusca (Haworth, 1809)	0.0	0.0	0.0	0.3
Bucculatricidae				
Bucculatrix ulmella Zeller, 1848	0.0	0.0	0.0	0.0
Gracillariidae				
Phyllonorycter sp.	0.0	0.0	0.6	0.0
Ypsolophidae				
Ypsolopha alpella (Denis et Schiffermüller, 1775)	1.8	0.0	0.0	0.9
Ypsolopha parenthesella (Linnaeus, 1761)	0.0	0.0	0.0	0.0
Ypsolopha ustella (Clerck, 1759)	0.3	1.1	0.0	0.3
Chimabachidae				
Diurnea fagella (Denis et Schiffermüller, 1775)	0.0	0.7	0.6	0.0
Diurnea lipsiella (Denis et Schiffermüller, 1775)	0.6	2.8	1.2	0.6
Peleopodidae				
Carcina quercana (Fabricius, 1775)	0.6	2.8	3.0	4.6
Coleophoridae				
Coleophora ibipennella Zeller, 1849	0.0	0.0	0.0	0.3
Coleophora kuehnella (Goeze, 1783)	0.0	0.0	0.0	0.0
Coleophora lutipennella (Zeller, 1838)	0.6	4.2	6.6	4.6
Coleophora siccifolia Stainton, 1856	0.0	0.0	0.0	0.3
Gelechiidae				
Anacampsis timidella (Wocke, 1887)	0.0	0.0	0.6	0.0
Carpatolechia decorella (Haworth, 1812)	0.0	0.4	0.0	0.0
Psoricoptera gibbosella (Zeller, 1839)	0.0	0.0	0.0	0.3
Stenolechia gemmella (Linnaeus, 1758)	0.0	0.4	0.0	0.0
Tortricidae				
Aleimma loeflingiana (Linnaeus, 1758)	7.7	0.0	12.0	3.4
Archips crataegana (Hübner, 1799)	0.6	0.0	0.0	0.3
Archips podana (Scopoli, 1763)	0.0	0.0	0.0	0.0
Eudemis profundana (Denis et Schiffermüller, 1775)	0.0	0.0	0.0	0.0
Pammene albuginana (Guenée, 1845)	0.0	0.0	0.0	0.0
Pandemis cerasana (Hübner, 1786)	0.0	0.7	2.4	0.0
Pandemis corylana (Fabricius, 1794)	0.0	0.4	0.0	0.3
Pandemis heparana (Denis et Schiffermüller, 1775)	0.0	0.4	0.0	0.3
Ptycholoma lecheana (Linnaeus, 1758)	0.0	0.0	0.0	0.3
Spilonota ocellana (Denis et Schiffermüller, 1775)	0.0	2.1	0.6	0.0
Tortricodes alternella (Denis et Schiffermüller, 1775)	1.8	1.4	3.6	0.9
Tortrix viridana (Linnaeus, 1758)	4.9	0.7	1.2	0.3
Zeiraphera isertana (Fabricius, 1794)	1.8	0.4	1.8	0.3

Apéndice 1. Lista de las especies de lepidópteros registradas en Q. dalechampii en las nueve parcelas del estudio ubicadas en los Pequeños Cárpatos con la dominancia (%), los meses de presencia de las larvas (MO), el grupo trófico (TG: S2. Oligófagas estrictas; S3. Especies oligófagas más amplias; G. Generalistas; U. Desconocido) y la especialización trófica de las larvas y los gremios estacionales (SG: FIF. Se alimentan durante la brotación; LSF. Se alimentan al final de la primavera; SF. Se alimentan en verano; AF. Se alimentan en otoño). (Para las abreviaturas de las parcelas del estudio, véase el apartado Material and methods).

HH	LL	LH	NA	NK	MO	TG	SG
0.0	0.0	0.0	0.0	0.0	5	G	FIF
0.0	0.0	0.0	1.5	0.4	6	G	LSF
0.0	0.0	0.0	0.0	0.0	7	U	SF
2.3	0.0	1.1	3.0	2.0	5–6	S2	FIF
0.0	0.0	1.1	0.0	0.0	5	G	FIF
2.3	0.9	1.1	0.8	0.4	5–6	G	FIF
4.5	0.5	0.0	3.8	0.4	6–9	G	SF
0.0	0.5	1.1	0.8	3.9	5–8	G	FIF
6.8	3.2	1.6	2.3	3.5	5–8	G	LSF
0.0	0.0	0.0	0.0	0.0	5	G	FIF
0.0	0.0	0.0	0.0	0.2	5	S2	FIF
0.0	6.0	2.7	3.8	5.0	4–6	S2	FIF
0.0	0.0	0.0	0.0	5.2	4–5	G	FIF
0.0	0.0	0.0	0.0	0.0	5	S2	FIF
0.0	0.0	0.0	0.0	0.0	5	G	FIF
2.3	0.0	0.5	0.0	0.0	5	G	FIF
0.0	0.0	1.1	0.0	0.0	5–6	S2	FIF
0.0	9.7	5.4	10.5	0.0	4–5	S2	FIF
0.0	0.0	1.1	0.0	0.4	5–6	G	FIF
0.0	0.5	0.0	0.0	0.0	5	G	FIF
0.0	0.5	0.0	0.0	0.0	5	S2	FIF
0.0	0.0	0.0	0.0	0.2	5	S2	FIF
0.0	0.0	3.8	1.5	0.7	5–7	G	FIF
0.0	0.5	0.0	0.0	0.0	5	G	FIF
0.0	1.4	0.5	0.0	0.7	5–6	G	FIF
0.0	0.0	0.0	0.8	0.2	5	G	FIF
0.0	0.0	0.0	0.0	0.0	5	G	FIF
4.5	1.4	1.1	1.5	0.9	5	G	FIF
0.0	9.3	0.0	2.3	1.3	4–5	S2	FIF
0.0	0.0	1.1	0.0	0.9	4–5	S2	FIF

Appendix 1. (Cont.)

Families and species	VI	CA	FU	LI
Lycaenidae				
Favonius quercus (Linnaeus, 1758)	0.3	0.0	1.2	0.3
Pyralidae				
Acrobasis repandana (Fabricius, 1798)	0.0	0.0	0.0	0.0
Acrobasis tumidana (Denis et Schiffermüller, 1775)	0.0	0.0	0.0	0.0
Phycita roborella (Denis et Schiffermüller, 1775)	0.9	0.4	0.0	0.9
Drepanidae				
Watsonalla binaria (Hufnagel, 1767)	0.0	0.4	0.6	0.0
Geometridae				
Agriopis aurantiaria (Hübner, 1799)	2.5	0.7	0.6	0.6
Agriopis leucophaearia (Denis et Schiffermüller, 1775)	3.1	1.4	1.8	6.8
Agriopis marginaria (Fabricius, 1776)	3.4	4.2	3.6	8.0
Alcis repandata (Linnaeus, 1758)	0.0	0.0	0.0	0.0
Alsophila aceraria (Denis et Schiffermüller, 1775)	0.3	0.4	0.0	1.5
Alsophila aescularia (Denis et Schiffermüller, 1775)	4.9	2.1	0.6	1.8
Apocheima hispidaria (Denis et Schiffermüller, 1775)	0.0	0.0	0.0	0.0
Biston betularia (Linnaeus, 1758)	0.0	0.4	1.2	0.0
Biston strataria (Hufnagel, 1767)	0.0	0.0	0.0	0.0
Campaea margaritaria (Linnaeus, 1761)	1.8	0.0	0.0	0.3
Colotois pennaria (Linnaeus, 1761)	1.5	0.4	0.0	1.8
Cyclophora linearia (Hübner, 1799)	0.3	3.2	8.4	2.5
Cyclophora punctaria (Linnaeus, 1758)	0.0	0.4	0.0	0.0
Ennomos autumnaria (Werneburg, 1859)	0.0	0.4	0.0	0.0
Ennomos erosaria (Denis et Schiffermüller, 1775)	0.9	0.0	0.6	0.3
Ennomos quercinaria (Hufnagel, 1767)	0.0	0.4	0.0	0.0
Epirrita dilutata (Denis et Schiffermüller, 1775)	1.2	7.7	1.2	0.9
Erannis defoliaria (Clerck, 1759)	1.8	2.1	0.0	0.6
Hypomecis punctinalis (Scopoli, 1763)	0.6	0.0	0.0	0.3
Lomographa temerata (Denis et Schiffermüller, 1775)	0.0	1.1	0.0	0.6
Lycia hirtaria (Clerck, 1759)	0.0	0.0	0.0	0.3
Operophtera brumata (Linnaeus, 1758)	11.4	22.9	10.2	8.0
Parectropis similaria (Hufnagel, 1767)	0.0	0.0	0.0	0.0
Peribatodes rhomboidaria (Denis et Schiffermüller, 1775)	0.0	0.0	0.0	0.0
Phigalia pilosaria (Denis et Schiffermüller, 1775)	0.0	0.0	0.0	0.0
Selenia lunularia (Hübner, 1788)	0.0	0.0	0.0	0.3
Selenia tetralunaria (Hufnagel, 1767)	0.0	0.0	0.0	0.0
Notodontidae				
Drymonia ruficornis (Hufnagel, 1766)	0.3	0.0	2.4	0.3
Phalera bucephala (Linnaeus, 1758)	0.0	0.0	0.0	0.0
Spatalia argentina (Denis et Schiffermüller, 1775)	0.0	0.0	0.0	0.0
Thaumetopoea processionea (Linnaeus, 1758)	0.0	0.0	0.0	0.6

HH	LL	LH	NA	NK	MO	TG	SG
0.0	0.0	0.5	0.0	0.6	5	S2	FIF
0.0	0.0	0.0	0.8	0.0	5	S2	FIF
0.0	2.3	1.6	0.0	0.7	5	S2	FIF
0.0	0.5	1.1	0.0	0.9	4–5, 9	S2	FIF
0.0	0.9	0.0	0.8	0.0	6, 8, 10	S3	AF
0.0	3.7	2.2	0.8	1.3	4–6	G	FIF
2.3	7.9	7.1	1.5	8.5	4–5	S3	FIF
2.3	0.9	7.6	1.5	5.7	4–5	G	FIF
0.0	0.0	0.5	0.0	0.0	9	G	AF
0.0	2.8	0.5	0.8	0.7	4–5	G	FIF
0.0	5.6	5.4	3.0	3.1	4–6	G	FIF
0.0	0.5	0.0	0.0	0.0	5	G	FIF
2.3	0.9	0.0	0.0	0.0	8–10	G	AF
0.0	0.0	0.5	0.0	0.0	5	G	FIF
0.0	1.4	2.2	3.8	1.3	4–9, 11	G	LSF
0.0	0.0	1.6	1.5	3.3	4–5	G	FIF
11.4	3.2	3.3	8.3	2.8	6–10	S3	LSF
0.0	0.0	0.5	0.0	0.0	6–7	S3	LSF
0.0	0.9	0.0	1.5	0.0	5–7	G	LSF
0.0	0.0	0.0	0.0	0.0	5, 9	S3	FIF
0.0	0.0	0.5	0.0	0.0	5–6	G	FIF
2.3	0.5	0.5	2.3	1.5	4–5	G	FIF
0.0	0.5	0.0	0.0	0.7	4–5	G	FIF
0.0	0.0	0.0	0.0	1.3	6–9	G	AF
0.0	0.0	0.0	0.0	0.4	7–8	G	SF
0.0	0.0	0.0	0.0	0.4	5	G	FIF
4.5	15.7	15.8	5.3	9.2	4–5	G	FIF
0.0	0.9	0.0	0.0	0.0	7, 10	G	AF
0.0	0.0	0.0	0.8	0.0	11	G	AF
0.0	0.0	0.0	0.0	0.2	5	G	FIF
0.0	0.0	0.0	0.0	0.0	6	G	LSF
0.0	0.0	0.5	0.0	0.0	6	G	LSF
0.0	0.0	0.0	0.0	0.9	5	S2	FIF
4.5	0.0	0.0	0.0	0.0	7	G	SF
0.0	0.0	0.5	0.0	0.2	6–7	G	SF
0.0	0.0	0.0	0.0	4.4	5–6	S2	FIF

Appendix 1. (Cont.)

Families and species	VI	CA	FU	LI
Erebidae				
Amata phegea (Linnaeus, 1758)	0.0	0.4	0.0	0.0
Calliteara pudibunda (Linnaeus, 1758)	1.2	0.0	0.0	0.9
Lymantria dispar (Linnaeus, 1758)	16.3	1.8	3.0	11.4
Orgyia antiqua (Linnaeus, 1758)	0.0	0.0	0.0	0.0
Nolidae				
Bena bicolorana (Fuessly, 1775)	0.3	0.4	0.0	0.3
Nycteola revayana (Scopoli, 1772)	0.0	0.0	0.6	0.0
Pseudoips prasinana (Linnaeus, 1758)	1.2	4.2	3.0	2.8
Noctuidae				
Acronicta auricoma (Denis et Schiffermüller, 1775)	0.0	0.7	0.0	0.3
Agrochola helvola (Linnaeus, 1758)	0.0	0.0	0.0	0.0
Amphipyra pyramidea (Linnaeus, 1758)	0.0	0.0	0.6	1.5
Colocasia coryli (Linnaeus, 1758)	0.0	0.4	0.0	0.0
Cosmia pyralina (Denis et Schiffermüller, 1775)	5.2	2.5	0.0	0.0
Cosmia trapezina (Linnaeus, 1758)	5.5	12.0	12.0	14.5
Dichonia convergens (Denis et Schiffermüller, 1775)	2.8	0.7	1.8	0.6
Dryobotodes eremita (Fabricius, 1775)	0.0	0.0	4.2	0.0
Eupsilia transversa (Hufnagel, 1766)	0.6	2.8	1.2	0.3
Lithophane ornitopus (Hufnagel 1766)	0.6	2.1	3.0	0.9
Moma alpium (Osbeck, 1778)	0.0	0.0	0.0	0.0
Noctuidae species 1	0.0	0.0	0.0	0.0
Noctuidae species 2	0.0	0.0	0.0	0.0
Noctuidae species 3	0.0	0.0	0.0	0.0
Noctuidae species 4	0.0	0.0	0.0	0.9
Noctuidae species 5	0.0	0.0	0.0	0.3
Noctuidae species 6	0.0	0.0	0.0	0.0
Noctuidae species 7	2.2	0.0	0.0	0.0
Orthosia cerasi (Fabricius, 1775)	5.5	2.1	0.0	3.4
Orthosia cruda (Denis et Schiffermüller, 1775)	2.2	1.1	1.8	2.5
Orthosia gothica (Linnaeus, 1758)	0.0	2.1	0.0	0.6
Orthosia incerta (Hufnagel, 1776)	0.0	0.4	0.0	0.0
Orthosia opima (Hübner, 1809)	0.0	0.0	2.4	3.4
No individuals	325	284	167	325
No species/taxons	38	46	35	53

HH	LL	LH	NA	NK	MO	TG	SG
2.3	0.0	0.0	0.0	0.0	4–5	G	FIF
0.0	0.0	0.0	0.8	0.0	6–8	G	SF
29.5	0.0	2.2	18.0	3.5	4–7	G	FIF
0.0	0.0	0.0	0.0	0.2	6	G	LSF
0.0	0.0	0.0	0.8	0.0	4, 8	S2	FIF
0.0	0.0	0.0	0.0	0.0	5	S3	FIF
6.8	0.9	1.1	1.5	1.1	6–10	G	LSF
0.0	0.0	0.0	0.0	0.0	5–6	G	LSF
0.0	0.0	0.5	0.0	0.0	5	G	FIF
0.0	0.0	0.5	0.0	0.2	5	G	FIF
0.0	0.5	0.0	0.8	0.0	6, 8	G	LSF
4.5	0.5	0.0	0.0	1.5	4–5	G	FIF
0.0	5.6	9.8	0.0	5.9	4–5	G	FIF
0.0	0.5	0.0	3.0	1.3	4–5	G	FIF
0.0	0.5	0.0	0.8	0.0	5	S2	FIF
4.5	2.3	0.5	3.0	0.7	4–6	G	FIF
0.0	0.9	0.5	2.3	0.2	4–6	G	FIF
0.0	0.5	0.0	0.8	0.4	7–8	G	SF
0.0	0.0	0.0	0.8	0.0	5	U	FIF
0.0	0.0	0.0	0.8	0.0	5	U	FIF
0.0	0.0	0.0	1.5	0.0	5	U	FIF
0.0	0.0	0.0	0.0	0.0	4	U	FIF
0.0	0.0	0.0	0.0	0.0	4	U	FIF
0.0	0.0	1.6	0.0	0.0	9	U	AF
0.0	0.0	0.0	0.0	0.0	4	U	FIF
0.0	3.2	6.5	0.0	1.1	4–7	G	FIF
0.0	1.4	0.0	0.0	7.0	4–6	G	FIF
0.0	0.0	0.0	0.8	0.4	5–6	G	FIF
0.0	0.0	0.0	0.0	0.0	5	G	FIF
0.0	0.0	0.5	0.0	2.8	4–6	G	FIF
44	216	184	133	462			
18	40	43	40	52			

Clear as daylight: analysis of diurnal raptor pellets for small mammal studies

M. Matos, M. Alves, M. J. Ramos Pereira, I. Torres, S. Marques & C. Fonseca

Matos, M., Alves, M., Ramos Pereira, M. J., Torres, I., Marques, S. & Fonseca, C., 2015. Clear as daylight: analysis of diurnal raptor pellets for small mammal studies. *Animal Biodiversity and Conservation*, 38.1: 37– 48.

Abstract

Clear as daylight: analysis of diurnal raptor pellets for small mammal studies.— Non–invasive approaches are increasingly investigated and applied in studies of small mammal assemblages because they are more cost–effective and bypass conservation and animal welfare issues. However, pellets of diurnal raptors have rarely been used for these purposes. We evaluated the potential of marsh harrier pellets (*Circus aeruginosus*) as a non–invasive method to sample small mammal assemblages, by comparing the results with those of sampling using Sherman live–traps and pitfalls. The three methods were applied simultaneously in an agricultural–wetland complex in NW Portugal. Estimates of species richness, diversity, evenness, abundance, and proportion of each species within the assemblage showed significant differences between the three methods. Our results suggest that the use of marsh harrier pellets is more effective in inventorying small mammal species than either of the two kinds of traps, while also avoiding any involuntary fatalities associated with the sampling of small non–volant mammals. Moreover, the analysis of pellets was the most cost–effective method. Comparison of the two trapping methodologies showed involuntary fatalities were higher in pitfalls than in Sherman traps. We discuss the advantages and flaws of the three methods, both from technical and conservational perspectives.

Key words: Animal welfare, *Circus aeruginosus*, Pitfalls, Pellets, Sherman traps, Small mammals

Resumen

Claro como el agua: análisis de las egagrópilas de aves rapaces diurnas para los estudios sobre pequeños mamíferos.— Los métodos no invasivos se investigan y se aplican cada vez más en los estudios de comunidades de pequeños mamíferos, ya que son más rentables en cuanto a sus costos y evitan los problemas relacionados con la conservación y el bienestar animal. Sin embargo, las egagrópilas de aves rapaces diurnas rara vez se han utilizado para estos fines. En este trabajo se evaluó el potencial que tienen las egagrópilas del aguilucho lagunero (*Circus aeruginosus*) como un método no invasivo para estudiar las comunidades de pequeños mamíferos, mediante la comparación de los resultados con los obtenidos en las trampas de tipo Sherman y las de caída (*pitfall*). Los tres métodos se utilizaron simultáneamente en un complejo formado por tierras agrícolas y humedales en el noroeste de Portugal. Las estimaciones de la riqueza, la diversidad, la uniformidad y la abundancia de especies y la proporción de cada una de ellas dentro de la comunidad mostraron diferencias significativas entre los tres métodos. Nuestros resultados sugieren que la utilización de las egagrópilas del aguilucho lagunero es más eficaz para inventariar las especies de pequeños mamíferos que cualquiera de los dos tipos de trampas, al mismo tiempo que evita la muerte involuntaria de animales asociada con el muestreo de pequeños mamíferos no voladores. Además, el análisis de las egagrópilas fue el método más rentable. Entre los dos métodos de captura, la muerte involuntaria de animales fue mayor en las trampas de caída que en las trampas de tipo Sherman. Se discuten las ventajas y los inconvenientes de los tres métodos tanto desde una perspectiva técnica como conservacionista.

Palabras clave: Bienestar animal, *Circus aeruginosus*, Trampas de caída, Egagrópilas, Trampas de tipo Sherman, Pequeños mamíferos

Milena Matos, Michelle Alves, Maria João Ramos Pereira, Inês Torres, Sara Marques & Carlos Fonseca, Dept. of Biology & CESAM, Univ. of Aveiro, 3810–193 Aveiro, Portugal.

Corresponding author: Michelle Alves. E–mail: michellealves@ua.pt

Introduction

Species inventories are usually the very first step towards biodiversity conservation in a certain region and the basis for integrative and effective management strategies (Begon et al., 2005).

Small, non–volant mammals are highly diverse and play a major role in ecosystem structure and function worldwide. With a wide range of reproductive, loco-motion and foraging strategies, small mammals are responsible for the maintenance of several interactions among wildlife communities, namely by promoting seed dispersal (Adler, 1995), or by constituting key prey for several groups of vertebrates (Carey & Johnson, 1995). Additionally, due to their sensitivity to environmental changes (Pardini et al., 2005), non-volant small mammals are excellent models for the study of ecosystem processes and patterns, and an important group to consider where the protection of ecological values is a concern (Converse et al., 2006).

Studying small mammals usually requires an effective capture plan to achieve a realistic assessment of the assemblies in accordance with the purpose of the work (Voss et al., 2001). Snap–trapping and live–trapping, with Sherman or Tomahawk traps, are the most commonly used methods to capture most small mammal species (Gurnell & Flowerdew, 2006), and supplementary surveys, such as pitfall trapping or active search are used for insectivorous or burrowing species (Voss et al., 2001). Differences in behavior, habitat use, diet, body size and use of vertical strata seem to significantly influence the effectiveness of traps (Sealander & James, 1958; Williams & Braun, 1983). It therefore seems that no single method will effectively yield an adequate sample of the species richness in an area (Voss et al., 2001). Besides efficiency and techniques, trapping small mammals also raises concerns related to ethics and animal welfare (Powell & Proulx, 2003; Putman, 1995), and to the ecological effects of involuntary or voluntary fatalities, the latter in removal–trappings. Acknowledging the usefulness of small mammals as bioindicators in terrestrial ecosystems, research and inventories of small mammal populations and assemblies have significantly increased in recent years. The potential associated fatality rates caused by hypothermia, discomfort or distress (Putman, 1995) may disrupt local populations and consequently, metapopulations (Sullivan & Sullivan, 2013), which taken to an extreme could result in conservation issues, such as monitoring in programmes dealing with sensitive or rare species.

Attending to all these constraints, and adding to the logistics and costs associated with trapping schemes, non–invasive approaches are increasingly investigated and applied (De Bondi et al., 2010; DeSa et al., 2012; Torre et al., 2013). When studies seek to examine aspects of assemblage composition, the most common non–invasive methods have long been the analysis of owl pellets, particularly from widespread and common species, such as *Tyto alba* (Torre et al., 2004, Rocha et al., 2011) and *Strix aluco* (Balčiauskienė, 2005; Petty, 1999), due to their generalist diets and close foraging ranges (Torre et

al., 2013). However, pellets of diurnal raptors have rarely been used for these purposes, perhaps due to the relative difficulty in finding suitable amounts of pellets or in identifying prey remains, as many raptor species decapitate their prey before ingestion (Balfour & Macdonald, 1970) or are able to digest the skeletal parts of mammalian prey (Glue, 1970). Pellets of diurnal raptors were used by Santos et al. (2009) and Scheibler & Christoff (2007) but in both cases only as a complementary method for the inventory of small mammals in their study areas, specifically salt ponds of Aveiro, Portugal, and in agricultural areas of southern Brazil.

The marsh harrier (*Circus aeruginosus*) is a diurnal, medium–sized bird of prey whose distribution ranges from Europe and central Africa to central Asia, the northern parts of the Middle East and the Indian subcontinent (BirdLife International and NatureServe, 2014). It occurs in a wide range of habitats, from wetlands to agricultural areas (Cardador et al., 2011) and other human–shaped environments (Vandermeer, 2010). Marsh harriers mostly roost (Moreno et al., 2014) and nest (González, 1991) in paludal vegetation. They also use perches (Kitowski, 2007), under which it is common to find pellets.

Marsh harriers usually forage in open agricultural areas, particularly on the edge of ponds of fresh or brackish water, using the raid as the main hunting technique (Clarke et al., 1993). However, they have a high foraging plasticity, allowing them to use habitats that are not accessible to other birds of prey (Kitowski, 2007). The diet of marsh harriers is usually characterized as generalist (Strandberg et al., 2008) and influenced by seasonal and local conditions (Witkowski, 1989; Cardador et al., 2012), but most studies list small mammals as their primary prey (González, 1991; Alves, 2013). Lagomorphs (Schipper, 1973) and birds (Mateo et al., 1999; Clarke et al., 1993) may also be important prey.

Here we aimed to evaluate the potential of the analysis of marsh harrier pellets as a non–invasive method to determine the composition of small mammal assemblages, by comparing the results with those of two other methods, Sherman live–trapping and pitfalls, applied simultaneously in the same mosaic of habitats. We discuss advantages and drawbacks of the three methods, both from technical and conservational perspectives. Although other authors have made some considerations about the viability of diurnal raptor pellets as a technique to sample small mammals (*e.g.*, Andrews, 1990), to our knowledge this is the first study specifically designed to compare the efficiency of this method with that of other widely used capture methods.

Methods

Study area

This study was developed in Baixo Vouga Lagunar, located approximately between 8° 32' 57" W and 8° 41' 32" W; and 40° 49' 43" N and 40° 41' 32" N, in NW Portugal. The study area occupies about 12,205 ha and encompasses important ecosystems integrated in the Natura 2000 Network (PTZPE0004, PTCON0061), featuring one of the most important Portuguese wetlands (Ria de Aveiro). The climate is Mediterranean with strong Atlantic influences, and it has an annual mean temperature of about 15ºC. The average annual humidity is 77% and the average annual rainfall is 1,387 mm, with a shortage of rainfall in the summer. The system represents a complex agriculture–wetland mosaic, integrating a variety of natural and human–altered habitats (fig. 1), such as pastures (2.35% of the area), rice fields *Oryza sativa* (0.98%) and maize *Zea mays* (28.13%); and 'Bocage' (7.09%), a typical and rare landscape unit which consists of hedgerows of trees (*e.g., Salix alba*), shrubs (*e.g., Rubus ulmifolius*) and ditches that compartmentalize farmlands and pastures. The wetlands are composed of reedbeds of *Phragmites australis* (4.50%), saltmarshes of *Spartina maritima* (12.79%) and rushes of *Juncus maritimus* (6.67%) (Alves et al., 2014). This region also houses high faunal richness (*e.g.* amphibians, birds (Special Protection Area for Birds PTZPE0004), and bats (Mendes et al., 2014). According to national bird censuses, the study area shelters 11 to 12 resident pairs of breeding marsh harriers, which corresponds to about 17% of the breeding population in the country (Rosa et al., 2006). The biological richness of Baixo Vouga Lagunar attracts ecotourists, and the region receives 25000+ visitors per year.

Sampling and identification of small non–volant mammals

Once a month, we collected marsh harrier pellets (*n* = 75) near nesting sites and under perches used by the species, during the breeding season of 2012, *i.e.* from February to August. The closest distance between a pellet collecting site and a nest was ca. 200 m. Our collecting procedures did not seem disturbing to the birds, especially considering the regular touristic visits to the area. Using a telescope, we spotted the collecting sites through direct observation of the birds from nine observation points each covering a circular area with a radius of 1.5 km. Observations took place monthly in the first three months and lasted for two hours per observation point. Regular flooding around perches and roosts prevented us from collecting pellets throughout the whole year. We oven–dried the pellets at 60°C for a day and the dry content was then separated after moisturizing. Food items were identified and quantified through the presence of non–digestible remains. Since harriers tend to rip the meat off their prey rather than swallow the entire animal (Hosking, 1943; Balfour & Macdonald, 1970), mammals were identified based on cranial structures described in the literature (Gállego & López, 1982; Gállego & Alemany, 1985; Blanco, 1998a; Blanco, 1998b) but also on detailed features of the fur: cuticular print, core and cross section (Teenrik, 1991; Quadros & Monteiro–Filho, 2006; Valente, 2012). *Talpa occidentalis* was the only species identified solely through cranial structures; all other species were identified by both cranial and hair structures. We used the 'minimum number of individuals' analysis in order to reduce possible erroneous counting of the number of prey (Lyman et al., 2003).

Simultaneously we sampled small non–volant mammals in the study area using Sherman and pitfall traps. For each habitat in the study area (reedbeds, rushes, saltmarshes, Bocage and rice and maize fields) and whenever possible, three replicate of small mammal sampling sites were randomly distributed within the nine 1.5 km radius harriers sampling areas, as long as 1 km of minimal distance between sampling sites was assured, to maintain spatial independence. Small mammals sampling sessions took place every two months, in a total of three sampling rounds for the study period. Each small mammal sampling site consisted of a line of 30 Sherman live traps (17.5 x 6 x 6 cm) separated 10 m from each other and baited with a mixture of canned sardines and hamster food, and a line of four pitfalls (buckets ca. 30 cm deep, 5 L capacity) connected with a drift fence buried to prevent animals from passing under it. In the Sherman traps, cotton was provided as nesting material. Whenever possible, traps were set under the cover of stones, shrubs or herbs to provide camouflage and some thermal insulation. Both standardized methodologies were applied simultaneously, in order to minimize the effect of the selectivity of each method and collect more representative data on the composition of the small mammal assemblage. At each sampling round, traps were active for five consecutive nights and visited every early morning as was previously tested (Gurnell & Flowerdew, 2006). At each trap check, we provided dry bedding material and a new food supply. We ringed the collected animals individually and released them after identification.

Statistical analysis

Since the trapping methods did not allow the identification of all small mammals to the species level, we used genera accumulation curves to assess patterns of genera richness in the incidence matrix obtained with each sampling method (Gotelli & Colwell, 2001). We calculated Mao Tau and Chao 1 richness estimators (Chao et al., 2009; Torre et al., 2013) using the software EstimateS 9.0 for Windows (Colwell, 2011). The completeness of the inventory made with each method was assessed by fitting the Clench equation to the observed genera accumulation curve, using the quasi–Newton method equation (Soberón & Llorente, 1993). We used the same procedures to estimate species richness with the pellet sampling method.

For each method we assessed the assemblage structure using the number of identified genera, abundance (measured as the number of individuals captured in 100 night–traps or in 100 pellets [Mills et al., 1991]), and the proportion of each taxa within the assemblage, measured through the percentage of occurrence of each taxa, calculated as $\%O_i = n_i / N_+ \times 100$, where n_i is the number of individuals of species i and N_+ is the total number of individuals identified or captured with a given method. We also calculated indices of evenness (Pielou index) and diversity (Shannon–Wiener and Simpson index) for all three methods. We calculated two diversity indices to ascertain possible effects of rarely recorded taxa: Simpson's index is less sensitive to presence, giving more weight to common taxa (Simpson, 1949); Shannon–Wiener's index is more sensitive to rare taxa (Magurran, 2004). To calculate evenness and diversity indices, due to a high number of zeros, pellet collecting sites ($n = 8$; see map in fig.1) and habitat replicates (for trappings; $n = 13$) were considered as samples.

We searched for differences between methods in assemblage composition and abundance of small mammals, as well as in evenness and diversity indices, using analysis of variance (ANOVA) and controlling for the effects of sample size (Rahbek, 1997).

Cost comparison

For each sampling method we calculated the associated costs for total working hours and expenses. Working hours for each trapping method considered two people working in the field, and included all steps of the monitoring programme (installation, checking, animal handling and trap removal), totaling on average 24 six–hour–days per person and per sampling round for Sherman traps and 24 three–hour–days per person and per sampling round for pitfall traps. As for pellets, field work was performed by two people, and included spotting of pellet collection sites (on average five five–hour–days per person and per month), and pellet collection (on average one four–hour–day per person and per month). Pellets (lab work) were analysed by the same person and took on average of 20 five–hour–days per month. Expenses included fuel for field trips and supplies, such as bait and cotton for the traps and cover slips, and microscope slides for pellet analysis. The price of traps and lab equipment was not included in the budget as these items were already available in the research facilities.

Results

In total, 429 small mammals of 11 species were recorded: seven rodents and four insectivore species (table 1). All eleven species were detected in marsh harrier pellets. Sherman traps captured five rodent and one insectivore taxa, and pitfall traps captured three rodent and two insectivore taxa. Six taxa were identified to species level only with pellets: *Arvicola sapidus*, *Mus musculus*, *Mus spretus*, *Sorex granarius*, *Sorex minutus* and *Talpa occidentalis*. Traps did not add any distinct species. Overall, pellets presented higher scores for the number of identified genera and species, evenness and diversity (table 1) than either of the two trapping techniques. The total number of captured individuals was highest with Sherman trap sampling (table 1).

The estimated species richness of small mammals using the pellet sampling was 11.33 ± 0.93; ($n = 75$; fig. 2A). The Clench equation showed strong adjustment to the species accumulation curve ($r^2 = 0.9999$), with a slope of 0.029, showing the proximity to an asymptote, and thus indicating that the sampling of small mammals with this method was quite complete and reliable (fig. 2A). Estimates of the

number of identified genera (Sherman 5.00 ± 0.45, $n = 39$; pitfalls 5.00 ± 0.17, $n = 39$; pellets 8.00 ± 0.25, $n = 75$), Clench model adjustment to the genera accumulation curve (Sherman $r^2 = 0.954$; pitfalls $r^2 = 1.000$; pellets $r^2 = 0.998$) and slopes of the obtained curves (Sherman 0.004; pitfalls 0.033; pellets 0.015) showed that further field efforts would not result in a relevant increase of detected genera (fig. 2B, 2C, 2D). All evaluated assemblage composition parameters showed significant differences between methods: number of genera ($F = 54.424$, $P < 0.0001$), abundance ($F = 30.548$, $P < 0.0001$), and the proportion of each genus within the assemblage ($F = 4.112$, $P = 0.017$). Tukey post–hoc tests showed there were differences between pairwise comparisons, in the number of genera and abundance between Sherman and pitfall traps and between Sherman traps and pellets; and percentage of occurrence between Sherman traps and pitfalls. Sherman traps detected a greater number of genera per sample (2.103 ± 1.046 genera/sample) than pellets (0.7475 ± 0.617 genera/sample) or pitfalls (0.462 ± 0.482 genera/sample) (results expressed as mean ± standard deviation).

Indices presented significant differences between methods: Pielou ($F = 15.685$, $P < 0.0001$), Shannon–Wiener ($F = 11.009$, $P < 0.0001$), and Simpson ($F = 3.742$, $P = 0.037$), with pellets consistently scoring the highest values.

Involuntary fatalities associated with trapping methods were 0.51 and 1.79 individuals per 100 trap–nights with Sherman traps and pitfalls, respectively. The species showing highest fatality rates were *Crocidura russula* and *Microtus lusitanicus*.

Cost estimations showed that pitfall trapping was the fastest method, although pellet analysis was the cheapest. Sherman trapping was the most time–consuming and the most expensive method (table 2).

Discussion

The combination of sampling methods used in this study identified 11 species of small non–volant mammals in the study area, where 12 species are known to be present (considering rodents and shrews, and excluding squirrels, hedgehogs and bats; Bandeira et al., 2013). Comparing our results with the independent and long–term mammal study of Bandeira et al. (2013), pellets only failed to detect *Rattus rattus*, a scavenger species that prefers to live around human settlements (Ewer, 1971). We did not sample within or around urban areas, but based on previous observations, it is plausible to assume that areas in such close proximity to humans are avoided by the marsh harriers (Alves et al., 2014).

In the pellets we found remains of small non–volant mammal species known to be less common or rare, such as *Sorex minutus* and *Arvicola sapidus*, an aquatic species. The two Iberian endemisms, *Sorex granarius* and *Talpa occidentalis*, were also only recorded in pellet samples. When comparing the three methods alone, pellet analysis seemed to be most cost effective and efficient method for inventorying small mammals

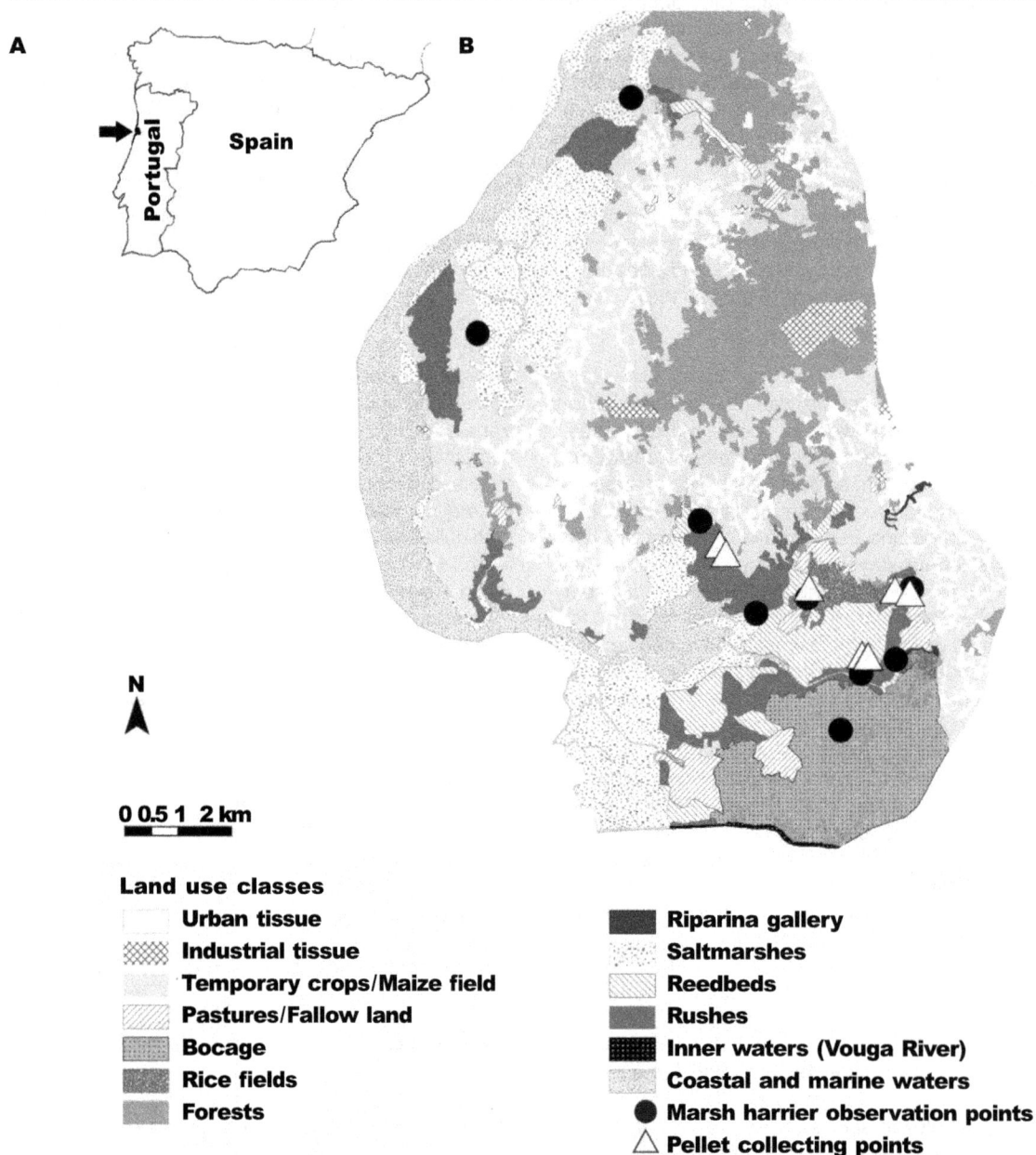

Fig. 1. Study area: A. Location of the study area in the Iberian peninsula; B. Land cover of the study area with marsh harrier observation points and pellet collecting sites.

Fig. 1. Área de estudio: A. Ubicación del área de estudio en la península ibérica; B. Cobertura del suelo en el área de estudio, con los puntos de observación de aguilucho lagunero y los puntos de recolección de las regurgitaciones.

in our study area. However, if we consider the two trapping schemes together the differences begin to fade. Nonetheless, in terms of species inventorying, our results indicate that even by analyzing a reduced number of pellets, information on the number of taxa detected can be significantly higher than that retrieved with large trapping efforts. Indeed, the species accumulation curve for pellet analysis indicates that 11 pellets were sufficient to reach five small mammals species, which is equivalent to the entire species count allowed by Sherman traps and pitfalls altogether throughout the whole study. At the genus level, trapping methods altogether yielded six genera, a score achieved with the analysis of 22 pellets.

Our field trapping did not always allow identification of some individuals to the species level, such

Table 1. Type of activity (N, nocturnal; D, diurnal, according to Blanco 1998a, 1998b), Portuguese conservation status (PT), international conservation status (IUCN), number of individuals (N), percentage of occurrence (%O), and abundance (A, measured as the number of individuals captured in 100 night–traps or in 100 pellets) of all species recorded with Sherman and pitfall trapping and with the analysis of marsh harrier pellets. Number of identified genera and species, number of captured individuals, abundance, Pielou's evenness index, Shannon–Wiener's diversity index and Simpson's diversity index are presented for each method. Indices were calculated with data to the genus level, allowing comparison between methods. Numbers in bold highlight the highest value obtained per considered parameter: * Iberian endemism.

*Tabla 1. Tipo de actividad (N, nocturna; D, diurna, de acuerdo con Blanco 1998a, 1998b), estado de conservación (PT); estado de conservación internacionañ (UICN), número de individuos (N), porcentaje de presencia (%O) y abundancia (A, medida como el número de individuos capturados en 100 trampas/noche o en 100 regurgitaciones), de todas las especies registradas con trampas de tipo Sherman y de caída y mediante el análisis de regurgitaciones de aguilucho lagunero. Para cada método evaluado se presentan el número de géneros y de especies identificados, el número de individuos capturados, la abundancia, el índice de uniformidad de Pielou, el índice de diversidad de Shannon–Wiener y el índice de diversidad de Simpson. Los índices se calcularon con los datos a nivel de género, lo que permitió comparar los métodos. Los números en negrita son los valores más altos obtenidos en los parámetros indicados: * Endemismo ibérico.*

Taxa	Activity	Conservation status PT	IUCN	Sherman N	%O	A	Pitfalls N	%O	A	Pellets N	%O	A
Rodentia												
Apodemus sylvaticus	N	LC	LC	79	22.44	1.35	1	5.26	0.13	8	13.79	10.67
Arvicola sapidus	N/D	LC	VU	–	–	–	–	–	–	2	3.45	2.67
Microtus agrestis	N	LC	LC	2	0.57	0.03	–	–	–	15	25.86	20.00
Microtus lusitanicus	N/D	LC	LC	33	9.38	0.56	–	–	–	7	12.07	9.33
Microtus sp.	–	–	–	–	–	–	9	47.37	1.15	–	–	–
Mus musculus	N	LC	LC	–	–	–	–	–	–	4	6.90	5.33
Mus spretus	N	LC	LC	–	–	–	–	–	–	6	10.34	8.00
Mus sp.	–	–	–	167	47.44	2.85	2	10.53	0.26	–	–	–
Rattus norvegicus	N	NA	LC	1	0.28	0.02	–	–	–	4	6.90	5.33
Eulipotyphla												
Crocidura russula	N/D	LC	LC	70	19.89	1.20	5	26.32	0.64	7	12.07	9.33
Sorex granarius	N/D	DD*	LC	–	–	–	–	–	–	3	5.17	4.00
Sorex minutus	N/D	DD	LC	–	–	–	–	–	–	1	1.72	1.33
Sorex sp.	–	–	–	–	–	–	2	10.53	0.26	–	–	–
Talpa occidentalis	N/D	LC*	LC	–	–	–	–	–	–	1	1.72	1.33

	Sherman	Pitfalls	Pellets
No. identified genera	5	5	**8**
No. identified species	5	2	**11**
Total captures	**352**	19	58
Abundance	6.02	2.44	**77.33**
Pielou eveness index	0.78	0.83	**0.84**
Shannon–Wiener index	1.26	1.33	**1.75**
Simpson diversity index	0.68	0.72	**0.80**

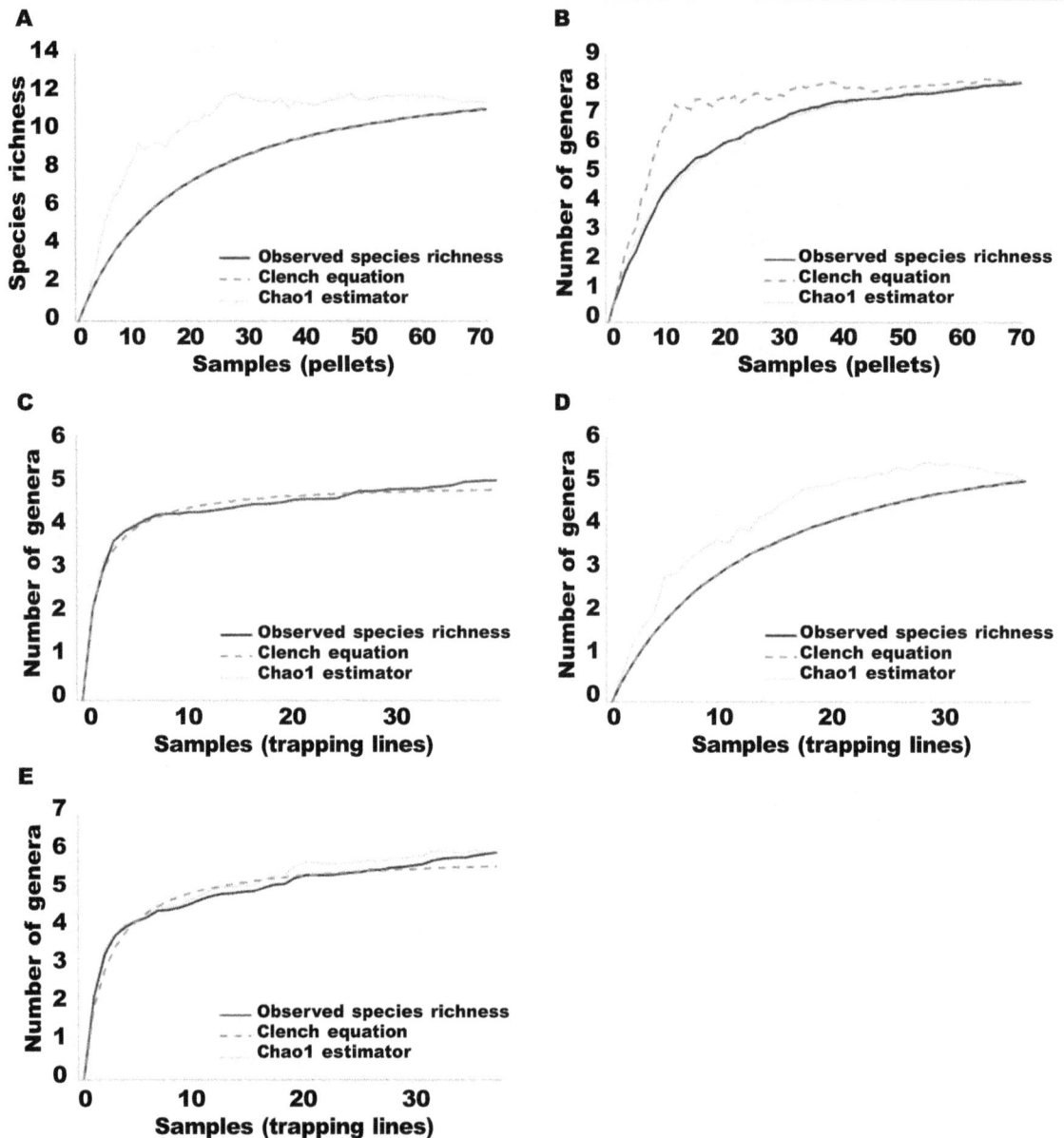

Fig. 2. A. Species accumulation curve and estimated number of species (Chao 1 estimator) for the small mammal assemblage preyed by marsh harriers: B–E. Genera accumulation curve and estimated number of genera identified through marsh harrier pellets (B), Sherman traps (C), pitfalls (D), Sherman and pitfall traps altogether (E). Observed data were fitted to the Clench equation to evaluate the completeness of the inventories.

Fig. 2. A. Curva de acumulación de especies y riqueza de especies estimada (estimador Chao 1) para el agregado de pequeños mamíferos cazado por los aguiluchos laguneros: B–E. Curva de acumulación de géneros y riqueza estimada de los géneros identificados a través de las regurgitaciones de aguilucho lagunero (B), las trampas de tipo Sherman (C), las trampas de caída (D) y las trampas de tipo Sherman y de caída juntas (E). Los datos observados se ajustaron a la ecuación de Clench para evaluar la exhaustividad de los inventarios.

as members of the genus *Crocidura*, which, when wet and anxious, may be difficult to carefully observe and distinguish. It is important to emphasize that no animals were intentionally sacrificed for later identification in the lab, so all morphometric analyses were done in the field and in live individuals, which were ultimately released. In fact, due to conservation constraints, the sacrifice of terrestrial vertebrates is

Table 2. Summary of the estimated cost for each sampling method (total working hours and total expenses): S. Sherman traps; Pt. Pitfall traps; Pl. Pellets.

Tabla 2. Resumen de la estimación de costos para cada método de muestreo (total de horas de trabajo y gastos totales): S. Trampas de tipo Sherman; Pt. Trampas de caída; Pl. Regurgitaciones.

	S	Pt	Pl
Total working (hours)	1,008	504	856
Total expenses (€)	1,596	1,211	365

neither ethically acceptable for the purpose of this study, nor legally permitted, so the identification of those individuals remained ambiguous. Conversely, all mammal remains present in marsh harrier pellets were identified to the species level, based on cranial and/or fur features.

Sherman traps tend to capture species with high capturability, oversampling these species and undersampling trap–shy species (Iriarte et al., 1989). Trap size and layout, among many other factors, influence their effectiveness (Smith et al., 1975). According to Boonstra & Krebs (1978), pitfalls are more efficient in sampling individuals of all ages among the small mammal assemblage but may fail to capture species with greater body size (M. Alves, personal observation). In our study area, it is possible that Sherman traps may have oversampled small mammal species that are more active on the surface, such as *Apodemus sylvaticus* and *Crocidura russula*, though presenting higher capturability rates than fossorial species, such as *Microtus* rodents, which may be undersampled. On the other hand, pellet sampling may be biased by the predator's ecological habits and preferences. Indeed, previous studies in the area have shown that during the breeding season, marsh harriers significantly prefer to forage on reedbeds (Alves et al., 2014), and to feed on *Microtus* species over other more abundant taxa (Alves, 2013). Also, marsh harriers may range over distances of up to 5,000 m from the nest during the breeding season (Cardador et al., 2009). Pellets may thus reflect a larger foraging area than that surveyed by trapping techniques; this may be useful when the objectives are to sample wide study areas.

Consistent differences in the estimates of species richness, abundance and proportion of species between the three methods suggest that supporting an assemblage study using only one method may lead to seriously biased results. Using various sampling methods combined is a way to overcome the biases of each method and obtain more complete information on the non–volant small mammal assemblage present in the study area. This is a well–established and recurring conclusion (Smith et al., 1975; Williams & Braun, 1983),

even in studies evaluating indirect and non–invasive sampling methods (Jaksić et al., 1981). Marsh harrier pellets, for instance, proved to be particularly efficient for inventorying species richness. However, population parameters such as abundance are probably more accurate when calculated with direct approaches, such as capture–recapture schemes (Hopkins & Kennedy, 2004). Also, due to the large home range and ecological preferences of raptors, pellets may fail to provide accurate information on the microhabitat preferences of small mammals.

Very few studies have used pellets of a diurnal raptor to study the assemblages of small non–volant mammals (but see Santos & Fonseca, 2009; Scheibler & Christoff, 2007). Most studies use pellets of common and widespread nocturnal birds of prey, such as *Tyto alba*, a generalist species that presents a relatively narrow home range, pellets that are very easy to find and collect, and prey remains that are easy to identify through cranial structures. However, owls mainly prey on small nocturnal mammals and may over or under sample some particular type of prey present in specific habitats that they may prefer or avoid, respectively (Torre et al., 2004). The same is true for marsh harriers, but these birds seem to forage on habitats that are inaccessible to other birds of prey, such as high crops and reedbeds (Kitowski, 2007). The predator's hunting time may bias the sampling results towards more nocturnal or diurnal species, but our data suggest this may not be a major issue, because though marsh harriers are mostly diurnal, their pellets contained relevant amounts of mammals described as predominantly nocturnal (*e.g., Apodemus sylvaticus* and *Microtus agrestis;* table 1). This may be due to the general activity patterns of small mammals, which are rarely exclusively nocturnal or diurnal. Even short periods of diurnal activity may represent hunting opportunities for fast raptors, such as the marsh harrier, that use the raid hunting technique. Furthermore, in wetlands and other open areas that lack nesting sites for owls —such as the Baixo Vouga Lagunar— owl pellets are not even an option. On the other hand, the marsh harrier is a widespread species in the region, easy to identify and spot for pellet deposit sites. Studies developed in the study area (Alves, 2013) confirmed the diet of marsh harrier as generalist and mostly (68%) constituted of small mammals. It was also confirmed that marsh harriers forage on reedbeds and saltmarshes —wetlands that can be quite difficult to sample with traps, due to regular flooding, tides (in salt marshes) and vegetation density— but also on crops, providing a general sampling of the small mammals of a vast number of habitats. Our results also show that marsh harrier pellets provided more complete information on small non–volant mammal richness, and potentially of evenness and diversity. Table 3 summarizes further advantages and disadvantages of the three methods assessed in our study.

Finally, it should be highlighted that when designing any study on the structure of animal assemblages the choice of methods must carefully consider not only the technical limitations, but also the purposes of the study and ethical and legal questions. The

Table 3. Comparative list of advantages and disadvantages of using Sherman traps, pitfalls and the analysis of marsh harrier pellets for small mammal studies, according to our study.

Tabla 3. Lista comparativa de las ventajas e inconvenientes de utilizar trampas de tipo Sherman, trampas de caída y el análisis de las regurgitaciones de aguilucho lagunero para los estudios sobre pequeños mamíferos, según nuestro estudio.

Advantages	Disadvantages
Marsh harrier pellets	
Low logistical requirements	Pellets are hard (or sometimes impossible) to collect, especially in flooded or hard–reaching areas
Cost–effective	
Less time–consuming than Sherman traps	
More positive identification of species	Collection is more time–consuming than pitfalls
Higher completeness and effectiveness in species inventory, especially in heterogeneous and wide areas	Not suitable to estimate density
	The quality of the results depends on the diet of the birds, which is influenced by environmental constraints (*e.g.* landscape features and prey availability)
Detection of species occupying habitats where trapping is not possible	
Higher potential to detect rare species	Does not provide information on the spatial ecology of small mammals.
Provides more accurate information at the species level, allowing better diversity and evenness calculations	May underestimate nocturnal and overestimate diurnal prey species
Non–invasive method for small mammals	
Sherman traps	
Lower mortality than pitfalls	Time–consuming and expensive. High logistical requirements
Allows the observation of gender, age and physical condition	Frequently sprung–but–empty
Allows capture–mark–recapture techniques	Not usable in all kind of habitats (*e.g.* wetlands)
Suitable for density estimation	
Provides information on the spatial ecology of small mammals	Lower potential to detect rare species
	Results biased towards trap–prone species
	May oversample species that live at the surface
	Trapping success largely depends on external conditions that influence animal activity
Pitfalls	
Less time–consuming than Sherman traps or pellet analysis	Time–consuming and expensive
	High logistical requirements
Allows simultaneous and sequential multiple captures	Very conspicuous apparatus, subject to vandalism or theft
Allows the observation of gender, age and physical condition	Lower potential to detect rare species
	Biased towards common species
Allows capture–mark–recapture techniques	May oversample species that live at the surface
Suitable for density estimation	May fail to capture larger animals
More successful in sampling trap–shy species than Sherman traps	High mortality rates, not suited for studies with endangered species
	Trapping success largely depends on external conditions that condition animal activity

efficiency of marsh harrier pellets —and also other non–invasive methods— showed that, in some cases, and depending on the objectives (for instance, presence–absence data), trapping is unnecessary, thereby avoiding disturbance and fatalities among small mammals (Powell & Proulx, 2003; Sullivan & Sullivan, 2013). Collecting pellets, in particular near nests, may disturb the birds (Fernández & Azkona, 1993), but if the study is well planned and takes the reproductive and spatial ecology and the habits of the species into consideration, impact can be minimized. In our study, pitfalls presented considerable involuntary fatality rates for small mammals, mostly due to the humidity in the buckets, causing the animals to die from hypothermia. This sampling approach cannot therefore be recommended for small non–volant mammals in wetlands or areas with high humidity levels. In regions that harbour endangered or rare species, we encourage researchers to seek for non–invasive methodologies, or, at least, to previously determine the least detrimental research protocol for wildlife, safeguarding animal and ecosystem welfare.

Acknowledgements

This work was co–supported by European Funds through COMPETE and by National Funds through the Portuguese Science Foundation (FCT) within project PEst–C/MAR/LA0017/2013. The authors would like to thank Câmara Municipal de Estarreja and OHM Estarreja for logistical and financial support. We also thank Eduardo Mendes for support in field work, Rita Rocha, Eduardo Ferreira and Victor Bandeira for their help identifying food items, and two anonymous reviewers for their helpful comments on a previous version of the manuscript. Milena Matos and Maria João Ramos Pereira were financed by post–doctoral grants from Fundação para a Ciência e Tecnologia (SFRH/BPD/74071/2010 and SFRH/BPD/72845/2010, respectively). All animals were captured and handled in accordance with Portuguese law (licenses 385/2011/ CAPT and 99/2012/CAPT issued by ICNF–Institute for the Conservation of Nature and Forests). Pellets were collected under the ICNF license 95/2012/ PERTURBAÇÃO.

References

Adler, G. H., 1995. Fruit and seed exploitation by Central American spiny rats, *Proechimys semispinosus*. *Studies on Neotropical Fauna and Environment*, 30: 237–244.

Alves, M., 2013. Foraging and spatial ecology of Marsh harrier in Baixo Vouga Lagunar. MSc thesis, University of Aveiro, Portugal.

Alves, M., Ferreira, J. P., Torres, I., Fonseca, C. & Matos, M., 2014. Habitat use and selection of the marsh harrier *Circus aeruginosus* in an agricultural–wetland mosaic. *Ardeola*, 61(2): 351–366.

Andrews, P., 1990. *Owls, Caves and Fossils: Predation, Preservation and Accumulation of Small Mammal Bones in Caves, with an Analysis of the Pleistocene Cave Faunas From Westbury–Sub–Mendip, Somerset*. University of Chicago Press, U.K.

Balčiauskienė, L., 2005. Analysis of Tawny Owl (*Strix aluco*) food remains as a tool for long–term monitoring of small mammals. *Acta Zoologica Lituanica*, 15: 85–89.

Balfour, E. & Macdonald, M., 1970. Food and feeding behaviour of the hen harrier in Orkney. *Scottish Birds*, 6: 57–66.

Bandeira, V., Azevedo, A. & Fonseca, C., 2013. *Guia de Mamíferos do BioRia*. Câmara Municipal de Estarreja, Estarreja.

Begon, M., Townsend, C. R. & Harper, J. L., 2005. Ecology: from individuals to ecosystems. Wiley–Blackwell, Oxford.

BirdLife International and NatureServe, 2014. *Bird Species Distribution Maps of the World*, 2013: Circus aeruginosus. The IUCN Red List of Threatened Species, version 2014.1

Blanco, J. C., 1998a. *Mamíferos de España I. Insectívoros, Quirópteros, Primates y Carnívoros de la Península Ibérica, Baleares y Canarias*. Editorial Planeta, Madrid.

– 1998b. *Mamíferos de España II. Cetáceos, Artiodáctilos, Roedores y Lagomorfos de la Península Ibérica, Baleares y Canarias*. Editorial Planeta, Madrid

Boonstra, R. & Krebs, C. J., 1978. Pitfall trapping of *Microtus townsendii*. *Journal of Mammalogy*, 59: 136–148.

Cardador, L., Carrete, M. & Mañosa, S., 2011. Can intensive agricultural landscapes favour some raptor species? The Marsh harrier in north–eastern Spain. *Animal Conservation*, 14: 382–390.

Cardador, L., Mañosa, S., Varea, A. & Bertolero, A., 2009. Short communication: Ranging behaviour of Marsh Harriers *Circus aeruginosus* in agricultural landscapes. *Ibis*, 151: 766–770.

Cardador, L., Planas, E., Varea, A. & Mañosa, S., 2012. Feeding behaviour and diet composition of marsh harriers *Circus aeruginosus* in agricultural landscapes. *Bird Study*, 59: 228–235.

Carey, A. B. & Johnson, M. L., 1995. Small mammals in managed, naturally young, and old–growth forests. *Ecological Applications*, 5: 336–352.

Chao, A., Colwell, R. K., Lin, C.–W. & Gotelli, N. J., 2009. Sufficient Sampling for Asymptotic Minimum Species Richness Estimators. *Ecology*, 90: 1125–1133.

Clarke, R., Bourgnje, A. & Casteljins, H., 1993. Food niches of sympatric Marsh Harriers *Circus aeruginosus* and Hen Harriers *Circus cyaneus* on the Dutch coast in winter. *Ibis*, 135: 424–431.

Colwell, R. K., 2011. *Estimates: statistical estimation of species richness and shared species from samples*. url: http://viceroy.eeb.uconn.edu/estimates/ (Accessed on March 2015)

Converse, S. J., Block, W. M. & White, G. C., 2006. Small mammal population and habitat responses to forest thinning and prescribed fire. *Forest Ecology and Management*, 228: 263–273.

De Bondi, N., White, J. G., Stevens, M. & Cooke, R., 2010. A comparison of the effectiveness of camera trapping and live trapping for sampling terrestrial small–mammal communities. *Wildlife Research,* 37: 456–465.

DeSa, M. A., Zweig, C. L., Percival, H. F., Kitchens, W. M. & Kasbohm, J. W., 2012. Comparison of Small–Mammal Sampling Techniques in Tidal Salt Marshes of the Central Gulf Coast of Florida. *Southeastern Naturalist,* 11: 89–100.

Ewer, R. F., 1971. The Biology and Behaviour of a Free–Living Population of Black Rats (*Rattus rattus*). *Animal Behaviour Monographs,* 4(3): 125–174.

Fernández, C. & Azkona, P., 1993. Human disturbance affects parental care of marsh harriers and nutritional status of nestlings. *The Journal of Wildlife Management,* 57: 602–608.

Gállego, L. & Alemany, A., 1985. *Vertebrados Ibéricos, vol. 6. Mamíferos Roedores y Lagomorfos.* Antiga Imprenta Soler, Palma de Mallorca.

Gállego, L. & López, S., 1982. *Vertebrados Ibéricos, vol. 5. Mamíferos Insectívoros.* Imprenta Sevillana, Palma de Mallorca.

Glue, D. E., 1970. Avian predator pellet analysis and the mammalogist. *Mammal Review,* 1: 53–62.

González, J. L., 1991. *El aguilucho lagunero Circus aeruginosus L. en España. Situación, biología de la reproducción, alimentación y conservación.* ICONA, CSIC, Madrid.

Gotelli, N. J. & Colwell, R. K., 2001. Quantifying biodiversity: procedures and pitfalls in the measurement and comparison of species richness. *Ecology letters,* 4: 379–391.

Gurnell, J. & Flowerdew, J., 2006. *Live trapping small mammals: a practical guide.* The Mammal Society London, London.

Hopkins, H. L. & Kennedy, M. L., 2004. An assessment of indices of relative and absolute abundance for monitoring populations of small mammals. *Wildlife Society Bulletin,* 32: 1289–1296.

Hosking, E. J., 1943. Some observations on the Marsh Harrier. *British Birds,* 37: 2–9.

Iriarte, J. A., Contreras, L. C. & Jaksić, F. M., 1989. A long–term study of a small–mammal assemblage in the central Chilean matorral. *Journal of Mammalogy,* 70: 79–87.

Jaksić, F. M., Yáñez, J. L. & Fuentes, E. R., 1981. Assessing a small mammal community in central Chile. *Journal of Mammalogy,* 62(2): 391–396.

Kitowski, I., 2007. Inter–sexual differences in hunting behaviour of Marsh Harriers (*Circus aeruginosus*) in southeastern Poland. *Acta Zoologica Lituanica,* 17: 70–77.

Lyman, R. L., Power, E. & Lyman, R. J., 2003. Quantification and sampling of faunal remains in owl pellets. *Journal of Taphonomy,* 1: 3–14.

Magurran, A. E., 2004. *Measuring biological diversity.* 2nd Ed. Blackwell, Oxford.

Mateo, R., Estrada, J., Paquet, J.–Y., Riera, X., Domínguez, L., Guitart, R. & Martínez–Vilalta., A.,1999. Lead shot ingestion by marsh harriers *Circus aeruginosus* from the Ebro delta, Spain. *Environmental Pollution,* 104: 435–440.

Mendes, E., Pereira, M., Marques, S. & Fonseca, C., 2014. A mosaic of opportunities? Spatio–temporal patterns of bat diversity and activity in a strongly humanized Mediterranean wetland. *European Journal of Wildlife Research,* 60: 651–664.

Mills, J. N., Ellis, B. A., McKee, K. T., Maiztegui, J. I. & Childs, J. E., 1991. Habitat associations and relative densities of rodent populations in cultivated areas of central Argentina. *Journal of Mammalogy,* 72: 470–479.

Moreno, C. A. T., Román, J. R. & Baticón, A. O., 2014. Excepcional dormidero invernal de Aguilucho Lagunero Occidental *Circus aeruginosus* en la laguna de Fes–el–Bali, región de Taza–Al Hoceima–Taounate (Marruecos). *Go–South Bull.,* 11: 14–16.

Pardini, R., de Souza, S. M., Braga–Neto, R. & Metzger, J. P., 2005. The role of forest structure, fragment size and corridors in maintaining small mammal abundance and diversity in an Atlantic forest landscape. *Biological Conservation,* 124: 253–266.

Petty, S. J., 1999. Diet of tawny owls (*Strix aluco*) in relation to field vole (*Microtus agrestis*) abundance in a conifer forest in northern England. *Journal of zoology,* 248: 451–465.

Powell, R. A. & Proulx, G., 2003. Trapping and marking terrestrial mammals for research: integrating ethics, performance criteria, techniques, and common sense. *Ilar Journal,* 44: 259–276.

Putman, R. J., 1995. Ethical considerations and animal welfare in ecological field studies. *Biodiversity and Conservation,* 4: 903–915.

Quadros, J. & Monteiro–Filho, E. L. A., 2006. Coleta e preparação de pêlos de mamíferos para identificação em microscopia óptica. *Revista Brasileira de Zoologia,* 23(1): 274–278.

Rahbek, C., 1997. The relationship among area, elevation, and regional species richness in neotropical birds. *The American Naturalist,* 149: 875–902.

Rocha, R. G., Ferreira, E., Leite, Y. L., Fonseca, C. & Costa, L. P., 2011. Small mammals in the diet of barn owls, *Tyto alba* (Aves: Strigiformes) along the mid–Araguaia river in central Brazil. *Zoologia (Curitiba),* 28: 709–716.

Rosa, G., Leitão, D., Mendes, C., Leão, F., Fernandes, C., Costa, H., Pacheco, C. & Pereira, J., 2006. Situação da Águia–sapeira *Circus aeruginosus* em Portugal: recenseamento dos efectivos nidificantes. *Airo,*16: 3–11.

Santos, J., Luís, A. & Fonseca, C., 2009. Mamíferos do sal. *Galemys,* 21 (nº especial): 81–99.

Scheibler, D. & Christoff, A., 2007. Habitat associations of small mammals in southern Brazil and use of regurgitated pellets of birds of prey for inventorying a local fauna. *Brazilian Journal of Biology,* 67: 619–625.

Schipper, W. J. A., 1973. A comparison of prey selection in sympatric harriers (*Circus*) in Western Europe. *Le Gerfaut,* 63:17–120.

Sealander, J. A. & James, D., 1958. Relative efficiency of different small mammal traps. *Journal of Mammalogy,* 39: 215–223.

Simpson, E. H., 1949. Measurement of diversity.

Nature, 163: 688.

Smith, M. H., Gardner, R. H., Gentry, J. B., Kaufman, D. W. & O'Farrell, M. J., 1975. Density estimations of small mammal populations. In: *Small mammals: their productivity and population dynamics*: 25–64 (F. B. Golley, K. Petrusewicz & L. Ryszkowski, Eds.). Cambridge University Press, London.

Soberón, J. & Llorente, B., 1993. The Use of Species Accumulation functions for the prediction of species richness. *Conservation Biology*, 7(3): 480–488.

Strandberg, R., Klaassen, R. H. G., Olofsson, P., Hake, M., Thorup, K. & Alerstam, T., 2008. Complex temporal pattern of Marsh Harrier *Circus aeruginosus* migration due to pre– and post–migratory movements. *Ardea*, 96: 159–171.

Sullivan, T. P. & Sullivan, D. S., 2013. Influence of removal sampling of small mammals on abundance and diversity attributes: scientific implications. *Human–Wildlife Interactions*, 7: 85–98.

Teenrik, B. J., 1991. *Hair of west European mammals*. Cambridge University Press, Cambridge.

Torre, I., Arrizabalaga, A. & Flaquer, C., 2004. Three methods for assessing richness and composition of small mammal communities. *Journal of Mammalogy*, 85: 524–530.

Torre, I., Arrizabalaga, A., Freixas, L., Ribas, A., Flaquer, C. & Díaz, M. 2013. Using scats of a generalist carnivore as a tool to monitor small mammal communities in Mediterranean habitats. *Basic and Applied Ecology*, 14: 155–164.

Valente, A., 2012. Estudo da microestrutura dos pelos de mamíferos terrestres ibéricos – Guia para identificação taxonómica de espécies. Relatório de Pesquisa. Licenciatura em Biologia, Univ. de Aveiro.

Vandermeer, J. H., 2010. *The ecology of agroecosystems*. Jones & Bartlett Publishers, Sudbury, MA.

Voss, R. S., Lunde, D. P. & Simmons, N. B., 2001. The mammals of Paracou, French Guiana: a Neotropical lowland rainforest fauna Part 2. Nonvolant species. *Bulletin of the American Museum of Natural History*, 263: 1–236.

Williams, D. F. & Braun, S. E., 1983. Comparison of pitfall and conventional traps for sampling small mammal populations. *The Journal of Wildlife Management*, 47: 841–845.

Witkowski, J., 1989. Breeding biology and ecology of the marsh harrier *Circus aeruginosus* in the Barycz Valley, Poland. *Acta Ornithologica*, 25: 223–320.

10

Exotic tree plantations and avian conservation in northern Iberia: a view from a nest–box monitoring study

I. de la Hera, J. Arizaga & A. Galarza

De la Hera, I., Arizaga, J. & Galarza, A., 2013. Exotic tree plantations and avian conservation in northern Iberia: a view from a nest–box monitoring study. *Animal Biodiversity and Conservation*, 36.2: 153–163.

Abstract

Exotic tree plantations and avian conservation in northern Iberia: a view from a nest–box monitoring study.— The spread of exotic tree plantations on the North Atlantic coast of the Iberian peninsula raises concern regarding the conservation of avian biodiversity as current trends suggest this region might become a monoculture of Australian Eucalyptus species. To shed more light on the factors promoting differences in avian communities between and within exotic tree (Monterey Pine *Pinus radiata* and *Eucalyptus* spp.) plantations and native forests in the Urdaibai area (northern Spain), this study aimed to explore (1) how the type of habitat and vegetation characteristics affect bird species richness and the settlement of some particular species during the breeding period, (2) if some reproductive parameters (*i.e.* egg–laying date and clutch size) vary among habitats in a generalist bird species (the Great Tit *Parus major*), and (3) the existence of differences among habitats in the abundance of a key food resource on which some insectivorous birds are expected to rely upon for breeding (*i.e.* caterpillars). Our results confirmed that Eucalyptus stands house the poorest bird communities, and identified understory development as an important determinant for the establishment of titmice species. Furthermore, we found that exotic trees showed lower caterpillar abundance than native Oak trees (*Quercus robur*), which might contribute to explain observed differences among habitats in bird abundance and richness in this region. However, we did not find differences among habitats in egg–laying date and clutch size for the Great Tit, suggesting that the potential costs of breeding in exotic tree plantations would occur in later stages of the reproductive period (*e.g.* number of nestlings fledged), a circumstance that will require further research.

Key word: Bird diversity, Planted forests, Land–use changes, Linear mixed models, MAB Biosphere reserve, Iberian peninsula.

Resumen

Plantaciones de árboles exóticos y conservación de la avifauna en el norte de la península ibérica: perspectiva de un estudio de seguimiento de cajas nido.— La expansión de plantaciones de árboles exóticos en la costa cantábrica de la península ibérica suscita preocupación por la conservación de la biodiversidad de aves, puesto que las tendencias actuales sugieren que esta región podría convertirse en un monocultivo de especies de eucalipto australiano. Para arrojar más luz sobre los factores que promueven las diferencias en las comunidades de aves entre y dentro de las plantaciones de árboles exóticos (pino de Monterrey *Pinus radiata* y *Eucalyptus* spp.) y los bosques nativos de la zona de Urdaibai (norte de España), el objetivo del presente estudio consistió en analizar (1) la forma en que el tipo de hábitat y las características de la vegetación afectan a la riqueza de especies de aves y el asentamiento de determinadas especies durante el período de cría; (2) si algunos parámetros reproductivos (p.ej. la fecha o el tamaño de puesta) varían entre los hábitats en una especie de ave generalista (el carbonero común, *Parus major*); y (3) la existencia de diferencias entre hábitats por lo que hace a la abundancia de una fuente clave de alimento de la que se prevé que las aves insectívoras dependan para la cría (las orugas). Nuestros resultados confirmaron que las poblaciones de eucalipto albergan las comunidades más pobres de aves y establecieron el desarrollo del sotobosque como un factor importante para el establecimiento de las especies de páridos. Asimismo, hallamos que los árboles exóticos presentaban una abundancia de orugas menor que la de los robles nativos (*Quercus robur*), lo que podría contribuir a explicar las diferencias observadas entre los hábitats en cuanto a la abundancia y la riqueza de aves de esta región. No obstante, no se hallaron diferencias entre los hábitats por lo que concierne a la fecha y el tamaño de puesta para el carbonero común, lo que sugiere que los posibles costes de criar en plantaciones de árboles exóticos se producirían en etapas posteriores del periodo reproductivo (p.ej. el número de pollos emplumados), una circunstancia que habrá que seguir investigando.

Palabras clave: Diversidad de aves, Bosques plantados, Cambio del uso de la tierra, Modelos lineales mixtos, Reserva de la biosfera del MAB, Península ibérica.

Iván de la Hera, Depto. de Zoología y Biología Celular Animal, Fac. de Farmacia, Univ. del País Vasco (UPV/ EHU), Paseo de la Universidad 7, 01006 Vitoria–Gasteiz, España (Spain).– Juan Arizaga, Urdaibai Bird Center. Sociedad de Ciencias Aranzadi, Orueta 7, 48314 Gautegiz–Arteaga, Bizkaia, España (Spain).– Aitor Galarza, Depto. de Agricultura, Diputación Foral de Bizkaia, 48014 Bilbao, España (Spain).

Corresponding author: I. de la Hera. E–mail: idelahera@bio.ucm.es

Introduction

Increasing human demands for wood and its by–products (e.g. paper) in contemporary time (Ajani, 2011), and greater concern over the loss of natural forests, particularly in tropical regions (Gibson et al., 2011), have favoured commercial forest plantations as a main source of timber supply (Barlow et al., 2007; Brockerhoff et al., 2008). As a consequence of their profitability, forest plantations are replacing other land uses of declining economic yield (such as pastures and agricultural lands; Sohngen et al., 1999). This shift in land use is dramatically changing the socio–economic context and the landscape of many regions, with potential effects on biodiversity worldwide (Foley et al., 2005; Bremer & Farley, 2010; Felton et al., 2010). Planted forests now cover more than 264 million hectares around the world (i.e. 7% of the global forest area), and they are expanding at a rate of five million hectares per year (FAO, 2010). In spite of the growing generalization of this land use, we still have a limited understanding of the consequences of plantations on biodiversity and other ecosystem services (Louzada et al., 2010). Such circumstances restrict our ability to design sustainable management policies to preserve native flora and fauna and improve habitat quality in these exploited areas (Hartley, 2002; Brockerhoff et al., 2008).

The spread of stands of the non–native Monterey Pine (Pinus radiata, of North–American origin) and Eucalyptus species (mainly Eucalyptus globulus, native of Australia and Tasmania) have transformed the landscape of the Atlantic coast of Northern Iberia, where the original view just a century ago showed a mosaic of farmlands interspersed with hedges and coppices of natural vegetation (Lautensach, 1964). In recent decades, exotic tree plantations have progressively replaced pastures and farmlands (most of them devoted to hay production) that constitute a declining traditional activity (GV, 2005; Santos et al., in press). From the description of the bird communities of the different habitats present in Northern Iberia, several studies have raised concern about the proliferation of plantations, since they house fewer, and more generalist bird species than farmlands and natural forests (Bongiorno, 1982; Tellería & Galarza, 1990; Proença et al., 2010). Such circumstances suggest that the above–mentioned landscape modifications are impoverishing regional avifauna and probably the communities of other taxonomic groups (Proença et al., 2010; Calvino–Cancela et al., 2012). This situation may be aggravated even more as a consequence of the current drop in the prices of pine wood, since many foresters now prefer Eucalyptus plantations, which are by far the poorest habitat with regard to avian communities (Pina, 1989; Tellería & Galarza, 1990).

Observed differences among habitats in avian richness suggest that exotic tree plantations could set limits to the local distribution of some bird species. Understanding the mechanisms responsible for these patterns could help us to design recommendations to increase biodiversity in plantations (Hartley, 2002). For this purpose, it is necessary to explore how variation in some important characteristics of plantations, which are susceptible to be managed (e.g. vegetation structure), can affect avian communities, an issue that has barely been tackled in this region. At a more detailed scale, another issue that remains to be addressed is to know how generalist bird species perform in commercial plantations and are able to persist in these habitats. The study of the ecology of these widely–distributed species, and the comparison among habitats of some relevant parameters (e.g. breeding performance, body condition) might help to understand the observed variation in their abundance, as well as the potential causes that are constraining the occurrence in plantations of other more ecologically–demanding species (Carrascal & Tellería, 1990; Tellería & Galarza, 1990; Proença et al., 2010).

In this study, we explored how habitat type and vegetation structure can affect species richness and the settlement of birds during reproduction in a study site located in northern Spain. We performed bird counts and monitored nest–boxes in seven localities representing the three main forest habitats of the area (natural Oak forests, and Pine and Eucalyptus plantations). We also analysed how some breeding parameters (egg–laying date and clutch size) varied in the Great Tit Parus major, a generalist species occurring at lower densities in plantations (Tellería & Galarza, 1990). Given that breeding success in the Great Tit and other species may depend on particular food resources (i.e. defoliating caterpillars; Visser et al., 2006; Wilkin et al., 2009), we also estimated caterpillar abundance, which is expected to differ between natural forests and exotic tree plantations, because a number of native phytophagous arthropods might be unable to thrive in the latter (Kolb, 1996). With this approach, our goal was to go one step forward in this research topic and shed additional light on the mechanistic factors that might be promoting variation in bird abundance and composition in plantations and natural forests in Northern Iberia.

Methods

Study area and nest–box study design

Fieldwork was carried out in the UNESCO–MAB Biosphere Reserve of Urdaibai (Bizkaia province, Basque Country, Spain) and some surrounding municipalities (i.e. Bakio, Ereño and Ea). The study site is located within the North Atlantic coast of Iberia, an area where it is possible to find farmlands, Pine and Eucalyptus plantations, and a few remnants of natural Oak Quercus robur forest. This latter formation constitutes the climax forest ecosystem that would develop in many of the areas currently occupied by farmlands and exotic trees (Loidi et al., 2009).

During mid–February 2012, we installed 212 nest–boxes in seven localities. These localities were not randomly selected but were previously identified as large enough to house at least 20 nest–boxes. Large patches of forest are rare in this region, which is characterized by the presence of a complex mosaic of small private plots that are normally devoted to different land–uses. Selected localities represented the three main wooded habitats of the region: three plots of Eucalyptus plantations in Jata (denoted as Euc–1 in the figures; 43° 24.855' N, 02° 51.778' W), Ea (Euc–2; 43° 22.063' N, 02° 35.649' W) and Mañu (Euc–3; 43° 24.744' N, 02° 47.041' W); three plots of Pine plantations in Arteaga (Pinus–1; 43° 21.595' N, 02° 39.633' W), Ereño (Pinus–2; 43° 20.739' N, 02° 36.783' W) and Matxitxako (Pinus–3; 43° 26.434' N, 02° 44.888' W); and one Oak forest in Arratzu (Oak; 43° 17.767' N, 02° 38.236' W), the only forest in the area that was large enough to hold a reasonable number of nest–boxes. Within each locality, nest–boxes were hung from a nail hammered into the tree trunk at about 3.5 m height and separated approximately 50 m from each other. However, we finally considered a reduced subset of nest–boxes (n = 186), because 26 disappeared during the course of the study, with between 18 and 42 nest–boxes remaining per site (see fig. 2A). Nest–boxes were checked regularly (at least once per week) from early April to late June to determine the laying date of the first egg (assuming a production of one egg per day) and clutch size of the nest–boxes occupied by birds. Although we were particularly interested in obtaining measurements of chicks' body condition, we failed in this purpose because anomalous bad weather conditions in May caused the death of recently hatched chicks or, less frequently, clutch desertion. In the end, only 32 chicks from eight Great Tit broods were able to fledge (from two to six fledglings per brood).

Bird counts, vegetation structure and estimation of caterpillar abundance

In mid–May, we also established 10 bird count stations in each of the seven localities. Each bird count station surveyed the proximity of a previously installed nest–box, all randomly selected from all the nest–boxes available within each locality. Surveys at bird count stations lasted five minutes and we annotated all the bird species that were detected (heard or seen) within a 25–m radius. All counts were conducted in the morning (between dawn and 11:00 h) and on non–rainy days without strong winds that could affect the reliability of our sampling protocol.

Variation in the structure of avian communities and bird reproductive performance among wooded formations are likely to be the result of variation in vegetation structure among habitats. With the purpose of separating effects of habitat and vegetation structure, we also characterized the vegetation around each nest–box (25 m of radio) using nine variables: (1) general cover of shrubs (%), (2) cover of deciduous shrubs (%), (3) average shrub height (m), (4) tree cover in the canopy (%), (5) deciduous tree cover (%), (6) average tree height (m), (7) number of tree stems with a diameter higher than 40 cm, (8) average tree trunk diameter (cm), and (9) number of tree and shrub species. Given that some of these variables were expected to be strongly correlated with each other, we performed a principal component analysis (PCA) to obtain a smaller number of uncorrelated variables

(the principal components, PCs), which were easier to interpret. For this purpose, we used the program STATISTICA 7.0 and a varimax rotation of factors. Such PCA yielded three independent components (table 1). PC1 values were associated with the age of the trees, PC2 was positively correlated with variables indicating a more developed shrub layer, and PC3 represented an index of tree cover development in the canopy (see factor loadings in table 1).

As stated above, each tree species may hold a different invertebrate community, a circumstance that could affect avian richness and breeding performance (Kolb, 1996; Hartley et al., 2010). We used specific sampling methods to roughly estimate among–habitat relative abundance of the favourite invertebrate prey item used by some insectivorous birds (particularly by the Great Tit) to feed their chicks (i.e. caterpillars, order Lepidoptera; Visser et al., 2006; Wilkin et al., 2009). For this purpose, we placed one plastic washbasin (diameter of 42 cm) on the ground, near the trunk of 23 trees (seven Oaks, eight Eucalyptus and eight Pines). Washbasins were partly filled with water and were also covered with a metallic mesh to avoid other animals (e.g. large mammals) having access to the water and affecting our caterpillar estimates. This method allowed us to collect drowned caterpillars which had descended from the canopy to the ground for pupation (see Zandt, 1994). Washbasins were checked approximately once per week between mid–April and late June, and the overall accumulated number of caterpillars found in each washbasin was used as a response variable in the statistical analyses.

Statistical analyses

First, we used linear mixed models (LMM) to explore the existence of differences between habitats in vegetation characteristics (PCs). Next, we used generalized linear mixed models (GLMMs) with Poisson errors to analyze whether bird species richness obtained from count stations varied among habitats, after controlling for vegetation characteristics (Zuur et al., 2009). A similar approach to the latter, but with a binomial error distribution, was performed to test habitat and vegetation effects on the probability of nest–boxes to be occupied (binary variable; empty nest–box = 0, occupied nest–box = 1). For the 25 first clutches detected for the Great Tit (see Results), we also tested for differences between habitats in egg–laying date (i.e. LMM) and clutch size (i.e. GLMM with Poisson errors). All these models were fitted in R using the package lme4 (Bates & Maechler, 2010). A Markov–Chain–Monte–Carlo sampling procedure (1×10^4 iterations) implemented in the package languageR was used to obtain the P–values for the fixed effects in the models analysing vegetation characteristics and egg–laying dates (Baayen, 2008). All previous analyses included locality as a random factor. Furthermore, we explored whether the number of collected caterpillars per washbasin differed among tree species. For this purpose, we performed a Kruskal–Wallis test.

Results

Variation in vegetation structure among habitats

Tree size or age (PC1) varied among habitats, with the Oak forest showing intermediate values of PC1 that did not differ significantly from the scores recorded in Pine (estimate = -1.148 ± 0.923, t [n = 99] = -1.25, P = 0.216) or Eucalyptus plantations (estimate = 0.601 ± 0.809, t [n = 117] = 0.74, P = 0.458), but PC1 was significantly higher in Pine than in Eucalyptus stands (estimate = 1.75 ± 0.572, t [n = 156] = 3.06, P = 0.003). Eucalyptus plantations did not differ significantly in shrub development (PC2) from the values recorded in Pine plantations (estimate = 0.064 ± 0.39, t [n = 156] = 0.17, P = 0.869). However, our Oak forest had a more developed understory than both types of exotic tree plantations, although this effect was only significant for the comparison with Pine plantations (Oak–Pine comparison: estimate = 0.845 ± 0.26, t [n = 99] = 3.26, P = 0.002; Oak–Eucalyptus comparison: estimate = 0.909 ± 0.545, t [n = 117] = 1.67, P = 0.097). For PC3, Pine and Eucalyptus plantations showed similar values (estimate = 0.021 ± 0.218, t [n = 156] = 0.1, P = 0.923), these being significantly lower than those observed in the Oak forest (Oak–Pine comparison: estimate = 2.124 ± 0.15, t [n = 99] = 14.18, P = < 0.001; Oak–Eucalyptus comparison: estimate = 2.146 ± 0.303, t [n = 117] = 7.09, P < 0.001).

Bird species richness

We recorded a total of 18 species (all passerines) after performing the 70 bird count station surveys (data are available from the authors upon request). Variation in bird species richness was better explained by habitat effects (see fig. 1) than by variation in vegetation structure (PC1 effect: estimate ± SE = 0.119 ± 0.109, Z [n = 70] = 1.09, P = 0.275; PC2 effect: estimate ± SE = 0.147 ± 0.084, Z [n = 70] = 1.75, P = 0.081; PC3 effect: estimate ± SE = -0.033 ± 0.135, Z [n = 70] = -0.25, P = 0.807). Thus, Eucalyptus plantations were the poorest habitat and differed significantly in species richness when compared to Pine plantations (estimate = 0.773 ± 0.269, Z [n = 60] = 2.87, P = 0.004) or the Oak forest (estimate = 1.205 ± 0.422, Z [n = 40] = 2.86, P = 0.004). Pine plantations and the Oak forest showed similar values of species richness (estimate ± SE = 0.27 ± 0.505, Z [n = 40] = 0.534, P = 0.593; fig. 1).

Nest–box occupancy rates

Out of the 186 nest–boxes considered in the study, 43 were occupied by birds for reproduction. We considered occupied nest–boxes as those in which eggs were laid. The Great Tit was the most common breeding species (n = 31) and the only species occurring in the seven localities (fig. 2A). Less frequently, we detected Coal Tits *Periparus ater* (n = 6) and Blue Tits *Cyanistes caeruleus* (n = 6). We arbitrarily distinguished between Great Tits' first and second clutches taking advantage of the fact that Coal Tits and Blue Tits are single–brooded

Table 1. Correlation coefficients (factor loadings) between the nine variables characterizing vegetation structure (VS) and the three principal components derived from the PCA. Eigenvalues and the percentage of variance explained by each component are also shown: 1. Overall shrub cover (in %); 2. Deciduous shrub cover (in %); 3. Average shrub height (in m); 4. Tree cover (in %); 5. Deciduous tree cover (in %); 6. Average tree height (in m); 7. Number of tree stems (d > 40 cm); 8. Average tree trunk diameter (in cm); 9. Number of tree and shrub species.

Tabla 1. Coeficientes de correlación (cargas factoriales) entre las nueve variables que caracterizan la estructura de la vegetación (VS) y los tres componentes principales derivados del análisis de componentes principales (ACP). También se muestran las raíces latentes y el porcentaje de la varianza explicados por cada componente: 1. Cubierta arbustiva total (en %); 2. Cubierta de arbustos caducifolios (en %); 3. Altura media de los arbustos (en m); 4. Cubierta arbórea (en %); 5. Cubierta de árboles caducifolios (en %); 6. Altura media de los árboles (en m); 7. Número de árboles (d > 40 cm); 8. Diámetro medio del tronco de los árboles (en cm); 9. Número de especies arbóreas y arbustivas.

VS	PC1	PC2	PC3
1	−0.22	0.80	0.27
2	−0.05	0.87	0.01
3	0.50	0.61	0.32
4	0.39	−0.11	0.76
5	−0.09	0.27	0.85
6	0.89	−0.01	0.10
7	0.88	−0.09	−0.06
8	0.93	0.08	0.17
9	0.11	0.80	−0.06
Eigenvalue	2.93	2.50	1.52
Explained variance	0.33	0.28	0.17

species (see appendix 1 for more details). Second clutches were only observed in six cases (their laying dates ranging from day 55 to day 68), all of them occurring in nest–boxes installed in Eucalyptus stands (fig. 2A), but they were not statistically more frequent in these plantations (results not shown). The overall percentage of occupied nest–boxes varied greatly among localities (ranging from four 4 to 40%; see fig. 2A). In our statistical model, shrub development (PC2) was the only significant factor affecting the probability of a nest–box to be occupied (PC2 effect: estimate = 0.493 ± 0.213,

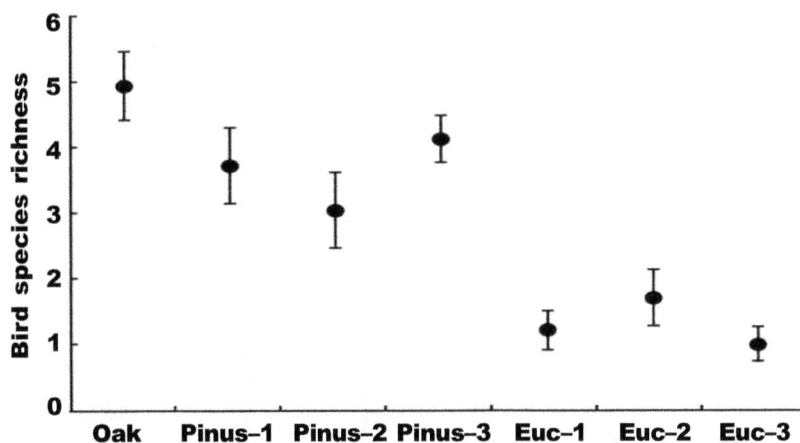

Fig. 1. Variation among localities in the number of bird species detected (*i.e.* species richness; mean ± SE) during five minute bird counts at stations (10 random bird counts per locality).

Fig. 1. Variación entre localidades en el número de especies de aves detectadas (riqueza de especies; media ± EE) durante recuentos de cinco minutos en las estaciones (10 recuentos aleatorios de aves por localidad).

Z [n = 186] = 2.32, P = 0.021; fig. 2B), while habitat type (post–hoc analysis Pine *vs.* Eucalyptus: estimate = 0.472 ± 0.67, Z [n = 156] = 0.71, P = 0.481; post–hoc analysis Oak *vs.* Eucalyptus: estimate = 1.421 ± 0.858, Z [n = 117] = 1.66, P = .098), PC1 (estimate = –0.163 ± 0.312, Z [n = 186] = –0.52, P = 0.602) and PC3 (estimate = –0.151 ± 0.312, Z [n = 186] = –0.48, P = 0.629) did not show significant effects.

Egg–laying date and clutch size in the Great Tit

For the 25 first clutches identified for the Great Tit, egg–laying date did not differ among Eucalyptus (mean date = 18 [18th of April] ± 5.3 d, n = 8), Pine (mean date = 24 [24th of April] ± 5 d, n = 9) and Oak stands (mean date = 21 [21st April] ± 5.3 d, n = 8; habitat effects: P > 0.05), but clutches were laid later in nest–boxes presenting older trees in their surroundings (PC1 effect: estimate = 10.17 ± 4.16, t [n = 25] = 2.45, P = 0.024).

Great Tit clutch size ranged between five and eight eggs. Clutch size was not affected by habitat (post–hoc analysis Pine *vs.* Eucalyptus: estimate = –0.01 ± 0.212, Z [n = 17] = –0.05, P = 0.964; post–hoc analysis Oak *vs.* Eucalyptus: estimate = –0.108 ± 0.393, Z [n = 16] = –0.27, P = 0.784), or by vegetation structure effects (PC1, PC2 and PC3 effects: P > 0.05).

Caterpillar abundance among habitats

The total number of caterpillars collected per washbasin differed among habitats (Kruskal–Wallis test: $H_{2,13}$ = 10.4, P = 0.006; fig. 3), with washbasins located under Oak trees containing a higher accumulated number of caterpillars than both Pine and Eucalyptus trees, where caterpillars were nearly absent.

Discussion

Our study confirms that the bird communities of Eucalyptus stands are significantly impoverished, with species richness during the breeding period being lower in commercial plantations than in natural forests. We also identified understory development as a main factor affecting the nest–box occupancy rate of titmice species in the study area. Thus, a more developed shrub layer increased the chances of a nest–box being occupied for breeding. Likewise, a significant difference in caterpillar abundance was observed between exotic and native trees. Although observed variation among habitats in this food resource might affect some reproductive parameters in bird species relying upon caterpillars for breeding, we did not detect differences in the breeding performance of Great Tits during the earliest stages of their reproductive process (*i.e.* egg–laying date and clutch size).

There is increasing concern about the burgeoning proliferation of exotic tree plantations around the world (Brockerhoff et al., 2008; Bremer & Farley, 2010; Putz & Redford, 2010), a trend that is also expected to have a pervasive impact in many areas of the Iberian Peninsula (Santos et al., 2006; Veiras & Soto, 2011). In coastal areas of northern Spain, the transformation of traditional land–uses (*i.e.* farmlands and pastures) into tree plantations seems to be an inexorable process that might imply the decline of many open–habitat bird species that normally would not occur in woodlands. Paradoxically, these open–habitat species were originally favoured by ancient human deforestation and farming (Tellería & Galarza, 1990; Williams, 2006). In order to maintain current regional avian biodiversity, conservation efforts should be channelled into pre-

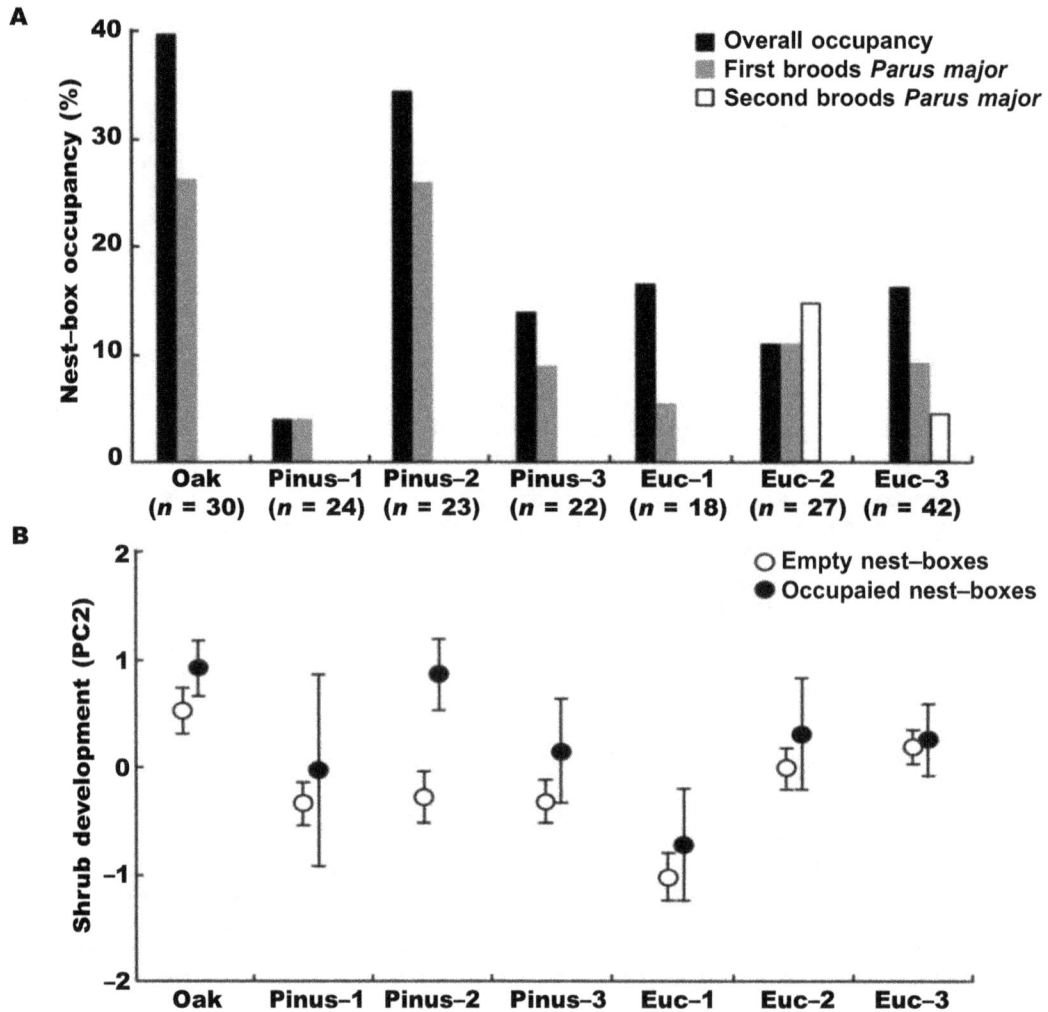

Fig. 2. A. Variation in the overall percentage of nest–boxes occupied by birds (black bars), and the percentage of Great Tit first (grey bars) and second clutches (open bars) among localities. Occupied nest–boxes were those in which eggs were laid. The number of available nest–boxes per locality is shown below the abscissa axis. B. Differences in understory development (mean ± SE of PC2) between empty and occupied nest–boxes for the seven study sites.

Fig. 2. A. Variación en el porcentaje total de las cajas nido ocupadas por aves (barras negras) y el porcentaje de primeras (barras grises) y segundas (barras blancas) puestas del carbonero común entre localidades. Las cajas nido ocupadas eran aquellas en las que se habían puesto huevos. El número de cajas nido disponibles por localidad se muestra a continuación en el eje de las abscisas. B. Diferencias en el desarrollo del sotobosque (media ± EE de PC2) entre las cajas nido vacías y ocupadas de los siete lugares del estudio.

serving conventional farmlands, which hold a singular avian breeding community and are also an important wintering destination of many European migratory populations (Tellería et al., 2008; Santos et al., in press). On the other hand, the generalization of tree plantations can be considered an opportunity to recover the woodland species that had been confined to the remnants of natural forest scattered throughout this region (Quine & Humphrey, 2010; Navarro & Pereira,

2012). However, our results confirm that exotic tree plantations are not able to fulfil the role of natural forests (Bongiorno, 1982; Tellería & Galarza, 1990; Proença et al., 2010) because they lack some bird species with high demands for old forest stands, such as the European Nuthatch *Sitta europaea* (only present in the Oak forest) and the Short–toed Treecreeper *Certhia brachydactyla* (common in American Pine formations but completely absent from Eucalyptus stands).

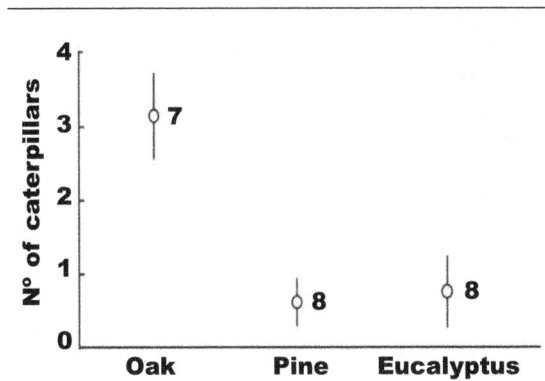

Fig. 3. Variation among Oak, Pine and Eucalyptus trees in the overall number of caterpillars collected. Graph shows means with standard errors and sample sizes.

Fig. 3. Variación entre el roble, el pino y el eucalipto en cuanto al número total de orugas recogidas. En el gráfico se muestran las medias con los errores estándar y los tamaños muestrales.

The previously–described scenario raises the need to develop management practices that help commercial plantations diversify their bird communities. A conventional solution to achieve this purpose would consist of promoting a well–developed natural shrub layer (López & Moro, 1997; Santos et al., 2006). This measure should be weighed in relation to wildfire risk, but it could be feasible in our study area given that plantations have a less complex understory than natural forests. This is probably a consequence of the regular removal of native scrublands in exotic tree stands (Veiras & Soto, 2011), which would be depicted in the lower values of PC2 in exotic tree plantations compared to the Oak forest. However, our results showed no clear association between understory development and bird richness obtained from bird counts ($P = 0.081$; see Results). We did detect, nevertheless, that shrub development may benefit the settlement of some hole–nesting sedentary species (*i.e.* Titmice species) known to attract other breeding (migratory) species that use year–round residents as cues for habitat selection (Forsman et al., 2009). At this point, we should point out that we provided birds with nest–boxes in all the study localities. Consequently, promoting a complex understory might be ineffective if nesting holes are a prerequisite for the settlement of birds during reproduction, because most exotic tree stands have very few natural cavities. This is particularly true for Eucalyptus plantations in the Basque Country, where trees are logged at a relatively early age (normally after 11 years of tree growth; Veiras & Soto, 2011), explaining the difficulty of finding Eucalyptus stands with high scores of PC1 in our study area.

In many European regions, Great Tits and other insectivorous forest bird species normally try to synchronize the hatching of their eggs with a short peak of tree defoliating caterpillars, which constitute an abundant and suitable food resource for feeding their chicks (Sanz et al., 2003; Visser et al., 2006). Although we used a very rough method based on washbasins to estimate caterpillar abundance (Zandt, 1994), our approach allowed us to corroborate that caterpillars are much rarer in exotic trees (both Pine and Eucalyptus) than in native Oaks. Such circumstance might affect the reproductive performance of Great Tits (Kolb, 1996). However, egg–laying date and clutch size did not differ among habitats, and only an effect of tree size/age (PC1) on egg–laying date emerged from our analyses, a finding that was difficult to interpret.

Although our sample size was relatively small ($n = 25$) and limited to only one year, the lack of variation between habitats in egg–laying date and clutch size agrees with the results obtained by Kolb (1996), who detected that exotic trees only had a negative effect on later stages of the reproductive period. Unfortunately, owing to bad meteorological conditions, we could not collect sufficient data from Great Tit chicks to explore this possibility. Kolb's study was carried out using a more Northern European population of Great Tits, which seems to be more dependent on caterpillar availability (Kolb, 1996; Wilkin et al., 2009). Consequently, it remains to be tested whether similar patterns will be detected in our population for which caterpillar availability is actually very low according to the data obtained from the use of washbasins, and also whether this food resource in the diet of the chicks will tend to be replaced by other invertebrates (*e.g.* spiders; Pagani–Nuñez et al., 2011).

In conclusion, our results further support the negative consequences of exotic tree stands for birds at community level (*i.e.* bird species richness), with these penalties being stronger in Eucalyptus than in Pine plantations. The study went one step further and explored the possibility that differences in the composition of bird species composition be explained by vegetation structure. Also we tested the existence of variation among habitats in the reproductive performance of a generalist bird species (the Great Tit) and found a marked difference between exotic and native trees in caterpillar abundance, two aspects that had not been considered before in our study area. Although limited and preliminary as a consequence of the reduced sample size and the use of only one year of data, the patterns we observed establish the basis for future research into the observed variation among habitats in bird abundance and composition in the North Atlantic coast of Iberian peninsula.

Our study also highlights the difficulty of uncoupling habitat from vegetation structure effects in the analyses, because management practices in this region (*e.g.* age at which trees are logged) differ notably depending on the exotic tree species considered (Eucalyptus *vs.* Pine). Together with the scarcity of Oak forests, these circumstances made it virtually impossible to find Eucalyptus, Pine and Oak stands with similar vegetation characteristics, preventing a realistic separation of the relative contribution of effects of habitat and vegetation structure on our response variables. Clarifying this issue could therefore help us to assess whether the conservation value of planted forests in Northern Iberia is constrained by exotic trees themselves or by the management practices they undergo (Sax, 2002).

Acknowledgements

We are grateful to all those who helped us in the fieldwork, particularly Álvaro Asteinza. We also want to thank the staff at the Urdaibai Bird Center from Aranzadi Society of Sciences for their support and suggestions in the course of the study, and Eriz Guerra for revising the English. Nest–boxes were provided by the Spanish Government (Ministerio de Medio Ambiente y Medio Rural y Marino), and ringing permissions by the Diputación Foral de Bizkaia. I. de la Hera was funded by the Department of Education, Universities and Research of the Basque Government (fellowship BFI. 09–13).

References

Ajani, J., 2011. The global wood market, wood resource productivity and price trends: an examination with special attention to China. *Environmental Conservation*, 38: 53–63.

Baayen, R. H., 2008. languageR: Data sets and functions with 'Analyzing linguistic data: a practical introduction to statistics'. R Package Version 0.953. http://CRAN.R–project.org/package=languageR.

Barlow, J., Gardner, T. A., Araujo, I. S., Ávila–Pires, T. C., Bonaldo, A. B., Costa, J. E., Esposito, M. C., Ferreira, L. V., Hawes, J., Hernandez, M. I. M., Hoogmoed, M. S., Leite, R. N., Lo–Man–Hung, N. F., Malcolm, J. R., Martins, M. B., Mestre, L. A. M., Miranda–Santos, R., Nunes–Gutjahr, A. L., Overal, W. L., Parry, L., Peters, S. L., Ribeiro–Junior, M. A., da Silva, M. N. F., da Silva Motta, C. & Peres, C. A., 2007. Identifying the biodiversity value of tropical primary, secondary, and plantation forests. *Proceedings of the National Academy of Sciences*, 104: 18555–18560.

Bates, D. & Maechler, M., 2010. lme4: Linear mixed–effects models using S4 classes. R package version 0.999375–33. http://CRAN.R–project.org/package=lme4.

Bongiorno, S. F., 1982. Land use and summer bird populations in Northwestern Galicia, Spain. *Ibis*, 124: 1–20.

Bremer, L. L. & Farley, K. A., 2010. Does plantation forestry restore biodiversity or create green deserts? A synthesis of the effects of land–use transitions on plant species richness. *Biodiversity & Conservation*, 19: 3893–3915.

Brockerhoff, E. G., Jactel, H., Parrotta, J. A., Quine, C. P. & Sayer, J., 2008. Plantation forests and biodiversity: oxymoron or opportunity? *Biodiversity & Conservation*, 17: 925–951.

Calvino–Cancela, M., Rubido–Bara, M. & van Etten, E. J. B., 2012. Do eucalypt plantations provide habitat for native forest biodiversity? *Forest Ecology & Management*, 270: 153–162.

Carrascal, L. M. & Tellería, J. L., 1990. Impacto de las repoblaciones de *Pinus radiata* sobre la avifauna forestal del norte de España. *Ardeola*, 37: 247–266.

FAO, 2010. Planted forests in sustainable forest management: a statement of principles. Food and Agriculture Organization of the United Nations, Rome, Italy.

Felton, A., Knight, E., Wood, J. Zammit, C. & Lindenmayer, D., 2010. A meta–analysis of fauna and flora species richness and abundance in plantations and pasture lands. *Biological Conservation*, 143: 545–554.

Foley, J. A., DeFries, R., Asner, G. P., Barford, C., Bonan, G., Carpenter, S. R., Chapin, F. S., Coe, M. T., Daily, G. C., Gibbs, H. K., Helkowski, J. H., Holloway, T., Howard, E. A., Kucharik, C. J., Monfreda, C., Patz, J. A., Prentice, I. C., Ramankutty, N. & Snyder, P. K., 2005. Global consequences of land use. *Science*, 309: 570–574.

Forsman, J. T., Hjernquist, M. B. & Gustafsson, L., 2009. Experimental evidence for the use of density based interspecific social information in forest birds. *Ecography*, 32: 539–545.

Gibson, L., Lee, T. M., Koh, L. P., Brook, B. W., Gardner, T. A., Barlow, J., Peres, C. A., Bradshaw, C. J. A., Laurance, W. F., Lovejoy, T. E. & Sodhi, N. S., 2011. Primary forests are irreplaceable for sustaining tropical biodiversity. *Nature*, 478: 378–381.

GV, 2005. Inventario Forestal de la Comunidad Autónoma del País Vasco. Departamento de Medio Ambiente, Planificación Territorial, Agricultura y Pesca, Gobierno Vasco.

Hartley, M. J., 2002. Rationale and methods for conserving biodiversity in plantation forests. *Forest Ecology and Management*, 155: 81–95.

Hartley, M. K., Rogers, W. E. & Siemann, E., 2010. Comparisons of arthropod assemblages on an invasive and native trees: abundance, diversity and damage. *Arthropod–Plant Interactions*, 4: 237–245.

Kolb, H., 1996. The reproductive biology of the Great Tit *Parus major* in small patches: exotic versus native tree species. *Journal of Ornithology*, 137: 229–242.

Lautensach, H., 1964. *Geografía de España y Portugal*. Vicens Vives, Barcelona.

Loidi, J., Biurrun, I., Campos, J. A., García–Mijangos, I. & Herrera, M., 2009. La vegetación de la CAPV. Leyenda del mapa de series de vegetación a escala 1:50.0000. Eusko Jaurlaritza–Gobierno Vasco.

López, G. & Moro, M. J., 1997. Birds of Aleppo pine plantations in south–east Spain in relation to vegetation composition and structure. *Journal of Applied Ecology*, 34: 1257–1272.

Louzada, J., Gardner, T., Peres, C. & Barlow, J., 2010. A multi–taxa assessment of nestedness patterns across a multiple–use Amazonian forest landscape. *Biological Conservation*, 143: 1102–1109.

Navarro, L. M. & Pereira, H. M., 2012. Rewilding abandoned landscapes in Europe. *Ecosystems*, 15: 900–912.

Pagani–Núñez, E., Ruiz, I., Quesada, J., Negro, J. J.

& Senar, J. C., 2011. The diet of Great Tit *Parus major* nestlings in a Mediterranean Iberian forest: the important role of spiders. *Animal Biodiversity and Conservation*, 34: 355–361.

Pina, J. P., 1989. Breeding bird assemblages in eucalyptus plantations in Portugal. *Annales Zoologici Fennici*, 26: 287–290.

Proença, V. M., Pereira, H. M., Guilherme, J. & Vicente, L., 2010. Plant and bird diversity in natural forests and in native and exotic plantations in NW Portugal. *Acta Oecologica*, 36: 219–226.

Putz, F. E. & Redford, K. H., 2010. The importance of defining 'forest': tropical forest degradation, deforestation, long–term phase shifts, and further transitions. *Biotropica*, 42: 10–20.

Quine, C. P. & Humphrey, J. W., 2010. Plantations of exotic tree species in Britain: irrelevant for biodiversity or novel habitat for native species? *Biodiversity and Conservation*, 19: 1503–1512.

Santos, T., Carbonell, R., Galarza, A., Pérez–Tris, J., Ramírez, A. & Tellería, J. L., in press. The importance of northern Spanish farmland for wintering migratory passerines: a quantitative assessment. *Bird Conservation International*.

Santos, T., Tellería, J. L., Díaz, M. & Carbonell, R., 2006. Evaluating the benefits of CAP reforms: can afforestations restore bird diversity in Mediterranean Spain? *Basic Applied & Ecology*, 7: 483–495.

Sanz, J. J., Potti, J., Moreno, J., Merino, S. & Frías, O., 2003. Climate change and fitness components of a migratory bird breeding in the Mediterranean region. *Global Change Biology*, 9: 461–472.

Sax, D. F., 2002. Equal diversity in disparate species assemblages: a comparison of native and exotic woodlands in California. *Global Ecology & Biogeography*, 11: 49–57.

Sohngen, B., Mendelsohn, R. & Sedjo, R., 1999. Forest management, conservation, and global timber markets. *American Journal of Agricultural Economics*, 81: 1–13.

Tellería, J. L. & Galarza, A., 1990. Avifauna y paisaje en el Norte de España: efecto de las repoblaciones con árboles exóticos. *Ardeola*, 37: 229–245.

Tellería, J. L., Ramírez, A., Galarza, A., Carbonell, R., Pérez–Tris, J. & Santos, T., 2008. Geographical, landscape and habitat effects on birds in Northern Spanish farmlands: implications for conservation. *Ardeola*, 55: 203–219.

Veiras, X. & Soto, M. A., 2011. *La conflictividad de las plantaciones de eucalipto en España (y Portugal). Análisis y propuestas para solucionar la conflictividad ambiental y social de las plantaciones de eucalipto en la península Ibérica*. Greenpeace, Madrid.

Visser, M. E., Holleman, L. J. M. & Gienapp, P., 2006. Shifts in caterpillar biomass phenology due to climate change and its impact on the breeding biology of an insectivorous bird. *Oecologia*, 147: 164–172.

Wilkin, T. A., King, L. E. & Sheldon, B. C., 2009. Habitat quality, nestling diet, and provisioning behaviour in great tits *Parus major*. *Journal of Avian Biology*, 40: 135–145.

Williams, M., 2006. *Deforesting the Earth: from prehistory to global crisis, an abridgment*. Univesity of Chicago Press, Chicago.

Zandt, H. S., 1994. A comparison of three sampling techniques to estimate the population size of caterpillars in trees. *Oecologia*, 97: 399–406.

Zuur, A., Ieno, E. N., Walker, N., Saveliev, A. A. & Smith, G. M., 2009. *Mixed Effects Models and Extensions in Ecology with R*. Springer, New York.

Appendix 1. Identifying second clutches of Great Tit.

Apéndice 1. Determinación de las segundas puestas de carbonero común.

Great tits are facultative multiple breeders and some pairs can undertake a second breeding attempt. Second clutches contain fewer eggs than first clutches and may be more frequent in some habitats than in others, possibly affecting the reliability of our between–habitat comparisons. We used the laying dates of two species known to be single–brooded (*i.e.* Blue Tit *Cyanistes caeruleus* and Coal Tits *Periparus ater*) to show the existence of second clutches in the Great Tit. According to the range of egg–laying dates in Blue Tits and Coal Tits (*i.e.* from day 4 to day 39 considering the 1st of April as day 1; see fig. A), we considered that Great Tit clutches laid later after May 11th (day 41) were second clutches, and they were consequently, excluded from the statistical tests that analysed nest–box occupancy rate, laying date, and clutch size.

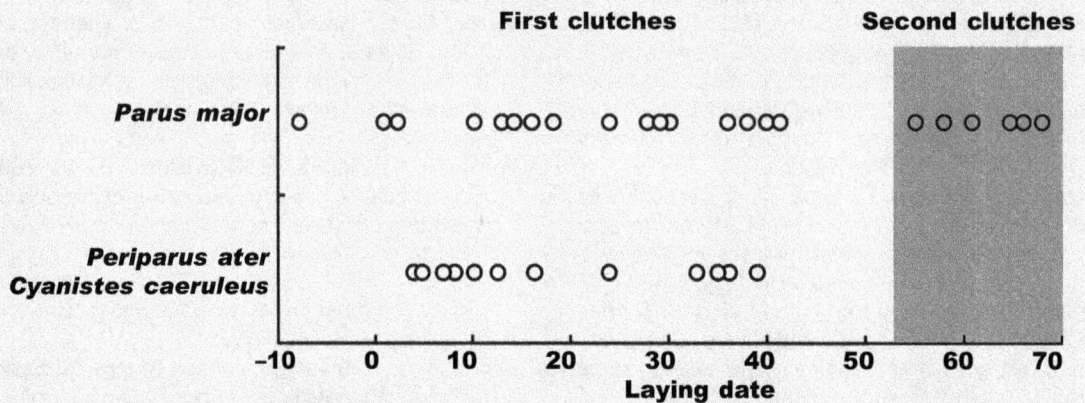

Fig. A. Comparison between the laying dates of Great Tits *Parus major* and the laying dates of two single–brooded species (*i.e.* Coal Tit *Periparus ater* and Blue Tit *Cyanistes caeruleus*). Note that there are some overlapping data points.

Fig. A. Comparación entre las fechas de puesta del carbonero común Parus major *y las de dos especies de puesta única (carbonero garrapinos* Periparus ater *y herrerillo común* Cyanistes caeruleus*). Nótese que algunos datos se superponen.*

11

Intensive monitoring suggests population oscillations and migration in wild boar *Sus scrofa* in the Pyrenees

M. Sarasa & J.–A. Sarasa

Sarasa, M. & Sarasa, J.–A., 2013. Intensive monitoring suggests population oscillations and migration in wild boar *Sus scrofa* in the Pyrenees. *Animal Biodiversity and Conservation*, 36.1: 79–88.

Abstract

Intensive monitoring suggests population oscillations and migration in wild boar Sus scrofa *in the Pyrenees.—* As few studies have analysed local variability in populations of wild boar *Sus scrofa* in Western Europe in recent years, our understanding of ecological processes currently affecting this species is limited. To analyse questions regarding local variability in wild boar abundance, we used information from 442 traditional drive hunts monitored throughout eight hunting periods in the Pyrenees mountain range (Urdués, N Spain). Results showed temporal oscillations in abundance, and a non–linear decrease of 23% in the number of wild boar seen per drive hunt between 2004 and 2011. Numbers of dogs and hunters per drive hunt also affected indexes of wild boar abundance. Inter–annual variations in bag size may cause overestimations of variations in boar abundance and may even deviate from the population dynamics inferred from the number of wild boars seen per drive hunt. The multimodal patterns of wild boar abundance during the hunting periods suggest migrations in the Pyrenees. Our findings highlight the limitations of hunting bag statistics in wild boar. Further studies are required to guarantee information–based sustainable management of wild boar populations.

Key words: Wild boar, *Sus scrofa*, Animal migration, Big game traditional hunting, Population dynamics, Wildlife management.

Resumen

*El seguimiento intensivo sugiere la existencia de oscilaciones demográficas y movimientos migratorios en las poblaciones de jabalí (*Sus scrofa*) en los Pirineos.—* Muy pocos estudios recientes han analizado la variabilidad local de las poblaciones de jabalí (*Sus scrofa*) en Europa occidental, lo que limita nuestra comprensión de los procesos ecológicos que en la actualidad afectan a esta especie. Usando la información recopilada mediante el seguimiento de 442 batidas durante ocho temporadas de caza en los Pirineos (Urdués, norte de España), se analizaron cuestiones relacionadas con la variabilidad local de la abundancia de jabalí. Los resultados revelaron oscilaciones temporales de la abundancia y una disminución discontinua del 23% en el número de jabalíes avistados por batida entre 2004 y 2011. El número de perros y de cazadores por batida también afectó a los índices de abundancia de jabalí. Las variaciones interanuales de animales abatidos pueden provocar que se sobreestimen las variaciones de la abundancia de jabalí e incluso pueden desviarse de la dinámica de poblaciones inferida del número de jabalíes avistados por batida. En los Pirineos, el patrón multimodal de la abundancia de jabalí durante las temporadas de caza sugiere la existencia de movimientos migratorios. Los resultados obtenidos destacan las limitaciones de las estadísticas de abundancia realizadas sobre el número de jabalíes abatidos y ponen de manifiesto la necesidad de llevar a cabo nuevos estudios que permitan gestionar las poblaciones de jabalí de forma sostenible y fundamentada.

Palabras clave: Jabalí, *Sus scrofa*, Migración, Caza mayor tradicional, Dinámica de poblaciones, Gestión de la fauna silvestre.

Mathieu Sarasa, Grupo Biología de las Especies Cinegéticas y Plagas (RNM–118), España (Spain).– Juan–Antonio Sarasa, Grupo de Caza Mayor de Urdués, España (Spain).

Corresponding author: M. Sarasa, Fédération Nationale des Chasseurs 13, Rue du Général Leclerc, F–92136 Issy les Moulineaux Cedex, France. E–mail: mathieusar@hotmail.com; msarasa@chasseurdefrance.com

Introduction

Over the last 30 years, most studies discussing or mentioning wild boar *Sus scrofa* abundance and densities in Western Europe have suggested that overall wild boar populations are increasing (Sáez–Royuela & Tellería, 1986; Melis et al., 2006; Marco et al., 2011). Wild boar populations have been associated with health problems in livestock and humans (Artois et al., 2002; Rossi et al., 2005; Meng et al., 2009), damage to crops and colonized ecosystems (Schley et al., 2008; Cuevas et al., 2010), and road accidents (Groot Bruinderink & Hazebroek, 1996; Lagos et al., 2012). It has also been suggested that wild boar might reduce the availability of summer grazing areas through soil disturbance (Bueno et al., 2010), although such issues have raised considerable controversy (Risch et al., 2010; Wirthner et al., 2011; Wirthner et al., 2012). In light of these concerns and of the predicted increase in wild boar populations as a response to global warming (Melis et al., 2006), management tools to control and reduce wild boar populations are of much interest (Massei et al., 2011). In recent years, however, few studies of wild boar population dynamics in Western Europe have been performed (Marco et al., 2011), limiting our understanding of current ecological processes in this species.

Generalized increases in wild boar densities are thought to be responsible for increasing presence of wild boar in agricultural ecosystems and even in urban environments (Jansen et al., 2007). However, very few studies have addressed whether these increases are the result of a source–sink gradient, sustained by woodland environments with increasing numbers of wild boars, or whether wild boars are locally adapting to agricultural and urban environments in which effective and perceived hunting pressure is low and opportunist foraging is facilitated by city dwellers (Cahill & Llimona, 2004).

Increasing interest is also being shown in the way in which wild boar use space, and a number of studies have revealed variable and complex movement patterns (Keuling et al., 2010). Whilst some authors suggest that wild boars are essentially sedentary animals (Saunders & Kay, 1996; Keuling et al., 2008; Mitchell et al., 2009), others indicate that wild boars might perform sex–specific habitat selection depending on their landscape of fear (Saïd et al., 2012) or even on local migrations (Andrzejewski & Jezierski, 1978; Singer et al., 1981; D'Andrea et al., 1995).

The essential parameters regarding population dynamics and space use in wild boar are therefore unclear, thus hindering the establishment of appropriate management practices.

In this study, we used hunting data collected over eight hunting periods in a locality of the Pyrenean mountain range (Urdués, N Spain) to analyse the population dynamics and space use in the wild boar. Historically, the Pyrenees have always been considered a propitious environment for this game species (Gortázar et al., 2000). Thus, if the increase in wild boar in agricultural and urban environments is the product of a source–sink gradient sustained by woodland environments, we would expect a positive population trend at our study site during the monitoring period (prediction 1). Given the overall increase in wild boar populations referred to in other studies (Sáez–Royuela & Tellería, 1986; Marco et al., 2011), we would also expect an increase in wild boar populations in the Pyrenees (prediction 1)

Secondly, we investigated the spatial ecology of wild boar by testing two mutually exclusive hypotheses. If the wild boar is a sedentary species (hypothesis 1), we would expect a decrease in wild boar abundance in our study site during the hunting period due to the population reduction caused by hunting pressure (prediction 2). Nevertheless, if the wild boar is migratory in the Pyrenees (hypothesis 2), we would expect to observe a pattern of abundance that, rather than corresponding to a simple model of population decrease during the hunting period, exhibits a bell–shaped or a multi–modal pattern of abundance indexes during the hunting period (prediction 3).

Thirdly, we also took into account previous studies of wild boar harvesting. Null or weak relationships were recorded between the numbers of dogs and hunters and bag size per drive hunt in Italy (Scillitani et al., 2010). Thus, we expected comparable results at our study site (prediction 4).

Material and methods

Study site

We analysed local wild boar abundance on the southern side of the Pyrenees (Hecho valley, Aragón, northern Spain). This area is characterized by extensive woodlands (mainly *Pinus sylvestris*, *Fagus sylvatica* and *Quercus sspp.*) and few open habitats (Acevedo et al., 2006). Human population density is low and traditional agricultural practices are mostly focused on animal husbandry (cows, sheep, goats, and horses). In the Pyrenees, local agriculture has been changing in recent decades and the natural reforestation of former open areas has led to a loss of diversity in the landscape mosaic (Ortigosa et al., 1990; García–Ruiz et al., 1996; Roura–Pascual et al., 2005). In this area, traditional hunting drives for wild boar are conducted by one or more beaters on foot with dogs and with hunters on stands. Despite apparent intensive harvesting, the global hunting pressure on the species in the region might be low, due to the abundance of shelter areas (Acevedo et al., 2006; Herrero et al., 2008). In this study we monitored the hunting group from the village of Urdués, which harvests wild boar in their local area (25 km²). In this study, the moderate scale of the area and the detailed information of our data set allowed a more precise monitoring of the local abundance of wild boar than in previous studies on wild boar in Aragón (detailed thereafter).

Data collection

Close collaboration with hunters allowed us to generate a database that included details of the drive hunts that were not included in previous studies on wild boar. Although reliable estimates of ungulate abundance can be made when the numbers of individuals seen on hunts are available (Ericsson & Wallin, 1999; Mysterud et al., 2007; Rönnegård et al., 2008), this information has rarely been available in previous studies on wild boars (Sáez–Royuela & Tellería, 1988; Tellería, 2004). Hunting statistics could also act as a good index of wild boar population abundance (Tellería, 2004; Imperio et al., 2010), even though official hunting statistics in Spain are incomplete and of questionable accuracy (Martínez–Jaúregui et al., 2011). We monitored the number of wild boars seen and culled during each drive hunt. The area in which the hunt took place and the number of dogs and hunters were also recorded. In total, data from 442 drive hunts were recorded during hunting periods (2.21 drive hunts per km^2 per hunting period) from 2004 to 2012. For this study our monitoring allowed a fine–resolution that is close to forty times greater than those of previous studies on wild boar in this region (2,657 drive hunts per hunting period for 47,669 km^2 in Aragón means close to 0.06 drive hunt per season per km^2 [Acevedo et al., 2006]). This underlines the difference of resolution between the data used in previous studies on wild boar in Aragón and the data set used in this study.

Analysis

We considered two indexes of wild boar abundance: the number of wild boars seen (index 1) and the number of wild boars culled (index 2) per drive hunt. To analyse the determining factors in these indexes of abundance, we used General Additive Models (GAMs) (Wood, 2006; Zuur et al., 2007). The explanatory variables considered were: (Y) the year in which the hunting period started; (D) the day of the hunting period: for the first day of each hunting period, we used the day number according to the Gregorian calendar and then added the number of days up to the end of the hunting period; (Nd) the number of dogs; (Nh) the number of hunters. In the Pyrenees, the number of hunters per drive hunt in traditional hunting groups (mean ± SE = 7 ± 2.8) rarely or never allows coverage of all the potential escape routes of the hunted patch. Also, data concerning the exact surface hunted by beaters and dogs (mean number of dogs per drive hunt ± SE = 9.8 ± 3.6) is usually unavailable because the courses of the dogs are not systematically recorded with telemetric tools. The exact hunted area is thus usually unknown. In this study, the approximate area potentially hunted during each drive hunt was close to 2.5 km^2. Furthermore, instead of using estimated surfaces characterized by overblown and unreliable accuracy, we tested the area in which the drive hunt took place as a potential co–factor to account for the potential effects of spatial heterogeneity on wild boar abundance.

We used an information–theoretic approach based on the Akaike's information criterion corrected for a small sample size (AICc; Burnham & Anderson, 2002). The analysis identified the most parsimonious model (lowest AICc) of possible subsets, ranging from the null model (M0, intercept only) to a model with all the considered explanatory variables. This analytical procedure selects the model that provides an accurate approximation to the structural information in the data at

hand, with the smallest possible number of parameters for adequate representation of the data (Burnham & Anderson, 2002). The Akaike weight of models (Wi) was presented —the weight of evidence in favour of the considered model being the best model for the situation at hand (Burnham & Anderson, 2002). The relative importance (RI) of the explanatory variables was estimated —by the sum of the Akaike weights over all models in which that variable appears– to highlight evidence for the importance of each variable within the set of models (Burnham & Anderson, 2002). Explained deviance values (Dev–expl), providing an estimate of the model fit (Wood, 2006), are also presented. All analyses were performed using the R statistical software (R Development Core Team, 2011).

Results

Variability in wild boar abundance indexes depended on temporal factors —the hunting period and the day in the hunting period— and on the characteristics of the drive hunt —the number of dogs and hunters and the area. Model selection suggests for both wild boars seen and wild boars culled that the best model for the data at hand includes as explanatory factors the year, the day of the season, the interaction between these two factors, the number of dogs and hunters, and the area (table 1 and 2). Over the considered period, the numbers of wild boars seen and culled per drive hunt showed non–linear trends (fig. 1). For the number of wild boars seen per drive hunt, the fitted model suggests an increase of 13% between 2004 and 2005, a decrease of 44% between 2005 and 2009, and an increase of 20% between 2009 and 2011. Between 2004 and 2011, this model suggests an overall reduction of 23% in the number of wild boars seen per drive hunt (fig. 1A, left). For the number of wild boars culled per drive hunt, the selected model suggests an oscillatory pattern with substantial increases (101% between 2004 and 2005; 57% between 2007 and 2008) and decreases (–46% between 2005 and 2007; –66% between 2008 and 2010). Between 2004 and 2011, this model suggests a 14% increase in the number of wild boars culled per drive hunt (fig. 1B, left). These inter–annual trends interact with a multimodal pattern that exhibits variations depending on the hunting period (fig. 2A).

The number of wild boars seen per drive hunt was highest at the beginning of the hunting period (early October), in early January, and in February in 2004–2006. However, this pattern changed over the study period and the number of wild boars seen was highest in December and February in 2006–2009. Since 2009, however, the periods with greatest numbers of wild boars seen were the same as in previous years but with the difference that the peaks of abundances in boar seen decreased in comparison with the period 2004–2009 (fig. 2A, left). The number of wild boars seen per drive hunt was positively associated with the number of hunters (at least up to ten hunters) (fig. 2B, left) and also increased strongly in drive hunts with 10–18 dogs (fig. 2B, left). A decrease in the number of wild boar

Table 1. Model selection for determining factors in the number of wild boar *Sus scrofa* seen per drive hunt: Y. Hunting period; D. Day of the hunting period; Nd. Number of dogs; Nh. Number of hunters; A. Area where the drive hunt took place; * interaction; K. Number of estimated parameters; AICc. Akaike's Information Criterion corrected for small sample size, lower values indicate a most–parsimonious model for the observed data; ΔAICc. Difference of AICc between the model and the most parsimonious model; the larger the ΔAICc, the less plausible it is that the fitted model is the best model given the data set; L(gi/x). Probability of the model being the best model given the data set; Wi. Akaike weight of the model; Dev–expl. Explained deviance of the fitted model; RI. Relative Importance of factors. Only the ten best models are reported (Burnham & Anderson, 2002; Wood, 2006).

*Tabla 1. Selección de modelos para determinar los factores que condicionan el número de jabalíes (Sus scrofa) avistados por batida: Y. Temporada de caza; D. Día de la temporada de caza; Nd. Número de perros; Nh. Número de cazadores; A. Área en la que tuvo lugar la batida; * Interacción; K. Número de parámetros estimados; AICc. Criterio de información de Akaike corregido para un tamaño muestral pequeño, los valores bajos indican un modelo principalmente parsimonioso para los datos observados; ΔAICc. Diferencia de AICc entre el modelo y el modelo más parsimonioso, cuánto mayor sea ΔAICc, menos plausible será que el modelo ajustado sea el mejor para el conjunto de datos; L(gi/x). Probabilidad de que el modelo sea el mejor para el conjunto de datos; Wi. Peso de Akaike del modelo; Dev–expl. Variabilidad explicacada del modelo ajustado; RI. Importancia relativa de los factores. Solo se muestran los diez modelos mejores (Burnham & Anderson, 2002; Wood, 2006).*

Model	K	AICc	ΔAICc	L(gi/x)	Wi	Dev–expl	RI	
Y+D+Y*D+Nd+Nh+A	55	1490.79	0.00	1.00	0.87	0.30	Y	1.00
Y+D+Y*D+Nd+A	48	1494.59	3.80	0.15	0.13	0.28	D	1.00
Y+D+Y*D+Nh+A	50	1509.94	19.15	0.00	0.00	0.28	Y*D	1.00
Y+D+Nd+Nh+A	34	1540.89	50.10	0.00	0.00	0.21	Nd	1.00
Y+D+Y*D+Nd+Nh	46	1548.35	57.56	0.00	0.00	0.25	Nh	0.87
D+Nd+A	21	1558.53	67.74	0.00	0.00	0.17	A	1.00
Y+D+Nh+A	29	1559.72	68.93	0.00	0.00	0.19		
Y+D+Nd+A	22	1560.71	69.92	0.00	0.00	0.17		
Y+Nd+A	14	1564.63	73.84	0.00	0.00	0.15		
D+Nh+A	25	1567.45	76.65	0.00	0.00	0.17		

seen was observed in drive hunts with more than 18 dogs, although this variation should be considered with caution due to its small sample size.

The number of wild boar culled per drive hunt revealed three key periods in the hunting periods, above all in the periods 2004–2005, 2007–2009, and 2011–2012 (fig. 2A, right): the end of December–early January and February, both characterized by the greatest number of wild boar culled per drive hunt, and lastly, the beginning of the hunting period (although the number of wild boars culled per drive hunt in this period was lower than in the other two periods). The number of wild boars culled per drive hunt also increased with the numbers of hunters and dogs, above all in drive hunts with 12–18 dogs (fig. 2B, right).

All the considered factors have very high relative importance (close to 1) in explaining the variability in the indexes of wild boar abundance (table 1 and 2). Yet, the explained deviance of the selected models was moderate (30% for wild boar seen and 23% for wild boar culled), which suggests that the considered factors only provide a partial understanding of the observed variability.

Discussion

Multimodal patterns in wild boar abundance indexes during hunting periods suggest that wild boar conduct seasonal migrations in our study site. Migrations are a more likely explanation than nomadism (Mueller & Fagan, 2008) because the environment is highly seasonal in the Pyrenees and because pulsations in wild boar abundance during the hunting period occur over years and in certain predictable times of the hunting period. The boar mating season at the end of December–early January (Delcroix et al., 1990), for instance, is one of the periods when high abundances of wild boar are most predictable. Thus, the observed variations in wild boar abundance may be linked —at least in part— to the behavioural ecology of the species in the area. The observed evidence of wild boar migrations in our area differs from the sedentary patterns reported in Germany and Australia (Keuling et al., 2008; Mitchell et al., 2009) but agrees with results from Poland and mountainous environments in Italy and in Tennessee, USA (Andrzejewski & Jezierski, 1978; Singer et al., 1981; D'Andrea et al., 1995). Patterns in the use of

Table 2. Model selection for determining factors in the number of wild boar *Sus scrofa* culled per drive hunt. Only the ten best models are reported. (Burnham & Anderson, 2002; Wood, 2006). (For abbreviations see table 1.)

Tabla 2. Selección de modelos para determinar los factores que condicionan el número de jabalíes Sus scrofa *abatidos por batida. Solo se han mostrado los diez modelos mejores (Burnham & Anderson, 2002; Wood, 2006). (Para las abreviaturas, ver tabla 1.)*

Model	K	AICc	ΔAICc	L(gi/x)	Wi	Dev–expl	RI	
Y+D+Y*D+Nd+Nh+A	26	758.41	0.00	1.00	0.94	0.23	Y	0.98
D+Nd+A	18	766.03	7.62	0.02	0.02	0.18	D	0.98
Y+Nd+A	14	767.11	8.70	0.01	0.01	0.16	Y*D	0.95
Y+D+Y*D+Nd+A	20	767.60	9.19	0.01	0.01	0.19	Nd	0.99
Y+D+Nd+A	19	767.84	9.43	0.01	0.01	0.18	Nh	0.95
Y+D+Nd+Nh+A	20	769.77	11.36	0.00	0.00	0.18	A	1.00
Y+D+Y*D+Nh+A	23	772.44	14.03	0.00	0.00	0.19		
Y+D+Nh+A	22	772.94	14.53	0.00	0.00	0.19		
Y+Nh+A	17	774.60	16.19	0.00	0.00	0.16		
Y+D+Y*D+Nd+Nh	18	782.29	23.88	0.00	0.00	0.17		

space in wild boar, therefore, appear to be highly dependent on the environment. As in other European ungulates (Albon & Langvatn, 1992; Mysterud, 1999; Ball et al., 2001) and as previously suggested for the wild boar (Andrzejewski & Jezierski, 1978), migrations may involve only part of the wild boar population (partial migration) and still require further study. The knowledge of wild boar migration in the Pyrenees may stimulate a reappraisal of significant variations in local populations. The sustainable management of migratory species requires an accurate understanding and familiarity with migratory routes (Thirgood et al., 2004; Bolger et al., 2008), and such knowledge would represent a substantial challenge for future management plans. Further studies should aim to characterize the life–history, the spatial scale, the phenology and the determining factors of migration (Ramenofsky & Wingfield, 2007) of wild boar in the Pyrenees. Previous studies on space use in wild boar suggested small home ranges at small time scales (< 1,000 ha in average; Massei et al., 1997; Keuling et al., 2008). The choice of temporal scale at which data are collected and the definition of home range can significantly influence biological inference (Börger et al., 2006). The size of our study site (2,500 ha) and our intense monitoring were key factors that allowed us to reach high–resolution analyses of variations in wild boar abundance. Further studies should use movement data at small temporal scale and take into account reproductive ecology and food availability, not just hunting period. Integrated and high–resolution monitoring is required to unravel the misunderstood complexity of space use in wild boar.

The number of wild boar seen and culled per drive hunt varied substantially during the monitoring period and the trends in these two abundance indexes differed. Inter–annual variations were greater for wild boars culled than for wild boars seen per drive hunt and on occasions the trends in abundance for each index were different. For instance, in the period 2004–2011, the results for the wild boars seen per drive hunt suggest a reduction of 23%, while the results for wild boars culled per drive hunt suggest an increase of 14%. The number of wild boars seen per drive hunt is probably a more reliable index of wild boar abundance than the number of wild boars culled because it is not dependent on shooting success and because indexes based on seen–individuals have previously been preferred in other ungulate species (Ericsson & Wallin, 1999; Mysterud et al., 2007). Variations in the migratory/resident ratio might also affect the relationship between the numbers of wild boar seen and culled through dilution effects on predation risk (Krause & Ruxton, 2002). Thus, strong inter–annual variations in the number of culled wild boar should be regarded with caution as this abundance index might overestimate or even deviate from true population dynamics. Further studies are required to unravel the relative importance of shooting success.

As seen above, the number of wild boars seen per drive hunt suggests a non–linear decrease of 23% in 2004–2011 in our study site. Indirect evidence of wild boar migration were observed and, therefore, further studies should analyse the spatial scale of this decreasing population trend. The population dynamics of wild boar in other areas should also be examined using indexes other than hunting bag alone. The observed inter–annual pattern disagrees with the results reported by Marco et al. (2011) that suggest —on the basis of official hunting statistics and for an area that included

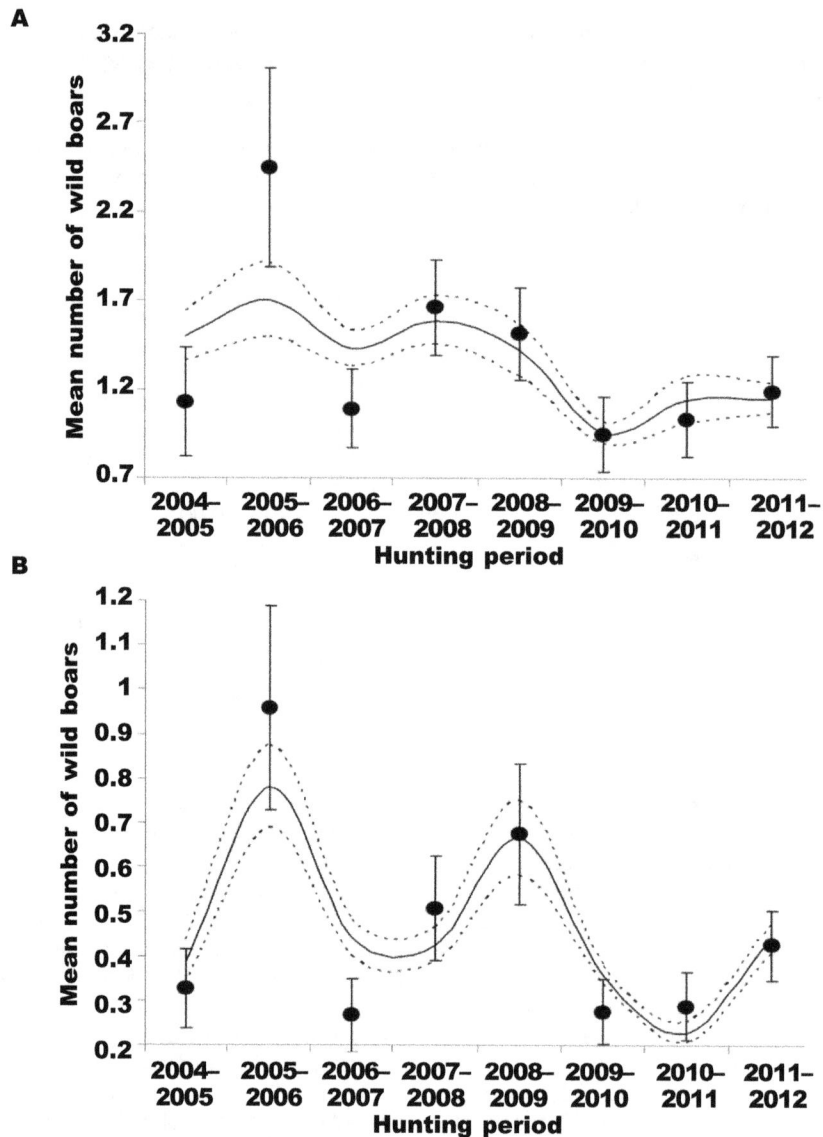

Fig. 1. Inter–annual variation in the indexes of wild boar abundance in Urdués (Pyrenees, northern Spain): A. Number of wild boars seen per drive hunt; B. Number of wild boars culled per drive hunt. Points and error bars represent mean values and related standard error. The solid lines represent the predicted patterns estimated by the best models and the dotted lines indicate the related standard error.

Fig. 1. Variación interanual en los índices de abundancia de jabalí en Urdués (Pirineos, norte de España): A. Número de jabalíes avistados por batida; B. Número de jabalíes abatidos por batida. Los puntos y las barras de error representan los valores medios y el error estándar relacionado. Las líneas continuas representan los patrones previstos estimados con los mejores modelos y las discontinuas indican el error estándar relacionado.

our study site— that wild boar populations increased in Aragón during this period. This incongruence might be caused by differences in the spatial scale considered. Nevertheless, as highlighted by Martínez–Jauregui et al. (2011), in Spain official hunting statistics can be incomplete and thus this discrepancy may be due to differences in the accuracy and in the completeness of the available information. Between 1997 and 2002, Acevedo et al. (2006) suggested a population increase in the Pyrenees (woodland habitat) and relative population stability or local decrease in central and south Aragón (which is characterized by a more developed agriculture than the Pyrenees). This might suggest that wild boar population dynamics might have changed in

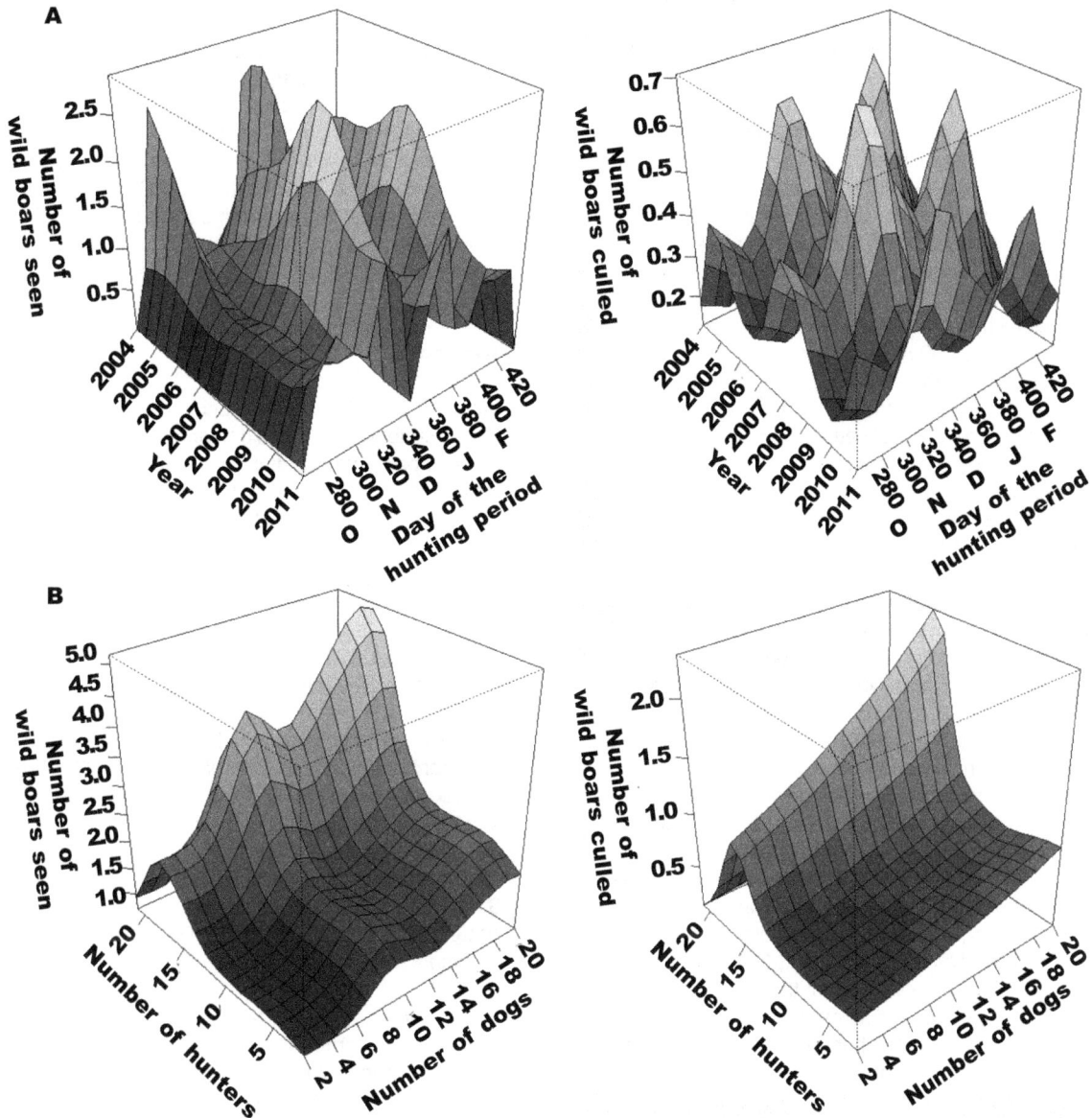

Fig. 2. Determining factors in the number of wild boar seen (left) and number of wild boar culled (right) per drive hunt: A. Effect of year and day of the hunting period (O. October; N. November; D. December; J. January; F. February); B. Effects of the numbers of dogs and hunters.

Fig. 2. Factores determinantes en el número de jabalíes avistados (izquierda) y número de jabalíes abatidos (derecha) por batida: A. Efecto del año y el día de la temporada de caza (O. Octubre; N. Noviembre; D. Diciembre; J. Enero; F. Febrero); B. Efectos del número de perros y de cazadores.

the Pyrenees, at least locally. However, this interpretation should be considered with caution because the results of Acevedo et al. (2006) were based on hunting bag size and their study had a much lower resolution than ours. We showed that inter–annual variations in bag size may deviate from the population dynamics inferred from the number of wild boars seen per drive hunt and, thus, results of previous studies should be considered with caution.

In Aragón, as in other European regions, wild boar can be hunted with no limit on bag size and the assumed population increase may have led authorities to advance the hunting period by two weeks since 2009 (Boletín Oficial de Aragón, 2008, 2009). Variations in wild boar abundance indexes observed during this study question the accuracy of the fine–tuning from one year to the next of the management of wild boar in Aragón. As in wild boar populations in Eastern Europe

(Danilov & Panchenko, 2012), wild boar populations in Western Europe oscillate —both increases and reductions occur— and this could be taken into account in further management plans. As for other game species, the sustainability of harvested wild boar populations —as a limited natural resource— will depend on the integration of results such as those we present here into future management plans to avoid population collapse (Fryxell et al., 2010). Past examples in Italy, Russia, Scandinavia and the United Kingdom already showed that severe population declines in wild boar or even a collapse were possible when the environmental conditions became too adverse (Apollonio et al., 1988; Leaper et al., 1999; Welander, 2000; Rosvold et al., 2010; Danilov & Panchenko, 2012). Thus, further studies are required to unravel the relative importance of down regulation (fructification, agricultural practices), characteristics of wild boar population (demographic structure, genetics) and of top regulation (pathogens, predators and hunting) in population dynamics of wild boar (e.g., Massolo & Mazzoni della Stella, 2006; Rosell et al., 2012).

Wild boar population dynamics at our study site does not support the hypothesis of a generalized increase in wild boar densities in woodland areas as the origin of the increasing presence of wild boars in urban environments (Jansen et al., 2007). The hypothesis suggesting that the species might be locally adapting to agricultural and urban environments —where hunting pressure is low and where opportunist foraging might even be facilitated by city dwellers (Cahill & Llimona, 2004)— should be analyzed more closely. Such an adaptation has already been observed in wild boar by other authors (Cahill et al., 2012; Rosell et al., 2012) and the importance of areas where hunting was banned in explaining crop damage was also highlighted (Amici et al., 2012).

In conclusion, as a result of the close collaboration with local hunters, our study was able to reveal that wild boar seen per drive hunt decreased by 23% in the period 2004–2011 in our study area in the Pyrenees. The observed patterns of wild boar abundance imply the existence of wild boar migrations and further studies should analyse population dynamics of wild boar in other areas using indexes other than hunting bag alone. Our study also highlighted the fact that the numbers of dogs and hunters affect the number of wild boars seen and culled per drive hunt, and that inter–annual variations in bag size might lead to overestimates and discrepancies with population dynamics inferred from the number of wild boars seen per drive hunt. Thus, there is still a need for further studies on the spatial ecology of wild boar and on the applied ecology of wild boar if we are to move towards sustainable and information–based management of wild boar populations.

Acknowledgements

We would like to thank Julia López, Agnès Sarasa, David Sarasa and Noelia Sánchez–López for logistic support during the study period. Thanks also to all the members of the Grupo de Caza Mayor de Urdués, and above all to César López, Jesús Jiménez, Francisco López, Eduardo López, Roberto Jiménez, Pablo López and Gabriel López for their collaboration during the monitoring period. We are grateful to Michael Lockwood and Agnès Sarasa for the English revision and to Cédric Girard–Buttoz and Sylvain Losdat for valuable comments on an earlier draft of this paper. The research activities of MS are partially supported by the Plan Andaluz de Investigación Desarollo e Innovación de la Junta de Andalucía (RNM–118). This work was conducted without specific financial support and complies with current Spanish laws.

References

Acevedo, P., Escudero, M. A., Munoz, R. & Gortazar, C., 2006. Factors affecting wild boar abundance across an environmental gradient in Spain. Acta Theriologica, 51: 327–336. DOI: 10.1007/BF03192685.

Albon, S. D. & Langvatn, R., 1992. Plant phenology and the benefits of migration in a temperate ungulate. Oikos, 65: 502–513.

Amici, A., Serrani, F., Rossi, C. M. & Primi, R., 2012. Increase in crop damage caused by wild boar (Sus scrofa L.): the 'refuge effect'. Agronomy for Sustainable Development, 32: 683–692.

Andrzejewski, R. & Jezierski, W., 1978. Management of a wild boar population and its effects on commercial land. Acta Theriologica, 23: 309–339.

Apollonio, M., Randi, E. & Toso, S., 1988. The systematics of the wild boar (Sus scrofa L.) in Italy. Bolletino di Zoologia, 55: 213–221.

Artois, M., Depner, K. R., Guberti, V., Hars, J., Rossi, S. & Rutili, D., 2002. Classical swine fever (hog cholera) in wild boar in Europe. Revue Scientifique et Technique de l'Office International des Epizooties, 21: 287–304.

Ball, J. P., Nordengren, C. & Wallin, K., 2001. Partial migration by large ungulates: characteristics of seasonal moose Alces alces ranges in northern Sweden. Wildlife Biology, 7: 39–47.

Boletín Oficial de Aragón, 2008. ORDEN de 10 de junio de 2008, del Departamento de Medio Ambiente, por al que se aprueba el Plan General de Caza para la temporada 2008–2009. 15277–15286.

Boletín Oficial de Aragón, 2009. ORDEN de 11 de junio de 2009, del Departamento de Medio Ambiente, por al que se aprueba el Plan General de Caza para la temporada 2009–2010. 16253–16263.

Bolger, D. T., Newmark, W. D., Morrison, T. A. & Doak, D. F., 2008. The need for integrative approaches to understand and conserve migratory ungulates. Ecology Letters, 11: 63–77. DOI: 10.1111/j.1461-0248.2007.01109.x.

Börger, L., Franconi, N., Ferretti, F., Meschi, F., De Michele, G., Gantz, A. & Coulson, T., 2006. An integrated approach to identify spatiotemporal and individual–level determinants of animal home range size. The American Naturalist, 168: 471–485.

Bueno, C. G., Barrio, I. C., García–González, R., Alados, C. L. & Gómez–García, D., 2010. Does wild boar rooting affect livestock grazing areas in alpine grasslands? European Journal of Wildlife Research,

56: 765–770. DOI: 10.1007/s10344–010–0372–2

Burnham, K. P. & Anderson, D. R., 2002. *Model selection and multimodel inference: a practical information–theoric approach.* 2nd edition. Springer–Verlag, New York.

Cahill, S. & Llimona, F., 2004. Demographics of a wild boar *Sus scrofa* Linnaeus, 1758 population in a metropolitan park in Barcelona. *Galemys,* 16: 37–52.

Cahill, S., Llimona, F., Cabañeros, L. & Calomardo, F., 2012. Characteristics of wild boar (*Sus scrofa*) habituation to urban areas in the Coillserola Natural Park (Barcelona) and comparison with other locations. *Animal Biodiversity and Conservation,* 35.2: 221–233.

Cuevas, M. F., Novillo, A., Campos, C., Dacar, M. A. & Ojeda, R. A., 2010. Food habits and impact of rooting behaviour of the invasive wild boar, *Sus scrofa*, in a protected area of the Monte Desert, Argentina. *Journal of Arid Environments,* 74: 1582–1585. DOI: 10.1016/j.jaridenv.2010.05.002.

D'Andrea, L., Durio, P., Perrone, A. & Pirone, S., 1995. Preliminary data of the wild boar (*Sus scrofa*) space use in mountain environment. *IBEX Journal of Mountain Ecology,* 3: 117–121.

Danilov, P. & Panchenko, D., 2012. Expansion and some ecological features of the wild boar beyond the northern boundary of its historical range in European Russia. *Russian Journal of Ecology,* 43: 45–51. DOI: 10.1134/S1067413612010043.

Delcroix, I., Mauget, R. & Signoret, J. P., 1990. Existence of synchronization of reproduction at the level of the social group of the European wild boar (*Sus scrofa*). *Journal of Reproduction and Fertility,* 89: 613–617.

Ericsson, G. & Wallin, K., 1999. Hunter observations as an index of moose *Alces alces* population parameters. *Wildlife Biology,* 5: 177–185.

Fryxell, J. M., Packer, C., McCann, K., Solberg, E. J. & Sæther, B. E., 2010. Resource management cycles and the sustainability of harvested wildlife populations. *Science,* 328: 903–906. DOI: 10.1126/science.1185802.

García–Ruiz, J. M., Lasanta, T., Ruiz–Flano, P., Ortigosa, L., White, S., González, C. & Martí, C., 1996. Land–use changes and sustainable development in mountain areas: a case study in the Spanish Pyrenees. *Landscape Ecology,* 11: 267–277.

Gortázar, C., Herrero, J., Villafuerte, R. & Marco, J., 2000. Historical examination of the status of large mammals in Aragón, Spain. *Mammalia,* 64: 411–422. DOI: 10.1515/mamm.2000.64.4.411.

Groot Bruinderink, G. W. T. A. & Hazebroek, E., 1996. Ungulate traffic collisions in Europe. *Conservation Biology,* 10: 1059–1067. DOI: 10.1046/j.1523–1739.1996.10041059.x.

Herrero, J., García–Serrano, A. & García–González, R., 2008. Reproductive and demographic parameters in two Iberian wild boar *Sus scrofa* populations. *Acta Theriologica,* 53: 355–364. DOI: 10.1007/BF03195196.

Imperio, S., Ferrante, M., Grignetti, A., Santini, G. & Focardi, S., 2010. Investigating population dynamics in ungulates: Do hunting statistics make up

a good index of population abundance? *Wildlife Biology,* 16: 205–214. DOI: 10.2981/08–051.

Jansen, A., Luge, E., Guerra, B., Wittschen, P., Gruber, A. D., Loddenkemper, C., Schneider, T., Lierz, M., Ehlert, D. & Appel, B., 2007. Leptospirosis in urban wild boars, Berlin, Germany. *Emerging Infectious Diseases,* 13: 739–742. DOI: 10.3201/eid1305.061302.

Keuling, O., Lauterbach, K., Stier, N. & Roth, M., 2010. Hunter feedback of individually marked wild boar *Sus scrofa* L.: dispersal and efficiency of hunting in northeastern Germany. *European Journal of Wildlife Research,* 56: 159–167. DOI: 10.1007/s10344–009–0296–x.

Keuling, O., Stier, N. & Roth, M., 2008. Annual and seasonal space use of different age classes of female wild boar *Sus scrofa* L. *European Journal of Wildlife Research,* 54: 403–412. DOI: 10.1007/s10344–007–0157–4.

Krause, J. & Ruxton, G. D., 2002. *Living in Groups.* Oxford Univ. Press, Oxford.

Lagos, L., Picos, J. & Valero, E., 2012. Temporal pattern of wild ungulate–related traffic accidents in northwest Spain. *European Journal of Wildlife Research,* 58: 661–668 DOI: 10.1007/s10344–012–0614–6.

Leaper, R., Massei, G., Gorman, M. L. & Aspinall, R., 1999. The feasibility of reintroducing wild boar (*Sus scrofa*) to Scotland. *Mammal Review,* 29: 239–258. DOI: 10.1046/j.1365–2907.1999.2940239.x.

Marco, J., Herrero, J., Escudero, M. A., Fernández–Arberas, O., Ferreres, J., García–Serrano, A., Giménez–Anaya, A., Labarta, J. L., Monrabal, L. & Prada, C., 2011. Veinte años de seguimiento poblacional de ungulados silvestres de Aragón. *Pirineos,* 166: 135–153. DOI: 10.3989/Pirineos.2011.166007.

Martínez–Jauregui, M., Arenas, C. & Herruzo, A. C., 2011. Understanding long–term hunting statistics: the case of Spain (1972–2007). *Forest Systems,* 1: 139–150. DOI: 10.5424/fs/2011201–10394.

Massei, G., Genov, P. V., Staines, B. W. & Gorman, M. L., 1997. Factors influencing home range and activity of wild boar (*Sus scrofa*) in a Mediterranean coastal area. *Journal of Zoology,* 242: 411–423.

Massei, G., Roy, S. & Bunting, R., 2011. Too many hogs? A review of methods to mitigate impact by wild boar and feral hogs. *Human–Wildlife Interactions,* 5: 79–99.

Massolo, A. & Mazzoni della Stella, R., 2006. Population structure variations of wild boar *Sus scrofa* in central Italy. *Italian Journal of Zoology* 73: 137–144.

Melis, C., Szafrańska, P. A., Jędrzejewska, B. & Bartoń, K., 2006. Biogeographical variation in the population density of wild boar (*Sus scrofa*) in western Eurasia. *Journal of Biogeography,* 33: 803–811. DOI: 10.1111/j.1365–2699.2006.01434.x.

Meng, X. J., Lindsay, D. S. & Sriranganathan, N., 2009. Wild boars as sources for infectious diseases in livestock and humans. *Philosophical Transactions of the Royal Society B: Biological Sciences,* 364: 2697–2707. DOI: 10.1098/rstb.2009.0086.

Mitchell, J., Dorney, W., Mayer, R. & McIlroy, J., 2009. Migration of feral pigs (*Sus scrofa*) in rainforests of north Queensland: fact or fiction? *Wildlife Re-*

search, 36: 110–116. DOI: 10.1071/WR06066.

Mueller, T. & Fagan, W. F., 2008. Search and navigation in dynamic environments – from individual behaviors to population distributions. *Oikos,* 117: 654–664. DOI: 10.1111/j.2008.0030–1299.16291.x.

Mysterud, A., 1999. Seasonal migration pattern and home range of roe deer (*Capreolus capreolus*) in an altitudinal gradient in southern Norway. *Journal of Zoology,* 247: 479–486. DOI: 10.1111/j.1469–7998.1999.tb01011.x.

Mysterud, A., Meisingset, E. L., Veiberg, V., Langvatn, R., Solberg, E. J., Loe, L. E. & Stenseth, N. C., 2007. Monitoring population size of red deer *Cervus elaphus*: an evaluation of two types of census data from Norway. *Wildlife Biology,* 13: 285–298. DOI: 10.2981/0909–6396(2007)13[285:MPSORD]2.0.CO;2.

Ortigosa, L. M., García–Ruiz, J. M. & Gil, E., 1990. Land reclamation by reforestation in the Central Pyrenees. *Mountain Research and Development,* 10: 281–288.

R Development Core Team, 2011. *R: a language and environment for statistical computing.* R Foundation for Statistical Computing (http://www.R–project.org/). Vienna.

Ramenofsky, M. & Wingfield, J. C., 2007. Regulation of migration. *Bioscience,* 57: 135–143. DOI: 10.1641/B570208.

Risch, A., Wirthner, S., Busse, M., Page–Dumroese, D. & Schütz, M., 2010. Grubbing by wild boars (*Sus scrofa* L.) and its impact on hardwood forest soil carbon dioxide emissions in Switzerland. *Oecologia,* 164: 773–784. DOI: 10.1007/s00442–010–1665–6

Rönnegård, L., Sand, H., Andrén, H., Månsson, J. & Pehrson, Å., 2008. Evaluation of four methods used to estimate population density of moose *Alces alces. Wildlife Biology,* 14: 358–371. DOI: 10.2981/0909–6396(2008)14[358:EOFMUT]2.0.CO;2.

Rosell, C., Navàs, F. & Romero, S., 2012. Reproduction of wild boar in a cropland and coastal wetland area: implications for management. *Animal Biodiversity and Conservation,* 35.2: 209–217.

Rossi, S., Fromont, E., Pontier, D., Cruciere, C., Hars, J., Barrat, J., Pacholek, X. & Artois, M., 2005. Incidence and persistence of classical swine fever in free–ranging wild boar (*Sus scrofa*). *Epidemiology and Infection,* 133: 559–568. DOI: 10.1017/S0950268804003553.

Rosvold, J., Halley, D. J., Hufthammer, A. K., Minagawa, M. & Andersen, R., 2010. The rise and fall of wild boar in a northern environment: evidence from stable isotopes and subfossil finds. *Holocene,* 20: 1113–1121. DOI: 10.1177/0959683610369505

Roura–Pascual, N., Pons, P., Etienne, M. & Lambert, B., 2005. Transformation of a rural landscape in the Eastern Pyrenees between 1953 and 2000. *Mountain Research and Development,* 25: 252–261. DOI: 10.1659/0276–4741(2005)025[0252:TOARLI]2.0.CO;2.

Sáez–Royuela, C. & Tellería, J. L., 1986. The increased population of the Wild Boar (*Sus scrofa* L.) in Europe. *Mammal Review,* 16: 97–101. DOI: 10.1111/j.1365–2907.1986.tb00027.x

– 1988. Las batidas como método de censo en especies de caza mayor: Aplicación al caso del jabalí (*Sus scrofa* L.) en la provincia de Burgos (Norte de España). *Doñana. Acta vertebrata,* 15(2): 215–223.

Saïd, S., Tolon, V., Brandt, S. & Baubet, E., 2012. Sex effect on habitat selection in response to hunting disturbance: the study of wild boar. *European Journal of Wildlife Research,* 58: 107–115. DOI: 10.1007/s10344–011–0548–4.

Saunders, G. & Kay, B., 1996. Movements and Home Ranges of Feral Pigs (*Sus Scrofa*) in Kosciusko National Park, New South Wales. *Wildlife Research,* 23: 711–719. DOI: 10.1071/WR9960711.

Schley, L., Dufrêne, M., Krier, A. & Frantz, A. C., 2008. Patterns of crop damage by wild boar (*Sus scrofa*) in Luxembourg over a 10–year period. *European Journal of Wildlife Research,* 54: 589–599. DOI: 10.1007/s10344–008–0183–x.

Scillitani, L., Monaco, A. & Toso, S., 2010. Do intensive drive hunts affect wild boar (*Sus scrofa*) spatial behaviour in Italy? Some evidences and management implications. *European Journal of Wildlife Research,* 56: 307–318. DOI: 10.1007/s10344–009–0314–z.

Singer, F. J., Otto, D. K., Tipton, A. R. & Hable, C. P., 1981. Home ranges, movements, and habitat use of European wild boar in Tennessee. *The Journal of Wildlife Management,* 45: 343–353.

Tellería, J. L., 2004. *Métodos de Censos en Vertebrados Terrestres.* Dpto. Biología. Animal I (Zoología de Vertebrados). Facultad de Biología, Univ. Complutense, Madrid.

Thirgood, S., Mosser, A., Tham, S., Hopcraft, G., Mwangomo, E., Mlengeya, T., Kilewo, M., Fryxell, J., Sinclair, A. R. E. & Borner, M., 2004. Can parks protect migratory ungulates? The case of the Serengeti wildebeest. *Animal Conservation,* 7: 113–120. DOI: 10.1017/S1367943004001404.

Welander, J., 2000. Spatial and temporal dynamics of wild boar (*Sus scrofa*) rooting in a mosaic landscape. *Journal of Zoology,* 252: 263–271. DOI: 10.1111/j.1469–7998.2000.tb00621.x.

Wirthner, S., Frey, B., Busse, M. D., Schütz, M. & Risch, A. C., 2011. Effects of wild boar (*Sus scrofa* L.) rooting on the bacterial community structure in mixed–hardwood forest soils in Switzerland. *European Journal of Soil Biology,* 47: 296–302. DOI: 10.1016/j.ejsobi.2011.07.003.

Wirthner, S., Schütz, M., Page–Dumroese, D. S., Busse, M. D., Kirchner, J. W. & Risch, A. C., 2012. Do changes in soil properties after rooting by wild boars (*Sus scrofa*) affect understory vegetation in Swiss hardwood forests? *Canadian Journal of Forest Research,* 42: 585–592. DOI: 10.1139/X2012–013.

Wood, S. N., 2006. *Generalized additive models, an introduction with R.* Chapman & Hall/CRC, Boca Raton.

Zuur, A. F., Ieno, E. N. & Smith, G. M., 2007. *Analysing ecological data.* Springer, New York.

12

Population estimates, density–dependence and the risk of disease outbreaks in the Alpine ibex *Capra ibex*

C. De Danieli & M. Sarasa

De Danieli, C. & Sarasa, M., 2015. Population estimates, density–dependence and the risk of disease outbreaks in the Alpine ibex *Capra ibex*. *Animal Biodiversity and Conservation*, 38.1: 101–119.

Abstract

Population estimates, density–dependence and the risk of disease outbreaks in the Alpine ibex Capra ibex.— Wildlife monitoring and the identification of factors associated with disease outbreaks are major goals in wildlife conservation. We reviewed demographic and epidemiological data for the Alpine ibex *Capra ibex* from 1975–2013 to characterize the species' abundance and distribution dynamics on a large scale. We also explored methodological bias in monitoring and analyzed the factors potentially associated with the risk of disease outbreaks. Our results revealed that the overall abundance and distribution of Alpine ibex appeared to be increasing at both national and international scales, in agreement with the IUCN's 'Least Concern' conservation status on the international scale and on the national scale for Italy, Switzerland and France. Our comparative analysis of common monitoring methods highlights the fact that abundance values from counts are underestimated and suggests that the Alpine ibex is more abundant than is usually reported. The appearance and persistence of disease outbreaks (*e.g.* sarcoptic mange, keratoconjunctivitis or brucellosis) are related to local ibex density and abundance. The observed correlation between the demographic growth of ibex populations and disease outbreaks suggests that the risk of epizooties may be increasing or might already be high in several populations of *Capra ibex*.

Key words: *Capra ibex*, Disease outbreak, Host density, Monitoring, Parasite transmission, Population dynamics

Resumen

Estimas de poblaciones, dependencia de la densidad y riesgo de aparición de brotes de enfermedades en el íbice de los Alpes, Capra ibex.— El seguimiento de la fauna silvestre y la identificación de los factores asociados con los brotes de enfermedades son algunos de los objetivos principales de la conservación de la fauna silvestre. En el presente estudio examinamos los datos demográficos y epidemiológicos del íbice de los Alpes, *Capra ibex*, entre los años 1975 y 2013 para caracterizar la dinámica de la distribución y la abundancia de la especie a gran escala. Asimismo, analizamos los sesgos metodológicos del seguimiento y estudiamos los factores que podrían estar relacionados con el riesgo de aparición y persistencia de brotes de enfermedades. Nuestros resultados revelaron que la abundancia y la distribución del íbice de los Alpes parecen estar aumentando tanto a escala nacional como internacional, de forma acorde con el estado de conservación de la Unión Internacional para la Conservación de la Naturaleza (UICN) de Preocupación Menor a escala internacional, y en el ámbito nacional para Italia, Suiza y Francia. Nuestro análisis comparativo de los métodos convencionales de seguimiento pone de relieve el hecho de que los valores de abundancia obtenidos a partir de los conteos son infravaloraciones y sugiere que el íbice de los Alpes es más abundante de lo que se suele registrar. La aparición y la persistencia de los brotes de enfermedades (p. ej. la sarna sarcóptica, la queratoconjuntivitis o la brucelosis) están relacionadas con la densidad y la abundancia del íbice a escala local. La correlación observada entre el crecimiento de las poblaciones de íbice y los brotes de enfermedades sugiere que el riesgo de padecer epizootias podría estar creciendo o ser ya elevado en varias poblaciones de *Capra ibex*.

Palabras clave: *Capra ibex*, Brote de enfermedades, Densidad del hospedador, Seguimiento, Transmisión de parásitos, Dinámica de poblaciones

Cédric De Danieli, Fédération Départementale des Chasseurs de la Haute–Savoie, Impasse des Glaises 74350 Villy–le–Pelloux.– Mathieu Sarasa, Fédération Nationale des Chasseurs, 13 rue du Général Leclerc 92136, Issy les Moulineaux, France.

Corresponding author: M. Sarasa. E–mail: msarasa@chasseurdefrance.com, mathieusar@hotmail.com

Introduction

The genus *Capra* includes flagship species living in rupicolous and mountain environments that were the subject of conservation, reintroduction and management programs during the past century (Stüwe & Nievergelt, 1991; Pérez et al., 2002). *Capra ibex,* known as the Alpine ibex due to its distribution (Sarasa et al., 2012), is a good example of this phenomenon. This species is present in the wild in at least six countries (Italy, Switzerland, France, Austria, Germany and Slovenia) and national reports from Italy and France reveal that its abundance and distribution is increasing (Apollonio et al., 2009; Corti, 2012). It was threatened by extinction at the beginning of the twentieth century but today is found in numerous colonies that are occasionally exposed to risk of disease outbreaks (Couturier, 1962; Gauthier et al., 1991; Stüwe & Nievergelt, 1991; Apollonio et al., 2009).

The improvement of wildlife monitoring and the identification of factors associated with disease outbreaks in animal populations are major goals in wildlife management and conservation (Lloyd–Smith et al., 2005; Putman et al., 2011). Nevertheless, the understanding of potential associations between the demography of host species and the causes of disease outbreaks (*e.g.* introduction, spread and persistence) is hampered by limited availability of data (Lloyd–Smith et al., 2005). The investigation of the potential correlation between host demography and epidemiology is of crucial interest for wildlife biologists aiming to conserve wild animal populations. Such research can aid in the identification of key factors regarding the compatibility —defined as a population's predisposition as a suitable environment for potential outbreaks (Combes, 2001)— to disease outbreaks on a population scale.

Several reviews of the recovery process and abundance of this ibex have been published (Couturier, 1962; Shackleton, 1997), and a number of national–wide reports on ibex populations have recently appeared highlighting increasing trends in Italy and France (Apollonio et al., 2009; Corti, 2012). Nevertheless, a novel synthesis of the management challenges facing the Alpine ibex populations on an international scale could help improve knowledge of the current status of the species and lead us to reassess the potential links between ibex demography and the risk of outbreaks of diseases such as sarcoptic mange, keratoconjunctivitis, and brucellosis.

Our first objective was to review the most recent demographic data to test the hypothesis that both the overall abundance and distribution of Alpine ibex are increasing on an international scale. In light of information contained in national reports (Apollonio et al., 2009; Corti, 2012), we also expected to observe an improvement in populations on a European scale.

The second objective was to assess the accuracy of abundance estimates and methodological limits. All evaluations of species abundance are conditioned by the inherent difficulties involved in monitoring wildlife populations. Although other methods such as capture–mark–recapture (CMR)

have been tested, counts (or censuses) performed in different seasons (depending on the population in question) are the most commonly used method for population estimates of the Alpine ibex (Toïgo et al., 2007; Apollonio et al., 2009; Guerra, 2010; Corti, 2012). Direct and indirect counts of ungulates have been reported in different environments (*e.g.* African forests and the boreo–nemoral zone on the west coast of Norway) but seem to be poor for predicting population changes below 10–50% (Plumptre, 2000; Mysterud et al., 2007). Nevertheless, as counts are still frequently used in the long–term monitoring of Alpine ibex (regardless of the population size and season) and, before any detailed analysis of the available data was carried out, we explored potential methodological biases. In line with previous estimates for this species (Gaillard et al., 2003; Largo et al., 2008; Giordano et al., 2012), we expected to find underestimated values in (1) census–based estimates *vs.* CMR estimates and in (2) summer censuses *vs.* winter censuses.

The third objective was to explore the factors potentially associated with the risk of disease outbreak in the Alpine ibex. Previous studies of *Capra* species have highlighted the fact that disease may be a strong destabilizing factor in population dynamics or even a conservation threat for *Capra* populations around the world (Couturier, 1962; Vyrypaev, 1985; Pérez et al., 2002). Epidemiological models predict that host density and local population size will be key factors controlling the transmission dynamics of infectious diseases (Anderson & May, 1979; Lloyd–Smith et al., 2005). The demographic characteristics of populations may determine host group size (Patterson & Ruckstuhl, 2013), for instance, and may set the threshold for successful parasite invasion and/or persistence (Lloyd–Smith et al., 2005; Jansen et al., 2012). We tested for potential links between host demography (density and population size) and the characteristics of disease outbreak (appearance and persistence).

Material and methods

Demographic data

Demographic and distribution data were compiled from scientific publications and official reports from institutions involved in the monitoring and management of Alpine ibex (table 1). We searched for data from all the countries in which Alpine ibex exist in the wild (Italy, Switzerland, France, Austria, Germany and Slovenia); nevertheless, most published data came from Italy, France and Switzerland. Comparisons between the different available sources enhance reliability and completeness of the compiled dataset. The name of the population, year of the population estimate, the estimated number of ibex and the estimation method used (censuses, CMR, monitoring season) were entered into a specially constructed database. The distribution of each colony or population was recorded using ArcGIS (ver. 10.1).

Table 1. Sources consulted for constructing the demographic dataset for Alpine Ibex *Capra ibex*.

Tabla 1. Fuentes consultadas para recopilar el conjunto de datos demográficos relativo al íbice de los Alpes, Capra ibex.

Italy
Gauthier et al. (1991); Bassano & Peracino (1992); Terrier & Rossi (1994); Weber (1994); Shackleton (1997); Mustoni et al. (2000); Dupré et al. (2001); Carlini (2004); CE.RI.GE.FA.S (2004); Dematteis et al. (2004); Giovo (2004); Jacobson et al. (2004); Parco Nazionale dello Stelvio (2004); Rosselli & Giovo (2004); Carmignola et al. (2005); Dotta & Meneguz (2006); Federazione Italiana Parchi e Riserve Naturali (2006); Gasparo & Borziello (2006); Giovo (2006); Parco Naturale Adamello Brenta (2006); Favalli (2007); Giovo (2007); Genero (2008); Giovo (2008); Maurino et al. (2008); Parco Naturale Adamello Brenta (2008); Von Hardenberg & Bassano (2008); Apollonio et al. (2009); Borgo (2009); Carnevali et al. (2009); Comprensorio Alpino, CN2(2009), TO2(2009), TO4(2009); Favalli (2009); Perrone & Cordero di Montezelomo (2009); Scillitani et al. (2009); Ufficio Faunistico (2009); Apollonio et al. (2010); Assessorato Agricoltura e Risorse naturali (2010); Genero & Favalli (2010); Giordano (2010); Guerra (2010); Mustoni et al. (2010); Servizio Foreste e Fauna (2010); Ufficio Faunistico (2010); Assessorato Agricoltura e Risorse naturali (2011); Attanasio & Pedrotti (2011); Favalli & Genero (2011); Federazione Italiana Parchi e Riserve Naturali (2011); Ferloni (2011); Giordano (2011); Giovo (2011); Maurino (2011); Scillitani (2011); Favalli (2012); Giordano (2012); Giordano et al. (2012); Maurino (2012); Ufficio Faunistico (2012); Giordano (2013); Giovo (2013); Parco Naturale Paneveggio–Pale di San Martino (2013); Servizio Foreste e Fauna (2013); Ufficio Faunistico (2013); Giordano (2014); Giovo (2014)

Switzerland
Weber (1994); Mayer et al. (1996, 1997); Shackleton (1997); Delétraz (2002); Apollonio et al. (2009); Biebach & Keller (2009); Willisch & Neuhｊaus (2009); Imesch–Bebié et al. (2010); Office fédéral de l'environnement (2010); Marreros et al. (2011); Aeschbacher et al. (2012); Office fédéral de l'environnement (2012, 2013b)

France
Pairaudeau et al. (1977); Reydellet (1984); Esteve & Villaret (1989); Gauthier et al. (1990, 1991); Michallet (1991); Michelot (1991); Blin et al. (1994); Huboux (1993); Darinot & Martinot (1994); Heuret & Coton (1994); Profit (1995); Anselme–Martin (1996); Toïgo et al. (1996); Shackleton (1997); Profit (1999); Anthoine & Delomez (2000); Girard (2000); Gardet (2001); Delétraz (2002); Toïgo et al. (2007); Corti (2008); Delorme (2008); Maillard et al. (2010); Delorme & Garnier (2011); Parc National du Mercantour (2011); Corti (2012); Delorme (2012); ONCFS (2012a, 2012b, 2012c); Papet (2012); Parc National du Mercantour (2012); Tardy et al. (2012); Hars et al. (2013b); Papet et al. (2013); Mick et al. (2014)

Austria
Shackleton (1997); Reimoser & Reimoser (2010)

Germany
Shackleton (1997); Wotschikowsky (2010)

Slovenia
Shackleton (1997); Adamic & Jerina (2010)

Density estimates

Density estimates were rarely available in articles and reports. Moreover, density estimates are scale– and method–dependent due to factors such as the variability in the spatial distribution of ungulates and the variability of the probability of detection during surveys (Wingard et al., 2011; Suryawanshi et al., 2012). Thus, to generate a proxy for density based on available information, we divided the estimated number of ibex in a population by the distribution area of the popu-

lation. This conservative approach assumes that the official reports of population estimates and distributions reflect the characteristics of the population in question despite the potential biases and poor data precision they contain (Largo et al., 2008; Wingard et al., 2011).

Disease data

Data on the occurrence of disease outbreaks were gathered from scientific publications and official reports from institutions monitoring and managing Alpine ibex populations (table 1–2). Previous studies have reported that macro– and micro–parasites have uneven and context–dependent impacts on ibex individuals and populations and may give rise to endemic or epidemic (*e.g.* outbreaks) interactions (Couturier, 1962; Hars & Gauthier, 1994). Thus, we only included disease outbreaks in our database when the host–parasite interaction was characterized as such by the authors of a publication. Moreover, the spread or the incidence of parasites does not necessarily predict the potential impact of parasites on host demography. Only a few diseases are ever associated with the occasional destabilization of ibex populations or actually have the potential —by affecting ecological, health and socio–economic factors— to jeopardize their futures. The most important diseases are sarcoptic mange (caused by *Sarcoptes scabiei*), pneumonia (for instance caused by *Mycoplasma agalactiae*) and keratoconjunctivitis (caused by *Mycoplasma conjonctivae*), although contagious agalactia, foot–rot, brucellosis and paratuberculosis are also of concern from health and socio–economic points of view (Couturier, 1962; Hars & Gauthier, 1994; Mick et al., 2014). Thus, we included in our dataset disease outbreaks that were both identified as such in previous studies and were concomitant with population decreases. We recorded the start of the outbreaks (pathogen invasion) as binary data. The persistence of the outbreaks (outbreak persistence) was the number of years it took for the epidemic to become inactive (based on previous studies) or to be no longer associated with any demographic decline. This proxy for outbreak persistence takes into account the fact that the causal agents of diseases might exhibit non–lethal or asymptomatic interactions that could be potentially widespread but not necessarily associated with any demographic impact on ibex populations (Ryser–Degiorgis et al., 2009).

Missing data inference

Previous reviews of the overall distribution and abundance of mountain ungulate have sometimes inferred missing data from previously reported estimates (Shackleton, 1997; Pérez et al., 2002). We also used this approach in the analyses focused on the descriptive characterization of the overall abundance and distribution of Alpine ibex populations. This conservative approach may underestimate population abundances and distributions in species whose populations are increasing but it can also provide robust estimates of minimum population size that are methodologically comparable with previous estimates.

We used a different approach in the analyses focused on the potential associations between demographic estimates and the occurrence of disease outbreaks. Values for all the considered factors were not available for every year. Thus, to avoid the loss of key information (in particular, of information on disease outbreaks), we performed a few (< 5%) missing–data inferences for population size and area estimates. Missing data were inferred using the predictions of linear models based on neighbouring available data. This approach is conservative and takes into account

the reported dynamics of Alpine ibex populations over the past half–century, which were mainly characterized by population increases (Darinot & Martinot, 1994; Girard et al., 1998).

Connectivity between populations

The identification and delimitation of independent population units is a complex task in population ecology. As a first proxy for population units, we took the population units that for practical reasons are used by the institutions that monitor and manage ibex populations (management units) (Apollonio et al., 2009; Corti, 2012). We also looked for information on reported connections between management units in scientific publications and in official reports to build a second proxy for meta–population units that group together connected management units. We considered a connected management unit to consist of populations (1) between which individuals are recorded to move or (2) with tangent/overlapping distributions. As the required information for testing potential associations between demography and epidemiology was only available for Italy and France (see below), population groups were only identified for these countries (table 2, fig. 1). We used a conservative approach, and when potential —but unconfirmed— connections were mentioned in reports, we distinguished between populations units (*e.g.* G4–G5, G11–G13; fig. 1).

Statistical analyses

In order to focus our study on the dynamic recovery of Alpine ibex on an international scale in recent decades and to reduce potential bias from unreliable former estimations we only used data from 1975 onwards.

Count (or census) data is the commonest form of population data for Alpine ibex. Thus, the overall population abundance and distribution of Alpine ibex were first predicted using count data and generalized additive models (GAM) (Wood, 2006). For some areas, population estimates inferred from CMR procedures were also available and the overall population abundance was also estimated using this data to quantify its effect on the total abundance estimates. Studies and reports on a local scale are essential for ibex management, and review analyses should support the dissemination of such studies. Thus, to maintain the large–scale focus of our analyses, to avoid pseudoreplication, and to encourage readers to refer directly to primary sources, local estimates are not presented and the references consulted for constructing the demographic dataset are given in table 1.

Paired abundance data (count *vs.* CMR estimates; end of spring–early summer counts vs. end of autumn–early winter counts) were analyzed with Student's *t*–test for paired samples.

Long–term monitoring of epidemiology, demography and distribution was only available from Italy and France. Thus, the association between demography and disease outbreaks could only be analysed for these two countries. We analyzed the potential association between the start of disease outbreaks and the demo-

Table 2. Disease outbreaks associated with population decreases taken from the literature: Fr. France; It. Italy.

Tabla 2. Brotes de enfermedades asociados con casos de disminución de poblaciones extraídos de la bibliografía: Fr. Francia; It. Italia.

Population	Year	Disease	Reference
Bargy (Fr)	1990	Keratoconjunctivitis	Huboux et al. (1992)
	2012–2013	Brucellosis	Hars et al. (2013a) Mick et al. (2014)
Parco Nazionale Stelvio (It)	2008	Contagious ecthyma	Dervaux (2012)
Antelao–Marmarole (It)	2001	Sarcoptic mange	Carmignola et al. (2006)
Croda Rossa–Croda del Becco (It)	2003	Sarcoptic mange	Carmignola et al. (2006)
Monzoni–Marmolada (It)	2004–2006	Sarcoptic mange	Carmignola et al. (2006), Guerra (2010)
Pale di San Martino (It)	2007–2008	Sarcoptic mange	Guerra (2010)
Sella (It)	2004–2005	Sarcoptic mange	Carmignola et al. (2006), Guerra (2010)
Dolomiti Friulane (It)	2010–2012	Sarcoptic mange	Favalli (2012)
Monte Canin (It)	2010	Sarcoptic mange	Favalli & Genero (2011)
Vanoise (Fr)	1976	Pneumonia	Pairaudeau et al. (1977), Deméautis (1982)
	1983	Keratoconjunctivitis	Gauthier et al. (1991), Hars & Gauthier (1994)
	2007–2011	Pneumonia/ /Keratoconjunctivitis	Delorme (2008), Parc National de la Vanoise (2009, 2010, 2011)
Parco Nazionale Gran Paradiso (It)	1976	Pneumonia	Pairaudeau et al. (1977), Gauthier et al. (1991)
	1981–1983	Keratoconjunctivitis	Gauthier et al. (1991), Hars & Gauthier (1994)
	1996–1997	Brucellosis	Ferroglio et al. (1998, 2007)
	2006–2009	Pneumonia	Delorme (2008)
Valli di Lanzo (It)	2006–2007	Pneumonia	Dotta (2009)
Tournette (Fr)	1990	Keratoconjunctivitis	Huboux (1990), Huboux et al. (1992), Hars & Gauthier (1994)

graphic factors characterizing Alpine ibex populations (density, abundance, year) using generalized additive models (GAM) and a model selection procedure based on Akaike's information criterion (Burnham & Anderson, 2002; Wood, 2006). We repeated this procedure to analyze the potential association between the persistence of disease outbreaks and factors characterizing Alpine ibex populations. In our models, we included spatial and temporal factors (population, meta–population, country, season and year).

Results

Estimated abundance

The analysis of the paired counts and CMR data ($n = 26$) revealed that counts underestimated Alpine ibex population abundance when compared to CMR protocols (mean absolute difference ± 95% confidence interval = –95 ± 27 ibex; mean relative difference = –53 ± 9%; paired t–test: $t = -7.22$, $P < 0.001$, df = 25).

Table 3. Meta–populations or groups of populations (G) included in the analyses of potential associations between demography and epidemiology in the Alpine ibex *Capra ibex*.

Tabla 3. Metapoblaciones o grupos de poblaciones (G) incluidas en los análisis de las posibles asociaciones entre la demografía y la epidemiología en relación con el íbice de los Alpes, Capra ibex.

G	Country	Colony name	References
G1	France	Mercantour Oriental Mercantour Occidental Cime de Tavels Aiguilles de Pélens L'Estrop Les Sagnes	Gauthier et al. (1991), Parc National Le Mercantour & Parco Naturale Alpi Marittime (2006), Apollonio et al. (2009), Corti (2012)
	Italy	Argentera Ciastella Valle Stura	
G2	France	Haute–Ubaye Saint–Ours Queyras Oriental	Rosselli & Giovo (2004), Giovo (2006), Apollonio et al. (2009), Corti (2012), Krammer (2013)
	Italy	Germanasca–Massello–Troncea Monviso–Val Pellice Parco Naturale Orsiera–Rocciavrè Valle Stura–Valle Maira	
G3	France	Cerces–Galibier	Corti (2012)
G4	France	Rochail–Muzelle	Corti (2012)
G5	France	Vieux Chaillol–Sirac	Corti (2012)
G6	France	Gorges de la Bourne Haut Plateau du Vercors	Gonin (2009)
G7	France	Belledonne	Corti (2012)
G8	France	Chartreuse	Corti (2012)
G9	France	Vanoise Encombres Champagny–Peisey Archeboc Sassière–Prariond Carro–Souces de l'Arc Dent d'Ambin	Couturier (1962), Gauthier et al. (1991), Weber (1994), Delorme (2008), Apollonio et al. (2009), Girard et al. (2009), Corti (2012)
	Italy	Gran Paradiso M. Levi–C. Vallonetto Rhêmes Rocciamelone–Lera Tersiva Valli di Lanzo	
G10	France	Tournette	Corti (2012)
G11	France	Aravis	Corti (2012)
G12	France	Sous–Dine	Corti (2012)
G13	France	Bargy	Corti (2012)
G14	France	Cornettes de Bise Arve–Giffre	Apollonio et al. (2009); Marreros et al. (2011), Office fédéral de l'environnement (2013b)
G15	France	Mont Blanc–Beaufortin	Gauthier et al. (1991), Apollonio et al. (2009), Marreros et al. (2011)
	Italy	Macugnaga–Valle Anzasca Val Veny–Gran San Bernardo Valle Antrona Valpelline–Valtournenche–Monte Rosa	

Table 3. (Cont.)

G	Country	Colony name	References
G15	Italy	Valsesia Formazza Monte Giove Premia Veglia–Devero	
G16	Italy	Alpi Lepontine V. Bregaglia–Cranna–Acqua Fraggia	Dupré et al. (2001), Apollonio et al. (2009)
G17	Italy	A. Orobie–P. 3 Signori–M. Legnone A. Orobie–Fiumenero–V. Seriana	Dupré et al. (2001), Apollonio et al. (2009)
G18	Italy	Val Malenco–Sasso di Fora–Sasso Moro Val Masino–Val di Mello	Dupré et al. (2001), Apollonio et al. (2009)
G19	Italy	Sperella–Viola–Redasco Parco Nazionale Stelvio Sesvenna Ultimo–Orecchia di Lepre	Dupré et al. (2001), Apollonio et al. (2009)
G20	Italy	C. Baitone–V. del Miller Parco Adamello Tredenus–Frisozzo	Dupré et al. (2001), Apollonio et al. (2009)
G21	Italy	Alto Garda–Tombea–Caplone	Dupré et al. (2001), Apollonio et al. (2009)
G22	Italy	Palla Bianca–Weisskugel Tessa–Senales Tribulaun	Dupré et al. (2001), Apollonio et al. (2009)
G23	Italy	Cima Dura–Durreck Ponte di Ghiaccio–Eisbruggspitze Tauri–Tauern Val di Vizze–Pfitschertal	Dupré et al. (2001), Apollonio et al. (2009)
G24	Italy	Antelao–Marmarole Croda Rossa–Croda del Becco Monzoni–Marmolada Pale di S. Martino Sella	Dupré et al. (2001), Apollonio et al. (2009)
G25	Italy	Dolomiti Friulane	Apollonio et al. (2009)
G26	Italy	Monte Canin Monte Plauris Tarvisio	Apollonio et al. (2009)

Analysis of paired–data from summer and winter counts (n = 24) suggested that Alpine ibex abundance in summer counts is underestimated when compared to winter counts (mean absolute difference = –15 ± 14 ibex; mean relative difference = –22 ± 13%; paired t–test: t = –2.22, P < 0.05, df = 23).

Using only the count data from countries with Alpine ibex populations (Italy, Switzerland, France, Austria, Germany, Slovenia), our GAM model predicted (estimate ± se) an overall population of 49,037 ± 1,012 individuals in 2013. Predicted values using only count data were also estimated for the three main countries with Alpine ibex (Italy, Switzerland and France; table 4, fig. 2). Data compiled for the other countries include several count estimations, but no long–term series that would have allowed us to build accurate and comparable models at a national scale for Austria, Germany and Slovenia. Nevertheless, in table 4 we summarize the population estimates reported in previous studies. Using CMR data when available and count data if not available, our GAM model predicted an overall population of 50,195 ± 1,012 individuals in 2013 (see table 4 for national predictions).

Estimated distribution

Using the available data on spatial distribution from the main countries harbouring Alpine ibex populations (Italy, Switzerland and France), our GAM model predicted (estimate ± se) a distribution of 5,058 ± 109 km² in

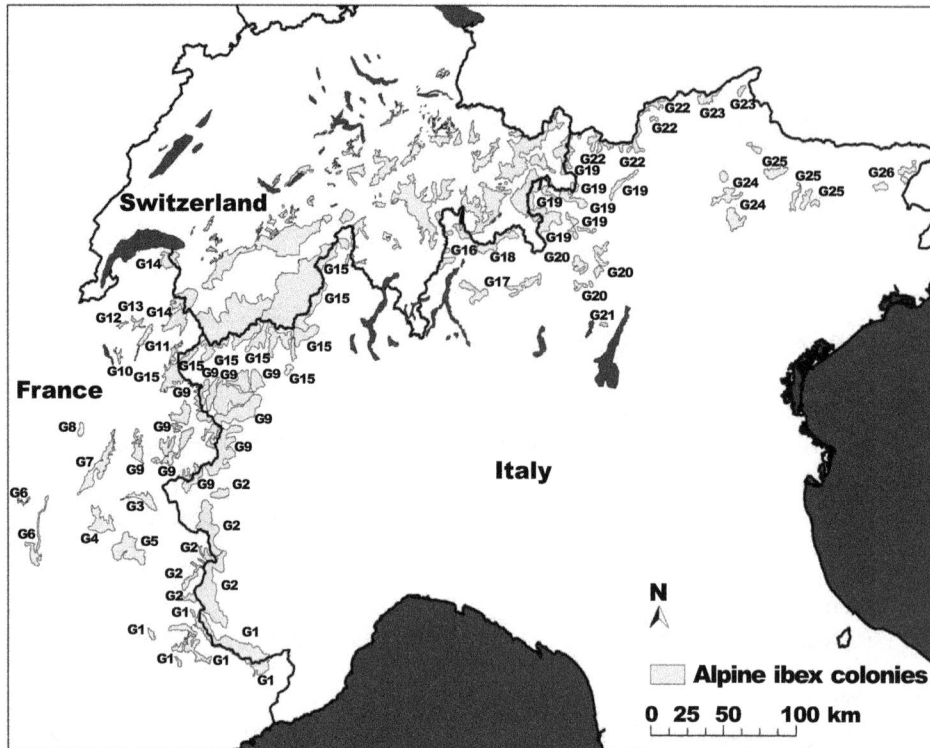

Fig. 1. Distribution of the Alpine ibex, *Capra ibex*. Lakes and the sea are depicted in dark grey, black lines are national boundaries. Redrawn from Apollonio et al. (2009), Corti (2012) and Office fédéral de l'environnement (2013a). See table 2 for the identity of the meta–populations.

Fig. 1. Distribución del íbice de los Alpes, Capra ibex. *El mar y los lagos se muestran en gris oscuro, las fronteras de los países son las líneas negras. Adaptado de Apollonio et al. (2009), Corti (2012) y Office fédéral de l'environnement (2013a). Véase la tabla 2 para consultar la identidad de las metapoblaciones.*

Italy and 2,568 ± 88 km² in France (table 4, fig. 3) for 2013. Data compiled for other countries (Switzerland, Austria, Germany and Slovenia) were too limited to be able to build reliable models or to predict estimated distributions for the whole range of the Alpine ibex.

Disease outbreaks

Several disease outbreaks associated with population decreases between 1975 and 2013 have been reported in the literature (table 2). The best model of the factors associated with the start of disease outbreak (AIC weight = 0.16) included only the factor 'density' (table 5, fig. 4A). Five other models were within 2 AIC units of the best model, all of which included the factor 'density' within other factors (year, abundance and country). The deviance explained by these six models ranges from 15.6% to 19.2% (table 5). The relative importance of variables underlined once more the fact that, of the tested variables, local density was the key factor associated with the appearance of a disease outbreak, although local abundance, year and country might also play role (table 5).

The best model for the persistence of disease out-break (AIC weight = 0.41) included four parameters: density, local abundance, year and meta–population. Two other models were within 2 AIC units of the best model and include (in addition to the previous four parameters) 'monitoring season' or 'country'. The deviance explained by these best models was about 70% (table 6, fig. 4B–4D).

Discussion

Abundance and distribution

The predicted values of the GAMs suggest that, in agreement with national reports, at both national and international scales, the Alpine ibex has increased in abundance (fig. 2) (Apollonio et al., 2009; Corti, 2012). Only the three countries with the largest ibex populations (Italy, Switzerland and France) have information from long–term monitoring schemes and up–to–date data from Austria, Germany and Slovenia would improve the overall estimates of ibex abundance in Europe. Analyses of paired data show that methodological designs have a major impact on the

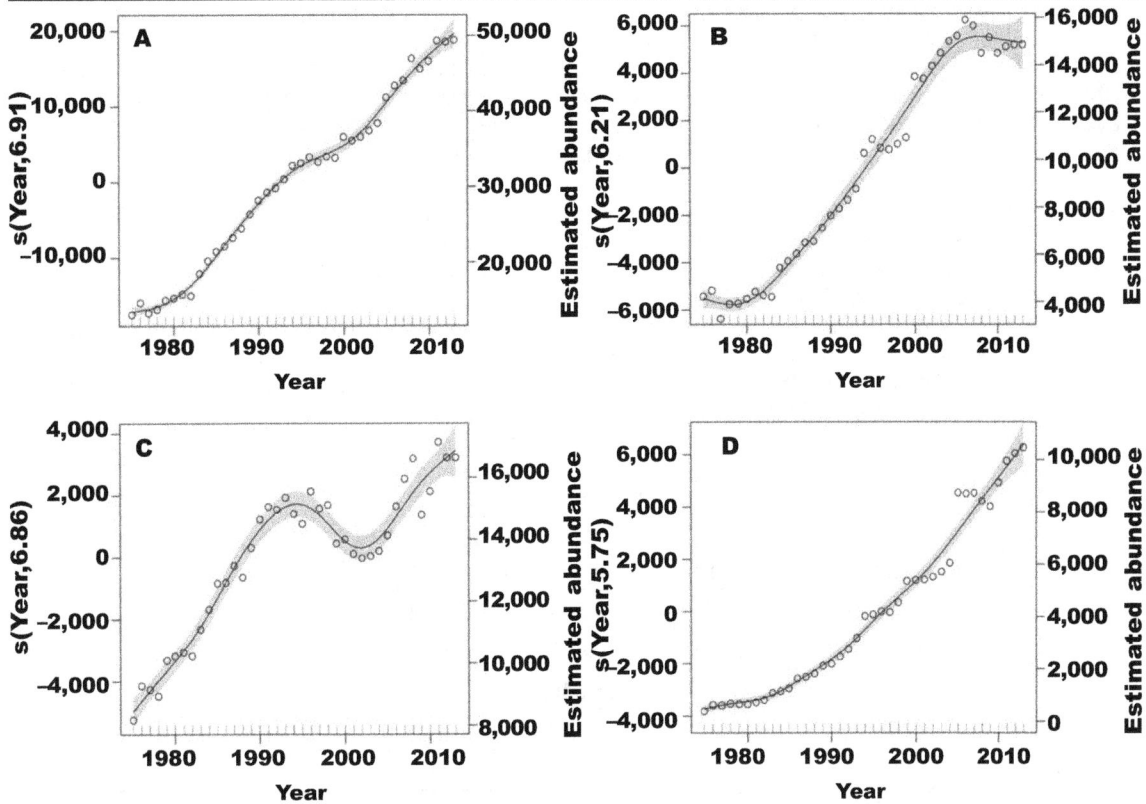

Fig. 2. Evolution of the abundance of the Alpine ibex, *Capra ibex*, between 1975 and 2013 in the whole of the Alps (A), Italy (B), Switzerland (C) and France (D). The solid lines represent the predicted patterns estimated by the generalized additive models, the grey shaded areas indicate the standard error, and the open circles are the observed values. The left–hand *y*–axis represents the centred values and specifies the smoothing factor 'Year', with the approximate degrees of freedom. The right–hand *y*–axis represents the estimated ibex abundance.

Fig. 2. Evolución de la abundancia del íbice de los Alpes, Capra ibex, *entre los años 1975 y 2013 en toda la cordillera de los Alpes (A), Italia (B), Suiza (C) y Francia (D). Las líneas continuas representan los patrones estimados por los modelos aditivos generalizados, las zonas sombreadas en gris indican el error estándar y los círculos son los valores observados. El eje de las Y de la izquierda representa los valores centrados y especifica el parámetro suavizado "Año", con los grados de libertad aproximados. El eje de las Y de la derecha representa la abundancia estimada de íbices.*

estimates of population abundance: abundances are underestimated in summer vs. winter counts and in census counts *vs.* CMR estimates. Although count data is the main source of population information in the literature, judging from our sample, on average they underestimate by half the Alpine ibex abundance when compared with results from CMR. Thus, most available information on ibex abundance (including the estimates in table 4) should be handled with care and is best thought of as an indicator of relative abundance rather than an accurate estimate of population size. The biases observed suggest that the Alpine ibex is probably more abundant that usually reported. Our total population estimate for Alpine ibex (about 50,200 individuals) is derived essentially from

counts, while the total estimate for the Iberian ibex *Capra pyrenaica* population (about 50,000 individuals according to Pérez et al. (2002)) are largely derived from line transects and distance sampling. Our results show that the mean bias of Alpine ibex counts leads to underestimations. However, the mean bias of Iberian ibex estimates —as for other mountain ungulates—is an issue that is still unresolved because the use of data truncation in distance sampling analysis can overestimate densities of mountain ungulates, and consequently, their abundance (Pérez et al., 2015). Thus, Alpine ibex may well be the most abundant ibex in Western Europe. Further studies will be required to refine monitoring methods and population estimates of mountain ungulate species.

Table 4. Abundance and distribution of Alpine ibex *Capra ibex* in 2013: Scd. Sum of count data only; P–GAM. Predicted by a GAM (generalized additive model) of count data; S–CMR. Sum of count plus CMR data when available; P–GAM–CMR. Predicted by a GAM of count data plus CMR data when available; Srp. Sum of reported population ranges; P–GAMrd. Predicted by a GAM of reported range data; * Reported from previous estimates; – Insufficient data available.

*Tabla 4. Abundancia y distribución del íbice de los Alpes, Capra ibex, en 2013: Scd. Suma de los datos obtenidos únicamente mediante conteo; P–GAM. Estimación mediante un GAM (modelo aditivo generalizado) de los datos de conteo; S–CMR. Suma de los datos de censos más los obtenidos con métodos de captura, marcaje y recaptura, si se dispone de ellos; P–GAM–CMR. Estimación mediante un GAM de los datos de censos más los obtenidos con métodos de captura, marcaje y recaptura, si se dispone de ellos; Srp. Suma de las áreas de distribución de la población registradas; P–GAMrd. Estimación mediante un GAM de los datos relativos a las áreas de distribución registradas; * Registrado en estimaciones anteriores; – Datos disponibles insuficientes.*

	Population size				Distribution range	
Country	Scd (ibex)	P–GAM (ibex ± se)	S–CMR (ibex)	P–GAM–CMR (individuals ± se)	Srp (km²)	P–GAMrd (km² ± se)
Italy	14,854	14,884 ± 588	14,854	14,879 ± 590	4,753	5,058 ± 109
France	9,302	9,819 ± 454	10,475	10,549 ± 453	2,509	2,568 ± 88
Switzerland	16,645	16,839 ± 436	16,645	–	–	–
Austria*	6,730*	–	6,730*	–	–	–
Germany*	400*	–	400*	–	–	–
Slovenia*	300*	–	300*	–	–	–
Total	48,231	49,037 ± 1,012	49,404	50,195 ± 1,012	–	–

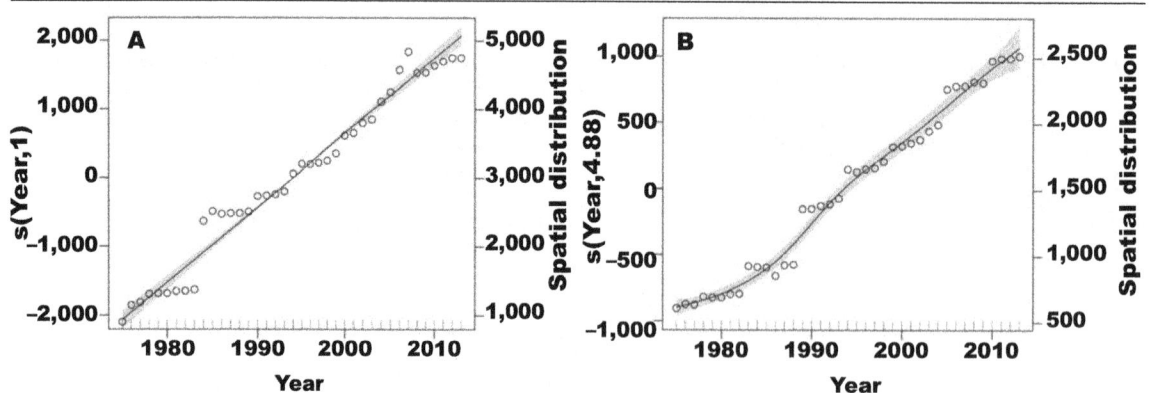

Fig. 3. Evolution of the spatial distribution of the Alpine ibex, *Capra ibex*, between 1975 and 2013 in Italy (A) and France (B). The solid lines represent the predicted patterns estimated by the generalized additive models, the grey shaded areas indicate the standard error, and the open circles are the observed values. The left–hand y–axis represents the centred values and specifies the smoothing factor 'Year', with the approximate degrees of freedom. The right–hand y–axis represents the estimated spatial distribution of ibex (km²).

Fig. 3. Evolución de la distribución espacial del íbice de los Alpes, Capra ibex, entre los años 1975 y 2013 en Italia (A) y Francia (B). Las líneas continuas representan los patrones estimados por los modelos aditivos generalizados, las zonas sombreadas en gris indican el error estándar y los círculos son los valores observados. El eje de las Y de la izquierda representa los valores centrados y especifica el parámetro suavizado "Año", con los grados de libertad aproximados. El eje de las Y de la derecha representa la distribución espacial estimada de íbices (km²).

Table 5. Model selection for factors associated with the appearance of disease outbreaks in Alpine ibex *Capra ibex* at population scale in Italy and France. D. Density; A. Abundance; Y. Year; G. meta–population; C. Country; S. Season of counts; P. population; *n*. Sample size; *K*. Number of estimated parameters; AIC. Akaike's information criterion; ΔAIC. Difference of AIC between the model and the most–parsimonious model; L(gi/x). Probability of the model being the best model given the data set; Wi. Akaike weight of the model; Dev–expl. Explained deviance of the fitted model; RI. Relative importance of factors. Only the ten best models are reported following Burnham & Anderson (2002) and Wood (2006).

Tabla 5. Selección de modelos para determinar los factores asociados con la aparición de brotes de enfermedades en el íbice de los Alpes, Capra ibex, a escala poblacional en Francia e Italia. D. Densidad; A. Abundancia; Y. Año; G. Metapoblación; C. País; S. Estación de los censos; P. Población; n. Tamaño muestral; K. Número de parámetros estimados; AIC. Criterio de información de Akaike; ΔAIC. Diferencia del AIC entre el modelo y el modelo de máxima parsimonia; L(gi/x). Probabilidad de que el modelo sea el mejor dado el conjunto de datos; Wi. Peso de Akaike del modelo; Dev–expl. Desviación explicada del modelo ajustado; RI. Importancia relativa de los factores. Solo se muestran los diez modelos mejores según Burnham & Anderson (2002) y Wood (2006).

Model	*n*	*K*	AIC	ΔAIC	L(gi/x)	Wi	Dev–expl	RI	
D	704	3	153.22	0.00	1.00	0.16	15.60	D	1.00
D+Y	704	5	153.67	0.44	0.80	0.13	18.20	Y	0.54
D+A	686	3	153.99	0.77	0.68	0.11	15.20	A	0.50
D+A+Y	686	5	154.14	0.92	0.63	0.10	17.70	C	0.34
D+Y+C	704	6	154.24	1.01	0.60	0.10	19.20	S	0.28
D+A+Y+C	686	6	155.01	1.78	0.41	0.07	19.10	G	0.00
D+A+C	686	5	155.33	2.10	0.35	0.06	15.90	P	0.00
D+A+Y+C+S	686	11	155.54	2.32	0.31	0.05	25.60		
D+S	704	5	155.60	2.38	0.30	0.05	17.00		
D+A+C+S	686	9	155.91	2.69	0.26	0.04	22.30		

In agreement with the predictions of ibex dynamics, information on the distribution of the Alpine ibex suggests that overall increases in ibex populations have occurred in Italy and France (fig. 3). The structure of the observed values (figs. 2–3) suggests that information on distribution has been updated less often than abundance estimates. Thus, further studies — particularly in ibex populations in Switzerland, Austria, Germany and Slovenia— would improve the present understanding of ibex distribution across Europe.

Abundance and distribution data, taken together, suggest that the IUCN 'Least Concern' conservation status is probably accurate at an international scale and at a national scale for Italy, Switzerland and France.

Population units

Analyses of epidemiological *vs.* demographic data show that the population units used for historical or practical reasons by institutions to monitor and manage ibex populations are excluded from the best models. However, the proxy for meta–populations that take into account the reported connections between populations is selected in the best models. Thus, the spatial structure and connectivity of ibex colonies must also be taken into account. These results also underline the relevance of trans–boundary monitoring and management, such as the programmes already underway in the Vanoise (France), Gran Paradiso (Italy) (Girard et al., 2009) and Mercantour (France) National Parks, and the Alpi Maritime Natural Park (Italy) (Parc National Le Mercantour & Parco Naturale Alpi Marittime, 2006).

Disease outbreaks

In agreement with theoretical models (Anderson & May, 1979; Lloyd–Smith et al., 2005), our results highlight the link between the local density of Alpine ibex and appearance and persistence of disease outbreaks (fig. 4), a finding that agrees with the results of recent studies of pneumonia epizootics in bighorn sheep (Sells et al., 2015). Moreover, ibex abundance and year were associated with at least the persistence of disease outbreaks (fig. 4C–4D). For the start of outbreaks, the deviance explained by the best models (15.6–19.2%) suggests that most of the variability in the appearance of disease outbreaks for

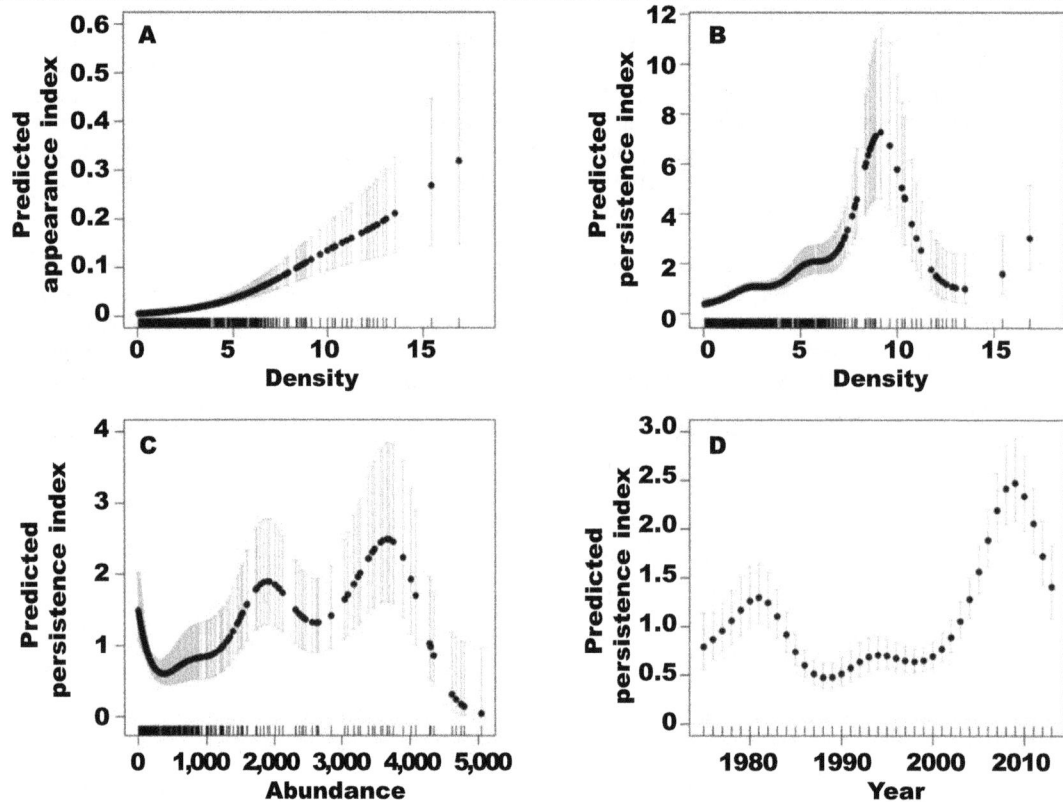

Fig. 4. Link between the appearance and the persistence of disease outbreaks in the Alpine ibex, *Capra ibex*, and density estimates (A, B), abundance (C) and year (D). The solid circles represent the predicted values estimated by the best generalized additive models for the available data, the grey bars indicate the standard error, and the thick lines represent the values with available observed data. The left–hand *y*–axis represents the relative probability of a disease outbreak (A) and the relative persistence of disease outbreaks (B, C, D).

Fig. 4. Relación entre la aparición y la persistencia de brotes de enfermedades en el íbice de los Alpes, Capra ibex, y las estimaciones de densidad (A, B), la abundancia (C) y el año (D). Los puntos representan los valores previstos estimados con los mejores modelos aditivos generalizados para los datos disponibles, las líneas grises indican el error estándar y las líneas gruesas representan los valores con los datos observados disponibles. El eje de las Y de la izquierda representa la probabilidad relativa de que se produzca un brote de enfermedad (A) y la persistencia relativa de los brotes de enfermedades (B, C, D).

the available data is not explained by local density and the other parameters included in our analyses. Thus, exposure to pathogens (encounter filter), defined as the probability of contact between the pathogen and the potential host population (Combes, 2001), rather than compatibility (compatibility filter), defined as the population predisposition as suitable environments for potential outbreaks, might be key to the appearance of disease outbreaks. In terms of outbreak persistence, the deviance explained by the best models was about 70%, which suggests that the considered factors of population compatibility to outbreaks explain most of the observed variability in the persistence of disease outbreaks. Thus, our results match the hypothesis of density–dependence for pathogen transmission and highlight the idea that disease outbreaks may

persist longer in high–density populations, at least in the areas with the commonest high–density values (7–10 ibex/km²). Nevertheless, our results also reveal a decrease in the persistence index at greatest density and abundance values (density greater than 10 ibex/km² and abundance over 4,000 individuals in the population), probably due to the low sample size for this range of values and the poorer accuracy of demographic and epidemiological data in the largest populations. The non–linear pattern linking the persistence of outbreaks and ibex abundance also suggests a localized decreasing pattern for small populations (0–350 ibex in the population). This pattern might be a localized artefact for this range of values due to the early intensive monitoring of reintroduced populations. However, it could also indicate an 'Allee

Table 6. Model selection for factors associated with persistence of disease outbreaks in Alpine ibex *Capra ibex* at population scale in Italy and France. D. Density; A. Abundance; Y. Year; G. meta–population; S. Season of counts; C. Country; P. population; *n*. Sample size; *K*. Number of estimated parameters; AIC. Akaike's information criterion; ΔAIC. Difference of AIC between the model and the most–parsimonious model; L(gi/x). Probability of the model being the best model given the data set; Wi. Akaike weight of the model; Dev–expl. Explained deviance of the fitted model; RI. Relative importance of factors. Only the ten best models are reported following Burnham & Anderson (2002) and Wood (2006).

Tabla 6. Selección de modelos para determinar los factores asociados con la persistencia de brotes de enfermedades en el íbice de los Alpes, Capra ibex, a escala de población en Francia e Italia. D. Densidad; A. Abundancia; Y. Año; G. Metapoblación; S. Estación de los conteos; C. País; P. Población; n. Tamaño muestral; K. Número de parámetros estimados; AIC. Criterio de información de Akaike; ΔAIC. Diferencia del AIC entre el modelo y el modelo de máxima parsimonia; L(gi/x). Probabilidad de que el modelo sea el mejor dado el conjunto de datos; Wi. Peso de Akaike del modelo; Dev–expl. Desviación explicada del modelo ajustado; RI. Importancia relativa de los factores. Solo se muestran los diez modelos mejores según Burnham & Anderson (2002) y Wood (2006).

Model	*n*	*K*	AIC	ΔAIC	L(gi/x)	Wi	Dev–expl	RI	
D+A+Y+G	686	49	306.97	0.00	1.00	0.41	70.12	D	1.00
D+A+Y+G+S	686	49	308.09	1.11	0.57	0.24	69.86	A	1.00
D+A+Y+G+C	686	49	308.13	1.16	0.56	0.23	70.20	Y	1.00
D+A+Y+G+C+S	686	50	309.45	2.47	0.29	0.12	69.99	G	1.00
A+Y+G+S	688	36	327.45	20.48	0.00	0.00	58.61	S	0.36
A+Y+G	688	34	328.17	21.19	0.00	0.00	57.82	C	0.35
D+A+Y+C+S	686	21	329.51	22.53	0.00	0.00	50.68	P	0.00
D+A+Y+S	686	21	336.17	29.19	0.00	0.00	49.04		
D+A+Y+C	686	22	337.96	30.99	0.00	0.00	48.96		
D+Y+G	704	40	338.01	31.04	0.00	0.00	58.49		

effect' linked to pathogens (Krkošek et al., 2013). The persistence index of outbreaks was greatest in very small and in large populations, that is, populations that are possibly at the edge and the core, respectively, of the ibex's range. Increased persistence in very small populations (edge of range) might be the result of repeated exposure or greater compatibility of colonies to pathogens, of contact or competition with livestock (Sells et al., 2015), or of high gregariousness in very small populations. Increased persistence in large populations (core of range) might be linked to areas with the high density of potential hosts that usually characterizes large populations. The observed links between outbreaks and density were non–linear and the patterns suggest relative threshold values (Lloyd–Smith et al., 2005) close to five and seven, respectively, for the appearance and persistence of disease outbreaks. As discussed in previous articles on wild species (Gortazar et al., 2006; Cross et al., 2010), environments favouring overabundant and aggregated populations (*e.g.* protected areas in the case of the Alpine ibex) sometimes also leave populations more prone to persistent disease outbreaks. Many threatened populations, species and rich ecosystems throughout the world are now protected. A new conservation challenge is emerging because protected populations on the increase may suffer from repeated disease outbreaks due to over–abundance facilitated by protection (Gortazar et al., 2006; Cross et al., 2010). Further studies will have to explore these conservation trade–offs (Leader–Williams et al., 2010) to ensure that management plans that are designed in one particular year but are then implemented year–after–year do not evolve into conservation threats.

Additionally, our results suggest that disease rose in the Alpine ibex in 2000–2010 despite its decrease in the 1980s. This decrease in the 1980s was probably the result of technical advances in diagnosis and increased investment in livestock prophylaxis (*e.g.* Fensterbank, 1986), coupled with research in eco–pathology and the management of disease risk at the wildlife–livestock interface during this period (*e.g.* Mayer et al., 1996). Nevertheless, the positive trends in Alpine ibex populations and the observed association between the demographic rise of ibex and disease outbreaks suggest that compatibility to epizooties may be increasing —or may already be high— in several populations of *Capra ibex*.

Proposal for further studies

The heterogeneity of Alpine ibex monitoring between populations prevented us from inferring abundance or distribution estimates for Switzerland, Austria, Germany or Slovenia. Further investment aimed at updating missing information would thus complete the data presently available for Alpine ibex.

Potential double counts in trans–boundary populations might also have affected abundance estimates. Nevertheless, the observed propensity to undercount probably compensates for double counts and further trans–boundary monitoring will probably minimize this potential bias.

Population structures are still too poorly documented, however, to be incorporated into our analyses (even though they may modulate demographic processes) (Yoccoz & Gaillard, 2006; Mignatti et al., 2012). Thus, further studies should explore the role of population structure as a potential modulating factor of population compatibility to disease outbreaks.

Outbreaks are relatively rare events and the available data do not allow for each disease to be analysed separately. In light of fresh outbreaks in the future, analysis should continue in the future, as should the search for potential variability in threshold values between pathogenic agents. Even so, our analysis permits us to explore factors that may determine associations with disease outbreaks at population scales.

In conclusion, our reappraisal of the available demographic and epidemiologic data on Alpine ibex highlights the methodological limitations, the increase in ibex populations, the increased risk of disease outbreaks and the links between host demography and disease outbreaks. A challenge for the future is how to integrate knowledge on density–dependent processes in wild species (e.g. disease outbreaks in the Alpine ibex) into the management of such species and ecosystems.

Acknowledgements

We would like to thank the Conseil Général de Haute Savoie, the Fédération Départementale des Chasseurs de Haute–Savoie (FDC–74) and the Fédération Nationale des Chasseurs (FNC) for supporting this project (n° FNC–PSN–PR8–2013). We are also particularly grateful to André Mugnier, Philippe Arpin, Eric Coudurier, Jean–Jacques Pasquier, and Yvon Crettenand for their help, advice and collaboration and to Michael Lockwood for the revision of the English.

References

Adamic, M. & Jerina, K., 2010. Ungulates and their management in Slovenia. In: *European Ungulates and Their Management in the 21st Century*: 507–526 (M. Apollonio, R. Andersen & R. Putman Eds.). Cambridge University Press, Cambridge.

Aeschbacher, S., Beaumont, M. A. & Futschik, A., 2012. A novel approach for choosing summary statistics in approximate Bayesian computation. *Genetics,* 192: 1027–1047.

Anderson, R. M. & May, R. M., 1979. Population biology of infectious diseases: Part I. *Nature,* 280: 361–367.

Anselme–Martin, S., 1996. *Compte–rendu du dénombrement bouquetin réalisé sur la population franco–suisse des Cornettes de Bise, Dent d'Oche, Grammont, le 27 juillet 1996*. Office National de la Chasse – Service Départemental de Garderie de Haute–Savoie.

Anthoine, F. & Delomez, L., 2000. *Comptage Bouquetin, septembre/octobre 1999, Réserves Naturelles et massif de Arve/Giffre, Compte–Rendu Technique*. APEGE – Réserves Naturelles de Haute–Savoie.

Apollonio, M., Ciuti, S., Pedrotti, L. & Banti, P., 2010. Ungulates and their management in Italy. In: *European Ungulates and Their Management in the 21st Century*: 475–506 (M. Apollonio, R. Andersen & R. Putman, Eds.). Cambridge University Press, Cambridge.

Apollonio, M., Giacometti, M., Lanfranchi, P., Lovari, S., Meneguz, P. G., Molinari, P., Pedrotti, L., Perco, F., Tosi, G., Toso, S. & Vigorita, V., 2009. *Piano di conservazione, diffusione e gestione dello stambecco sull'arco alpino italiano*. Settore Agricoltura e Risorse Ambientali, Provincia di Sondrio.

Assessorato Agricoltura e Risorse naturali, 2010. *Piano Regionale Faunistico–Venatorio 2008–2012*. Regione Autonoma Valle d'Aosta.

– 2011. *Piano Regionale Faunistico–Venatorio 2008–2012, Verifica intermedia*. Regione Autonoma Valle d'Aosta.

Attanasio, G. & Pedrotti, L., 2011. *Ulteriori informazzioni sullo stambecco nel Parco Nazionale dello Stelvio*. Corpo Forestale dello stato, Coordinamento Territoriale per l'Ambiente del Parco Nazionale dello Stelvio, Bornio (SO), Italy.

Bassano, B. & Peracino, A., 1992. Annual average increases of Alpine ibex (*Capra ibex ibex* L.) populations reintroduced from Gran Paradiso National Park. In: *Ongulés/Ungulates 91*: 579–581 (F. Spitz, G. Janeau, G. Gonzalez & S. Aulagnier, Eds.). S.F.E.P.M., I.R.G.M., Paris, Toulouse.

Biebach, I. & Keller, L. F., 2009. A strong genetic footprint of the re–introduction history of Alpine ibex (*Capra ibex ibex*). *Molecular ecology,* 18: 5046–5058.

Blin, L., Bonardi, J. & Michallet, J., 1994. La population du bouquetin des Alpes du massif de Belledonne – Sept Laux, dix ans déjà!. In: S*peciale IBEX, Atti dell'incontro del Gruppo Stambecco Europa 24-26 janvier 1993, Grenoble*: 63–72 (V. Peracino, B. Bassano & C. Mann, Eds.). Gruppo Stambecco Europa, Parco Nazionale Gran Paradiso, Torino.

Borgo, C., 2009. *Piano di Programmazione per la Gestione degli Ungulati*. Comprensorio Alpino TO3, Bassa Valle Susa e Val Sangone.

Burnham, K. P. & Anderson, D. R., 2002. *Model selection and multimodel inference: a practical information–theoric approach*. Springer, New York.

Carlini, E., 2004. *Lo stambecco delle Alpi (*Capra [ibex]

ibex *Linnaeus, 1758) nel Parco Naturale Adamello Brenta, status e indicazioni per il monitoraggio*. Parco Naturale Adamello Brenta, Strembo (TN).

Carmignola, G., Stefani, P. & Gerstgrasser, L., 2005. *Rapporto rogna sarcoptica*. Ufficio Caccia e Pesca, Ripartizione Foreste, Provincia autonoma di Bolzano.

– 2006. *Rapporto rogna sarcoptica*. Ufficio Caccia e Pesca, Ripartizione Foreste, Provincia autonoma di Bolzano.

Carnevali, L., Pedrotti, L., Riga, F. & Toso, S., 2009. Banca Dati Ungulati: Status, distribuzione, consistenza, gestione e prelievo venatorio delle popolazioni di Ungulati in Italia. Rapporto 2001–2005. *Biologia e Conservazione della Fauna*, 117: 1–168.

CE.RI.GE.FA.S, 2004. *Resoconto monitoraggio 2004 sulla cheratocongiuntivite dei bovini selvatici alpini in Valle Po e in Valle Varaita*. Centro Ricerche sulla Gestione della Fauna Selvatica – Fondazione Universitaria, Sampeyre (CN).

Combes, C., 2001. *Parasitism. The Ecology and Evolution of Intimate Interactions*. University of Chicago Press, Chicago.

Comprensorio Alpino CN2–2009. *Piano di programmazione per la gestione degli ungulati – CA CN2, 2009–2013*. Comprensorio Alpino CN2.

TO2–2009. *Piano Programmazione Gestione Ungulati 2009–2013 – Versione modificata*. Comprensorio Alpino – TO2 Alta Val Susa.

TO4–2009. *Piano di Programmazione per la Gestione degli Ungulati*. Comprensorio Alpino – TO4 Valli di Lanzo, Ceronda e Casternone.

Corti, R., 2008. Le bouquetin des Alpes. In: *Inventaire des populations françaises d'ongulés de montagne – Mise à jour 2006*: 40–45. Réseau Ongulés Sauvages ONCFS–FNC–FDC.

– 2012. Le bouquetin des Alpes. In: *Inventaire des populations françaises d'ongulés de montagne – Mise à jour 2011*: 44–50. Réseau Ongulés Sauvages ONCFS–FNC–FDC.

Couturier, M., 1962. *Le bouquetin des Alpes (Capra aegagrus ibex ibex L.)*. Arthaud, Grenoble.

Cross, P., Cole, E., Dobson, A., Edwards, W., Hamlin, K., Luikart, G., Middleton, A., Scurlock, B. & White, P., 2010. Probable causes of increasing brucellosis in free–ranging elk of the Greater Yellowstone Ecosystem. *Ecological Applications*, 20: 278–288.

Darinot, F. & Martinot, J.–P., 1994. Les populations de bouquetins des Alpes (*Capra ibex ibex* L.) dand le Parc national de la Vanoise: bilan de trente années de protection. *Travaux scientifiques du Parc national de la Vanoise*, 18: 177–203.

Delétraz, C., 2002. *Le piétin chez les ongulés sauvages: étude clinique et épidémiologique chez le bouquetin des Alpes*. D. V. M., Université Claude–Bernard Lyon 1, Lyon.

Delorme, M., 2008. *Recensement des populations de bouquetin des Alpes dans le Parc national de la Vanoise – Été 2008*. Parc National de la Vanoise.

– 2012. *Recensement des populations de bouquetin des Alpes dans le Parc national de la Vanoise – Été 2012*. Parc National de la Vanoise.

Delorme, M. & Garnier, A., 2011. *Recensement des populations de bouquetin des Alpes dans le Parc national de la Vanoise – Été 2011*. Parc National de la Vanoise.

Dematteis, A., Menzano, A., Tizzani, P., Craveri, P. & Meneguz, P. G., 2004. Updating of ibex distribution in Southern Italian Cotian Alps. *Second International Conference on Alpine ibex*. 2–3 December 2004, Cogne, Aosta, Italy.

Deméautis, G., 1982. Enzooties 1974–1976: Pleuropneumonies des chamois et bouquetins en Vanoise. *Acta Biologica Montana*, 1: 217–245.

Dervaux, J., 2012. *Cheptel domestique et grande faune sauvage de montagne: risques liés à la transmission d'agents pathogènes et proposition de mesures de prévention dans le Parc national des Écrins*. D. V. M. Ecole Nationale Vétérinaire de Lyon – Université Claude–Bernard – Lyon I, Lyon.

Dotta, R., 2009. *Piano di Programmazione per la Gestione degli Ungulati per il quinquennio 2009–2014*. Comprensorio Alpino Torino 4 Valli di Lanzo, Ceronda e Casternone.

Dotta, R. & Meneguz, P. G., 2006. The "swarming" of the ibex of Gran Paradiso towards the Lanzo valley (Alpi Graie, Torino). *Proceedings of the 3rd International Conference on Alpine ibex, 12–14 October 2006*: 12–14. Pontresina, Switzerland.

Dupré, E., Pedrotti, L. & Arduino, S., 2001. *Alpine Ibex Conservation Strategy – The Alpine ibex in the Italian Alps: status, potential distribution and management options for conservation and sustainable development*. WWF International, Italy.

Esteve, R. & Villaret, J., 1989. Le Bouquetin *Capra ibex* L. en Haute–Savoie: un premier bilan après 10 ans de réintroduction et de protection. *Le Bièvre*, 10: 23–38.

Favalli, M., 2007. *Censimento Camoscio–Stambecco 2007*. Parco Naturale Regionale delle Dolomiti Friulane, Cimolais (PN).

– 2009. *Censimenti stambecchi–camoscio e controllo sanitario sul camoscio nel Parco Naturale Dolomiti Friulane. Ricerca studio e monitoraggio degli ungulati della Riserva Naturale Forra del Cellina*. Reserva Naturale Forra del Cellina – Parco Naturale Dolomiti Friulane, Cimolais (PN).

– 2012. *Monitoraggio della colonia di stambecco (Capra ibex) nel Parco Dolomiti Friulane – Applicazione protocolli fanALP. Anno 2011–2012*. Parco Naturale Dolomiti Friulane, Cimolais (PN).

Favalli, M. & Genero, F., 2011. *Monitoraggio stambecco – Resultati dei censimenti estivi 2011*. Interreg IV, project fanAlp.

Federazione Italiana Parchi e Riserve Naturali, 2006. *Alpi Marittime, crescono stambecchi e camosci*. http://www.parks.it/federparchi/rassegna.stampa/dettaglio.php?id=2359. [Accessed on June 2014]. Europarc, Sezione Italiana.

– 2011. *Parco dello Stelvio o degli stambecchi?* http://www.parks.it/federparchi/rassegna.stampa/dettaglio.php?id=23178. [Accessed on June 2014]. Europarc, Sezione Italiana.

Fensterbank, R., 1986. La Brucellose des bovins et des petits ruminants: diagnostic, prohylaxie, vaccination. *Revue scientifique et technique – Office international des épizooties*, 5(3): 587–603.

Ferloni, M., 2011. *Piano faunistico venatorio.* Provincia di Sondrio, Sondrio.

Ferroglio, E., Gennero, M. S., Pasino, M., Bergagna, S., Dondo, A., Grattarola, C., Rondoletti, M. & Bassano, B., 2007. Cohabitation of a Brucella melitensis infected Alpine ibex (*Capra ibex*) with domestic small ruminants in an enclosure in Gran Paradiso National Park, in Western Italian Alps. *European Journal of Wildlife Research,* 53: 158–160.

Ferroglio, E., Tolari, F., Bollo, E. & Bassano, B., 1998. Isolation of *Brucella melitensis* from alpine ibex. *Journal of wildlife Diseases,* 34: 400–402.

Gaillard, J.–M., Loison, A. & Toïgo, C., 2003. Variation in life history traits and realistic population models for wildlife management: the case of ungulates. In: *Animal Behavior and Wildlife Conservation:* 115–132 (M. Festa–Bianchet & M. Apollonio, Eds.). Island Press, Washington.

Gardet, P., 2001. *Le bouquetin des Alpes, Bilan des connaissances dans les réserves naturelles de Haute–Savoie 1986–2000.* ASTERS–Réserves Naturelles de Haute–Savoie, Pringy.

Gasparo, D. & Borziello, G., 2006. *Sentieri ritrovati: il ritorno dei grandi animali sulle alpi orientali, Atti del Convegno.* Club Alpino Italiano, Milano.

Gauthier, D., Bouvier, M., Choisy, J.P., Estève, R., Martinot, J.–P., Michallet, J., Terrier, G. & Villaret, J.–C., 1990. Bilan sur le statut du Bouquetin dans les Alpes françaises en 1986. In: *Actes de la conférence internationale. Le bouquetin des Alpes: réalité actuelle et perspectives. 17–19 septembre 1987, Valdieri, Cuneo, Italia:* 25–37 (T. Balbo, D. De Meneghi, P. Meneguz & L. Rossi, Eds.). Dipartimento di Patologia Animale dell'Università di Torino. Cattedra di Malattie Parassitarie, Torino, Italia.

Gauthier, D., Martinot, J.–P., Choisy, J.–P., Michallet, J., Villaret, J.–C. & Faure, E., 1991. Le bouquetin des Alpes. *Revue d'écologie,* Suppl. 6: 233–275.

Genero, F., 2008. *Piano pluriennale di gestione della Fauna, Triennio 2008–2010.* Parco Naturale delle Prealpo Giulie, Regione Autonoma Friuli–Venezia Giulia.

Genero, F. & Favalli, M., 2010. Camoscio, stambecco e cervo nel Parco, un'interessante evoluzione. *LA VOCE del Parco Prealpi Giulie,* Anno X n°1 nueva serie: 8–9.

Giordano, O., 2010. *1° censimento invernale stambecco Capra ibex, Valle Varaita.* Comprensorio Alpino CN2 "Valle Varaita", Melle, Italy.

– 2011. *2° censimento invernale stambecco Capra ibex, Valle Varaita.* Comprensorio Alpino CN2 "Valle Varaita", Melle, Italy.

– 2012. *3° cencimento invernale stambecco Capra ibex, Valle Varaita.* Comprensorio Alpino CN2 "Valle Varaita", Melle, Italy.

– 2013. *4° censimento invernale stambecco Capra ibex, Valle Varaita.* Comprensorio Alpino CN2 "Valle Varaita", Melle, Italy.

– 2014. *5° censimento invernale stambecco Capra ibex Valle Varaita.* Comprensorio Alpino CN2 "Valle Varaita", Melle, Italy.

Giordano, O., Ficetto, G. & Giovo, M., 2012. *Summer and winter ibex censuses in the Varaita Valley (Cuneo, Italy): comparison of two monitoring techniques.* XXIIth Incontro del gruppo Stambecco Europa (GSE–AIESG), 26–28 October 2012, Zernez, Switzerland.

Giovo, M., 2004. *La cheratocongiuntivite nel camoscio e nello stambecco nelle valli Pellice, Chisone e Germanasca.* Comprensorio Alpino TO1 – Valli Pellive, Chisone e Germanasca, Bricherasio.

– 2006. *2° censimento invernale stambecco (Capra ibex) – Valli Pellice e Germanasca.* Comprensorio Alpino TO1 Valle Pellice, Chisone e Germanasca, Bricherasio, Italy.

– 2007. *3° censimiento invernale stambecco (Capra ibex) – Valli Pellice e Germanasca.* Comprensorio Alpino TO1 Valli Pellice, Chisone e Germanasca, Bricherasio, Italy.

– 2008. *4° censimiento invernale stambecco (Capra ibex) – Valli Pellice e Germanasca.* Comprensorio Alpino TO1 Valli Pellice, Chisone e Germanasca, Bricherasio, Italy.

– 2011. *5° – 6° – 7° censimenti invernali stambecco (Capra ibex) – Valli Pellice e Germanasca.* Comprensorio Alpino TO1, Valli Pellice, Chisone e Germanasca, Bricherasio, Italy.

– 2013. *9° censimento invernale stambecco (Capra ibex) – Valli Pellice e Germanasca.* Comprensorio Alpino TO1, Valli Pellice, Chisone e Germanasca, Bricherasio, Italy.

– 2014. *10° censimento invernale stambecco (Capra ibex) – Valli Pellice e Germanasca.* Comprensorio Alpino TO1 – Valli Pellice, Chisone e Germanasca, Bricherasio, Italy.

Girard, I., 2000. *Dynamique des populations et expansion geographique du bouquetin des Alpes (Capra ibex ibex, L.) dans le Parc National de la Vanoise.* Ph. D. Thesis, Université de Savoie.

Girard, I., Adrados, C., Bassano, B. & Janeau, G., 2009. Application de la technologie GPS au suivi du déplacement de bouquetins des Alpes (*Capra ibex ibex*, L.) dans les Pacs Nationaux de la Vanoise et du Gran Paradiso (Italie). *Travaux scientifiques du Parc national de la Vanoise,* XXIV: 105–126.

Girard, I., Gauthier, D. & Martinot, J. P., 1998. Evolution démographique des populations de bouquetin des Alpes (*Capra ibex ibex*) présentes dans le Parc National de la Vanoise ou réintroduites à partir de celui-ci. In: *Proceedings of the XXIIIrd Congress of the International Union of Game Biologists (IUGB), 1–6 Sept 1997, Lyon, France:* 417–431 (P. Havet, E. Taran & J. C. Berthos, Eds.), Gibier Faune Sauvage, 15 (Hors série Tome 2).

Gonin, O. B., 2009. Suivi des bouquetins du Royans. *LPO Info Drôme,* 2: 21.

Gortazar, C., Acevedo, P., Ruiz–Fons, F. & Vicente, J., 2006. Disease risks and overabundance of game species. *European Journal of Wildlife Research,* 52: 81–87.

Guerra, L., 2010. *Analisi dell'andamento post–epidemico di una colonia di stambecchi (Capra ibex) affetta da rogna sarcoptica.* D. V. M., Università degli studi di Padova.

Hars, J., Anselme–Martin, S., Da Silva, E., Game, Y., Gauthier, D., Gibert, P., Guyonnaud, B., Le Horgne, J.–M., Losinger, I., Maucci, E., Pasquier, J.–J., Rautureau, S., Rossi, S., Toïgo, C. & Garin–Bastuji, B., 2013a. Un foyer de brucellose a Brucella melitensis chez le bouquetin et le chamois du massif du Bargy (Haute–Savoie/France). *31èmes Rencontres du GEEFSM*, 14–16 June 2013. Canillo, Principality of Andorra.

Hars, J. & Gauthier, D., 1994. Pathologie du bouquetin des Alpes: bilan sanitaire des populations françaises. *Travaux scientifiques du Parc national de la Vanoise*, 18: 53–98.

Hars, J., Rautureau, S., Jaÿ, M., Game, Y., Gauthier, D., Herbaux, J.–P., Le Horgne, J.–M., Maucci, E., Pasquier, J.–J., Vaniscotte, A., Mick, V. & Garin–Bastuji, B., 2013b. Un foyer de brucellose chez les ongulés sauvages du massif du Bargy. *Bulletin épidémiologique, santé animale et alimentation,* 60: 2–6.

Heuret, J. & Coton, C., 1994. *Compte–rendu du comptage bouquetins Mont–Blanc du 25 octobre 1994.* APEGE, Sixt-Fer-à-Cheval.

Huboux, R., 1990. *Suivi de l'épizootie de kérato–conjonctivite infectieuse du bouquetin dans le Département de la Haute–Savoie.* FDC/ONC, CNERA FM.

– 1993. *Compte rendu succinct du comptage bouquetin du 17 juillet 1993. Secteur France: massif Cornettes de Bise, Dent Doche, Dent du Velan. Secteur Suisse du Valais: Mont Gardy Les Jumelles, Mont Granon.* Service Départemental de la Garderie Nationale de la Chasse et de la Faune Sauvage.

Huboux, R., Mogeon, G., Gay, J., Gervason, J., Gruffat, A., Lambrech, M., Revillard, C., Sallaz, J., Hanscotte, D., Oudar, J., Bellon, B. & Gauthier, D., 1992. Suivi de l'épizootie de kérato–conjonctivite infectieuse du bouquetin (*Capra ibex ibex*) dans le département de la Haute–Savoie en 1990 et 1991. *Bulletin Mensuel de l'Office National de la Chasse,* 168: 38–44.

Imesch–Bebié, N., Gander, H. & Shnidrig–Petrig, R., 2010. Ungulates and their management in Switzerland. In: *European Ungulates and Their Management in the 21st Century*: 357–391 (M. Apollonio, R. Andersen & R. Putman, Eds.). Cambridge University Press, Cambridge.

Jansen, P. A., Kristoffersen, A. B., Viljugrein, H., Jimenez, D., Aldrin, M. & Stien, A., 2012. Sea lice as a density–dependent constraint to salmonid farming. *Proceedings of the Royal Society B: Biological Sciences*, 279: 2330–2338.

Jacobson, A. R., Provenzale, A., Von Hardenberg, A., Bassano, B. & Festa-Bianchet, M., 2004. Climate forcing and density dependence in a mountain ungulate population. *Ecology,* 85: 1598–1610.

Krammer, M., 2013. Le bouquetin des Alpes (*Capra ibex ibex*) en Provence–Alpes–Côte d'Azur: passé, présent et avenir. *Faune–PACA Publication,* 30.

Krkošek, M., Ashander, J., Frazer, L. N. & Lewis, M. A., 2013. Allee Effect from Parasite Spill–Back. *The American Naturalist,* 182: 640–652.

Largo, E., Gaillard, J. M., Festa-Bianchet, M., Toïgo, C., Bassano, B., Cortot, H., Farny, G., Lequette, B., Gauthier, D. & Martinot, J. P., 2008. Can ground counts reliably monitor ibex *Capra ibex* popula-

tions? *Wildlife Biology,* 14: 489–499.

Leader–Williams, N., Adams, W. M. & Smith, R. J., 2010. *Trade–offs in Conservation: Deciding what to Save.* John Wiley & Sons, Oxford.

Lloyd–Smith, J. O., Cross, P. C., Briggs, C. J., Daugherty, M., Getz, W. M., Latto, J., Sanchez, M. S., Smith, A. B. & Swei, A., 2005. Should we expect population thresholds for wildlife disease? *Trends in Ecology & Evolution,* 20: 511–519.

Maillard, D., Gaillard, J., Hewison, A., Ballon, P., Duncan, P., Loison, A., Toïgo, C., Baubet, E., Bonenfant, C. & Garel, M., 2010. Ungulates and their management in France. In: *European ungulates and their management in the 21st century*: 441–474 (M. Apollonio, R. Andersen & R. Putman, Eds.). Cambridge University Press, Cambridge.

Marreros, N., Hüssy, D., Albini, S., Frey, C. F., Abril, C., Vogt, H.–R., Holzwarth, N., Wirz–Dittus, S., Friess, M., Engels, M., Borel, N., Willisch, C. S., Signer, C., Hoelzle, L. E. & Ryser–Degiorgis, M.–P., 2011. Epizootiologic investigations of selected abortive agents in free–ranging Alpine ibex (*Capra ibex ibex*) in Switzerland. *Journal of Wildlife Diseases,* 47: 530–543.

Maurino, L., 2011. *Gestione faunistica nel Parco Naturale Val Troncea 2010.* Parco Naturale Val Troncea, Bricherasio, Italy.

– 2012. *Gestione faunistica nel Parco Naturale Val Troncea 2011.* Parco Naturale Val Troncea, Bricherasio, Italy.

Maurino, L., Alberti, S., Boetto, E., Fornero, C., Peyrot, V., Rosselli, D. & Usseglio, B., 2008. Lo stambecco *Capra ibex* nel Parco Naturale Val Tronceo: metodologie di conteggio e risultati. *XXI Incontro del Gruppo Stambecco Europa, Alpine Ibex European Specialist Group*. 11–12 december, Ceresole Reale, Italy.

Mayer, D., Degiorgis, M.–P., Meier, W., Nicolet, J. & Giacometti, M., 1997. Lesions associated with infectious keratoconjunctivitis in alpine ibex. *Journal of Wildlife Diseases,* 33: 413–419.

Mayer, D., Nicolet, J., Giacometti, M., Schmitt, M., Wahli, T. & Meier, W., 1996. Isolation of Mycoplasma conjunctivae from conjunctival swabs of Alpine Ibex (*Capra ibex ibex*) affected with infectious keratoconjunctivitis. *Journal of Veterinary Medicine Series B,* 43: 155–161.

Michallet, J., 1991. Inventaire des populations de bouquetin des Alpes en France. *Bulletin Mensuel de l'Office National de la Chasse,* 159: 20–27.

Michelot, J., 1991. Réintroductions et introductions de vertébrés sauvages dans la région Rhône–Alpes. *Le Bièvre,* 12: 71–99.

Mick, V., Le Carrou, G., Corde, Y., Game, Y., Jay, M. & Garin–Bastuji, B., 2014. *Brucella melitensis* in France: Persistence in Wildlife and Probable Spillover from Alpine Ibex to Domestic Animals. *PLoS ONE,* 9, e94168.

Mignatti, A., Casagrandi, R., Provenzale, A., Von Hardenberg, A. & Gatto, M., 2012. Sex–and age-structured models for Alpine ibex *Capra ibex ibex* population dynamics. *Wildlife Biology,* 18: 318–332.

Mustoni, A., Cali, T. & Tosi, G., 2000. *La reintroduzione*

dello stambecco in Val di Genova – Rapporto finale. Parco Naturale Adamello Brenta, Strembo (TN).

Mustoni, A., Chirichella, R., Chiozzini, S., Liccioli, S. & Zibordi, F., 2010. *Ruolo ecosistemico degli ungulati selvatici nel Parco Naturale Adamello Brenta.* Parco Naturale Adamello Brenta, Ufficio Faunistico, Strembo (TN).

Mysterud, A., Meisingset, E. L., Veiberg, V., Langvatn, R., Solberg, E. J., Loe, L. E. & Stenseth, N. C., 2007. Monitoring population size of red deer *Cervus elaphus*: an evaluation of two types of census data from Norway. *Wildlife Biology,* 13: 285–298.

Office fédéral de l'environnement, 2010. *Inventaire des colonies de bouquetins. Etat 2007.* OFEV, division Gestion des espèces. fichiers SIG, version du 14/01/2002 http://www.bafu.admin.ch/gis/02911/07403/index.html?lang=fr

– 2012. *Statistique fédérale de la chasse.* BAFU/OFEV/UFAM 2012, http://www.wild.uzh.ch/jagdst/

– 2013a. *Colonies de bouquetin. Etat 2007.* http://www.bafu.admin.ch/tiere/09262/09401/index.html?lang=fr, CH, 3003 Berne.

– 2013b. *Statistique fédérale de la chasse.* BAFU/OFEV/UFAM 2013, http://www.wild.uzh.ch/jagdst/

ONCFS, 2012a. *Répartition des ongulés de montagne, bouquetin 1994.* http://carmen.carmencarto.fr/38/ongules_montagne.map. Réseau Ongulés Sauvages – ONCFS/FNC/FDC.

– 2012b. *Répartition des ongulés de montagne, bouquetin 2005.* http://carmen.carmencarto.fr/38/ongules_montagne.map. Réseau Ongulés Sauvages – ONCFS/FNC/FDC.

– 2012c. *Répartition des ongulés de montagne, bouquetin 2010.* http://carmen.carmencarto.fr/38/ongules_montagne.map. Réseau Ongulés Sauvages – ONCFS/FNC/FDC.

Pairaudeau, D., Moulin, A., Prave, M., Gastellu, J., Hars, J. & Joubert, L., 1977. Sur deux enzooties ayant sévi dans le Parc National de la Vanoise: kérato–conjonctivite infectieuse du chamois et pleuropneumonie enzootique du chamois et du bouquetin. *Travaux scientifiques du Parc national de la Vanoise,* 8: 157–172.

Papet, R., 2012. *Bilan annuel 2011: 16 ans après sa réintroduction, la colonie de boquetins (Capra ibex) amorce une étape de diminution de croissance.* Parc National des Ecrins, Gap.

Papet, R., Bouche, M. & Farny, G., 2013. *22ème Colonie de bouquetin des Alpes «Vieux Chaillol/Sirac» – Bilan annuel 2012.* Parc National des Ecrins, Gap.

Parc National de la Vanoise, 2009. *Rapport d'activité 2008.* Parc National de la Vanoise, Chambery.

– 2010. *Rapport d'activité 2009.* Parc National de la Vanoise, Chambery.

– 2011. *Rapport d'activité 2010.* Parc National de la Vanoise, Chambery.

Parc National du Mercantour, 2011. *Bilan annuel 2010 du contrat d'objectifs – Rapport d'activité de l'établissement pour l'année 2010.* Ministère de l'Ecologie du Développement durable, des Transports et du Logement, République Française.

– 2012. *Secteur Roya–Bévéra – Brèves secteur 2012.*

Parc National Le Mercantour & Parco Naturale Alpi Marittime, 2006. *Répartition spatiale du bouquetin des Alpes.*

Parco Naturale Adamello Brenta, 2006. *Progetto di ricerca e conservazione dello stambecco delle Alpi (Capra [ibex] ibex Linnaeus, 1758) nel Parco Naturale Adamello Brenta 2005/2006.*

– 2008. *Stambecco – Monitoraggio radiotelemetrico (2005–2007).* http://www.pnab.it/natura–e–territorio/stambecco/monitoraggio.html. Strembo (TN), Italy.

Parco Naturale Paneveggio – Pale di San Martino, 2013. *Rilasciati sulle Pale di San Martino altri due stambecchi.* http://www.parcopan.org/it/news/rilasciati–sulle–pale–di–san–martino–altri–due–stambecchi–n273.html.

Parco Nazionale dello Stelvio, 2004. *Stambecco, grafico dei censimenti 1984–1994.* http://www.stelviopark.it/italiano/Fauna/Mammiferi/Stambecco.html. Consorzio del Parco Nazionale dello Stelvio.

Patterson, J. E. & Ruckstuhl, K. E., 2013. Parasite infection and host group size: a meta–analytical review. *Parasitology,* 140: 803–813.

Pérez, J. M., Granados, J. E., Soriguer, R. C., Fandos, P., Márquez, F. J. & Crampe, J. P., 2002. Distribution, status and conservation problems of the Spanish Ibex, *Capra pyrenaica* (Mammalia : Artiodactyla). *Mammal Review,* 32: 26–39.

Pérez, J. M., Sarasa, M., Moço, G., Granados, J. E., Crampe, J. P., Serrano, E., Maurino, L., Meneguz, P. G., Afonso, A. & Alpizar–Jara, R., 2015. The effect of data analysis strategies in density estimation of mountain ungulates using distance sampling. *Italian Journal of Zoology,* 82(2): 262–270. Doi: 10.1080/11250003.2014.974695

Perrone, A. & Cordero di Montezelomo, N., 2009. *Piano di Programmazione per la gestione degli ungulati 2009–2013.* Comprensorio alpino CN1 – Valle Po.

Plumptre, A. J., 2000. Monitoring mammal populations with line transect techniques in African forests. *Journal of Applied Ecology,* 37: 356–368.

Profit, C., 1995. *Compte–rendu du dénombrement des bouquetins (Capra–ibex ibex,L), Landron – Sous–Dine – Roche Parnale.* Office National des Forêts, Annecy.

– 1999. *Compte–rendu du dénombrement des bouquetins (Capra–ibex–ibex) réalisé sur les massifs de Landron – Sous–Dine – Roche Parnal.* Office National des Forêts, Annecy.

Putman, R., Apollonio, M. & Andersen, R., 2011. *Ungulate management in Europe: problems and practices.* Cambridge University Press, Cambridge.

Reimoser, F. & Reimoser, S., 2010. Ungulates and their management in Austria. In: *European Ungulates and Their Management in the 21st Century*: 338–356 (M. Apollonio, R. Andersen & R. Putman, Eds.). Cambridge University Press, Cambridge.

Reydellet, M., 1984. *Le bouquetin en France.* Fédération des Chasseurs des Hautes–Alpes.

Rosselli, D. & Giovo, M., 2004. Stato della colonia di stambecco (*Capra ibex*) della Val Troncea e della Val Germanasca (Torino, Italia). *2nd International Conference on Alpine ibex – XIX Meeting of the Alpine ibex European Specialist Group (GSE–*

AIESG). 2–3 December 2004, Parco Nazionale Gran Paradiso, Cogne, Aosta, Italy.

Ryser–Degiorgis, M.–P., Bischof, D. F., Marreros, N., Willisch, C., Signer, C., Filli, F., Brosi, G., Frey, J. & Vilei, E. M., 2009. Detection of *Mycoplasma conjunctivae* in the eyes of healthy, free–ranging Alpine ibex: Possible involvement of Alpine ibex as carriers for the main causing agent of infectious keratoconjunctivitis in wild Caprinae. *Veterinary Microbiology,* 134: 368–374.

Sarasa, M., Alasaad, S. & Pérez, J. M., 2012. Common names of species, the curious case of *Capra pyrenaica* and the concomitant steps towards the 'wild–to–domestic' transformation of a flagship species and its vernacular names. *Biodiversity and Conservation,* 21: 1–12.

Scillitani, L., 2011. *Ecology of Alpine ibex (*Capra ibex ibex, *Linnaeus 1758) in relation to management actions in the Marmolada Massif, Italy.* Ph. D. Thesis, Dipartimento di Scienze Animali, Università degli studi di Padova.

Scillitani, L., Sturaro, E. & Ramanzin, M., 2009. *Il progetto "Stambecco Marmolada", Relazione finale sul monitoraggio: maggio 2006 – maggio 2009.* Safari Club International – Italian Chapter, Amministrazione provinciale di Belluno, Corpo Forestale dello Stato, Regione Friuli – Venezia Giulia, Dipartimento di Produzioni Animali Epidemiologia Ecologia di Torino, Dipartimento di Scienze Animali di Padova.

Sells, S. N., Mitchell, M. S., Nowak, J. J., Lukacs, P. M., Anderson, N. J., Ramsey, J. M., Gude, J. A., Krausman, P. R., 2015. Modeling risk of pneumonia epizootics in Bighorn sheep. *Journal of Wildlife Management,* 79(2): 195–210. Doi: 10.1002/jwmg.824

Servizio Foreste e Fauna, 2010. *Rapporto sullo stato delle Foreste e della Fauna – 2010.* Provincia Autonoma di Trento.

– 2013. *Relazione sull'attività svolta dal Servizio Foreste e fauna nel 2012.* Provincia Autonoma di Trento.

Shackleton, D. M., 1997. *Wild sheep and goats and their relatives: status survey and conservation action plan for Caprinae.* IUCN, Gland, Switzerland and Cambridge, UK.

Stüwe, M. & Nievergelt, B., 1991. Recovery of alpine ibex from near extinction: the result of effective protection, captive breeding, and reintroductions. *Applied Animal Behaviour Science,* 29: 379–387.

Suryawanshi, K. R., Bhatnagar, Y. V. & Mishra, C., 2012. Standardizing the double–observer survey method for estimating mountain ungulate prey of the endangered snow leopard. *Oecologia,* 169: 581–590.

Tardy, F., Baranowski, E., Nouvel, L.–X., Mick, V., Manso–Silvàn, L., Thiaucourt, F., Thébault, P., Breton, M., Sirand–Pugnet, P. & Blanchard, A., 2012. Emergence of atypical *Mycoplasma agalactiae* strains harboring a new prophage and associated with an alpine wild ungulate mortality episode. *Applied and Environmental Microbiology,* 78: 4659–4668.

Terrier, G. & Rossi, P., 1994. Le bouquetin (*Capra ibex ibex*) dans les Alpes Maritimes Franco–Italiennes: occupation de l'espace, colonisation et régulation naturelle. *Travaux scientifiques du Parc national de la Vanoise,* 18: 271–287.

Toïgo, C., Blanc, D., Michallet, J. & Couilloud, F., 2007. La survie juvénile comme moteur des fluctuations des populations de grands herbivores: l'exemple du bouquetin des Alpes. *ONCFS Rapport scientifique,* 2007: 11–14.

Toïgo, C., Gaillard, J. & Michallet, J., 1996. La taille des groupes: un bioindicateur de l'effectif des populations de bouquetin des Alpes (*Capra ibex ibex*)? *Mammalia,* 60: 463–472.

Ufficio Faunistico, 2009. *Relazione attivita'2008 del Gruppo di Ricerca e Conservazione dell'Orso Bruno del Parco.* Parco Naturale Adamello Brenta.

– 2010. *Relazione attivita'2009 del Gruppo di Ricerca e Conservazione dell'Orso Bruno del Parco.* Parco Naturale Adamello Brenta.

– 2012. *Relazione attivita'2011 del Gruppo di Ricerca e Conservazione dell'Orso Bruno del Parco.* Parco Naturale Adamello Brenta.

– 2013. *Relazione attivita'2012 del Gruppo di Ricerca e Conservazione dell'Orso Bruno del Parco.* Parco Naturale Adamello Brenta.

Von Hardenberg, A. & Bassano, B., 2008. Modificazioni temporali nei parametri demografici della popolazione di stambecco *Capra ibex* nel Parco Nazionale Gran Paradiso. In: Hystrix, *(n. s.) Supp 2008, Atti del VI Congresso Italiano di Teriologia, 16-18 Aprile 2008, Cles (Trento)*: 20 (C. Prigioni, A. Meriggi & E. Merli, Eds.), Associazione Teriologica Italiana.

Vyrypaev, V., 1985. The influence of an epizootic of *Sarcoptes scabiei* infection on a population of the central Asiatic mountain ibex (*Capra sibirica*) in Tien–Shan. *Parazitologiya,* 19: 190–194.

Weber, E., 1994. *Sur les traces des bouquetins d'Europe.* Delachaux et Niestlé.

Willisch, C. S. & Neuhaus, P., 2009. Alternative mating tactics and their impact on survival in adult male Alpine ibex (*Capra ibex ibex*). *Journal of Mammalogy,* 90: 1421–1430.

Wingard, G. J., Harris, R. B., Amgalanbaatar, S. & Reading, R. P., 2011. Estimating abundance of mountain ungulates incorporating imperfect detection: argali *Ovis ammon* in the Gobi Desert, Mongolia. *Wildlife Biology,* 17: 93–101.

Wood, S. N., 2006. *Generalized additive models, an introduction with R.* Chapman & Hall/CRC, Boca Raton.

Wotschikowsky, U., 2010. Ungulates and their management in Germany. In: *European Ungulates and Their Management in the 21st Century:* 201–222 (M. Apollonio, R. Andersen & R. Putman, Eds.). Cambridge University Press, Cambridge.

Yoccoz, N. G. & Gaillard, J.–M., 2006. Age structure matters for Alpine ibex population dynamics: comment on Lima and Berryman (2006). *Climate Research,* 32: 139–141.

13

Assessing the response of exploited marine populations in a context of rapid climate change: the case of blackspot seabream from the Strait of Gibraltar

J. C. Báez, D. Macías, M. de Castro, M. Gómez–Gesteira, L. Gimeno & R. Real

Báez, J. C., Macías, D., De Castro, M., Gómez–Gesteira, M., Gimeno, L. & Real. R., 2014. Assessing the response of exploited marine populations in a context of rapid climate change: the case of blackspot seabream from the Strait of Gibraltar. *Animal Biodiversity and Conservation*, 37.1: 35–47.

Abstract

Assessing the response of exploited marine populations in a context of rapid climate change: the case of blackspot seabream from the Strait of Gibraltar.— There is a growing concern over the decline of fisheries and the possibility of the decline becoming worse due to climate change. Studies on small–scale fisheries could help to improve our understanding of the effect of climate on the ecology of exploited stocks. The Strait of Gibraltar is an important fishery ground for artisanal fleets. In this area, blackspot seabream (*Pagellus bogaraveo*) is the main species targeted by artisanal fisheries in view of its relevance in landed weight. The aims of this study were to explore the possible effects of two atmospheric oscillations, the North Atlantic Oscillation (NAO) and the Arctic Oscillation (AO), on the capture of blackspot seabream in the Strait of Gibraltar, to determine their association with oceanographic conditions, and to improve our knowledge about the possible effects of climate change on fisheries ecology so that fishery management can be improved. We used two types of data from different sources: (i) landings per unit of effort reported from a second working group between Morocco and Spain on *Pagellus bogaraveo* in the Gibraltar Strait area, for the period 1983–2011, and (ii) the recorded blackspot seabream landings obtained from the annual fisheries statistics published by the *Junta de Andalucía* (Andalusian Regional Government). Our results indicate that the long–term landing of blackspot seabream in the Strait of Gibraltar is closely associated with atmospheric oscillations. Thus, prolonged periods of positive trends in the NAO and AO could favour high fishery yields. In contrast, negative trends in NAO and AO could drastically reduce yield.

Key words: Arctic Oscillation, Blackspot seabream, Climate, Fisheries collapse, North Atlantic Oscillation, Oceanography.

Resumen

Evaluación de la respuesta de las poblaciones marinas explotadas en un contexto de cambio climático rápido: el caso del besugo de la pinta en el estrecho de Gibraltar.— Existe una creciente preocupación por la disminución de la pesca y la posibilidad de que esta disminución se acelere debido al cambio climático. Los estudios sobre la pesca a pequeña escala podrían ayudar a mejorar nuestra comprensión de los efectos del clima en la ecología de las poblaciones explotadas. El estrecho de Gibraltar es una importante zona de pesca para la flota artesanal. En esta zona, el besugo de la pinta (*Pagellus bogaraveo*) es la especie más importante para la pesca artesanal en vista de su volumen de descarga. Los objetivos de este estudio consisten en estudiar los posibles efectos de dos oscilaciones atmosféricas: la oscilación del Atlántico Norte (NAO) y la oscilación del Ártico (AO), en la captura del besugo de la pinta en el estrecho de Gibraltar con objeto de determinar su relación con las condiciones oceanográficas, y mejorar nuestro conocimiento sobre los posibles efectos del cambio climático en la ecología de la pesca, para poder mejorar la gestión de la actividad pesquera. Utilizamos dos tipos de datos de diferentes fuentes: (i) los desembarques por unidad de esfuerzo registrados por un segundo grupo de trabajo entre Marruecos y España sobre el besugo de la pinta en la zona del estrecho de Gibraltar, para el período 1983–2011, y (ii) los desembarques registrados de besugo de la pinta obtenidos de las estadísticas anuales de pesca publicadas por la Junta de Andalucía. Nuestros resultados indican que el desembarque a largo plazo del besugo de la pinta en el estrecho de Gibraltar está íntimamente relacionado con las oscilaciones atmosféricas. Por lo tanto, los períodos prolongados de tendencias positivas en la NAO y la AO podrían favorecer altos rendimientos pesqueros. En contraste, las tendencias negativas de la NAO y la AO reducen drásticamente el rendimiento pesquero.

Palabras clave: Oscilación del Ártico, Besugo de la pinta, Clima, Colapso pesquero, Oscilación del Atlántico Norte, Oceanografía.

José Carlos Báez, Inst. Español de Oceanografia (IEO), Centro Oceanográfico de Málaga, Puerto pesquero de Fuengirola s/n., 29640 Fuengirola, Málaga, España (Spain); investigador asociado de la Fac. de Ciencias de la Salud, Univ. Autónoma de Chile, Chile.– David Macías, Inst. Español de Oceanografia (IEO), Centro Oceanográfico de Málaga, Puerto pesquero de Fuengirola s/n., 29640 Fuengirola, Málaga, España (Spain).– Maite de Castro, Moncho Gómez–Gesteira & Luis Gimeno, Ephyslab, Fac. de Ciencias de Ourense, Univ. de Vigo, Ourense, España (Spain).– Raimundo Real, Depto. de Biología Animal, Fac. de Ciencias, Univ. de Málaga, 29071 Málaga, España (Spain).

*Corresponding author: J. C. Báez. E–mail: granbaez_29@hotmail.com

Introduction

Fisheries are an important source of food and income for many local communities, and their value as a source of animal protein was recently emphasized in a Food and Agriculture Organization report (FAO, 2010). Several studies (*e.g.* Thurstan et al., 2010) have suggested that over the last decade, 88% of monitored marine fish stocks in EU waters have been overfished, and some authors have predicted a global collapse of fisheries within the next few decades (Worm et al., 2006, 2009). The observed decline in fisheries is mainly due to overfishing at an industrial scale (Worm & Myers, 2004; Pitcher, 2005). However, this situation could be aggravated by the response of fish populations to climate change (*e.g.* see Brandt & Kronbak, 2010). Thus, with the aim of integrating fisheries within sustainable ecosystems, Pitcher (2005) proposed studying the effect of climate parameters and their temporal variability on global fisheries. Some fisheries have been shown to respond to multi–decadal oscillations, such as the oscillation of El Viejo (The Old Man), or La Vieja (The Old Woman), in the Pacific (Chavez et al., 2003), and decadal oscillations, such as the North Atlantic Oscillation (Báez et al., 2011; Báez & Real, 2011).

The North Atlantic Oscillation (NAO) is a dominant pattern of coupled ocean–climate variability in the North Atlantic and Mediterranean basin (Hurrell, 1995). Many authors have observed a relationship between the NAO and changes in fishery abundance (Graham & Harrod, 2009; Báez et al., 2011; Báez & Real, 2011) and recruitment (Fromentin, 2001; Borja & Santiago, 2002; Mejuto, 2003). The NAO reflects fluctuations in atmospheric pressure at sea–level between the Icelandic Low and the Azores High. The NAO is associated with many meteorological variations in the North Atlantic region, affecting wind speed and direction and differences in temperature and rainfall (Hurrell, 1995). Recent studies (*e.g.* Overland et al., 2010) have discussed the effect of large–scale climate variability on several marine ecosystems and suggest that marine ecosystems could respond to climate change. Straile & Stenseth (2007) have suggested that the NAO can be used to explain inter–annual variability in ecological series, citing the following reasons: (1) a strong relationship between the NAO and weather conditions during the winter season; (2) qualitative changes in environmental conditions in response to winter weather conditions, especially temperature; and (3) the great importance of these environmental conditions in the distribution and population dynamics of species in temperate and boreal regions.

Nevertheless, the dominant mode of variability in atmospheric circulation variability in the Northern Hemisphere is determined by the Arctic Oscillation (AO). The AO is characterized by a meridional dipole in atmospheric sea level pressure between the northern Polar Regions and mid–latitudes (Thompson & Wallace, 1998). The NAO and AO are closely correlated (Thompson et al., 2000). The AO has been attributed to stratosphere–troposphere coupling. According to Thompson et al. (2000), this includes the NAO, which may be considered a different view of the same phenomenon. Thus, the AO and the NAO both tend to be in a positive phase during winters when the stratospheric vortex is strong (Douville, 2009). Few studies have analyzed the possible effect of the AO on fisheries ecology, for example Gancedo (2005) and Yatsu et al. (2005).

The possible effects of global climate change on the ecology of exploited stocks are difficult to study due to the multitude of other factors affecting these stocks, such as overfishing, coastal development, and pollution. Regional studies focused on small–scale fisheries could help to understand the effect of global climate change on the ecology of exploited stocks due to the reduction of the number of other variables (*e.g.* Meynecke et al., 2012; Pranovi et al., 2013).

The Strait of Gibraltar connects the western Mediterranean Sea with the Atlantic Ocean, providing an important fisheries ground for artisanal fleet (Silva et al., 2002). Because of the high frequency of maritime traffic in the Strait of Gibraltar, the largest Spanish fishing boats do not operate in this area; thus, the fishery is carried out by small numbers of artisanal boats working near the coast (Báez et al., 2009). According to Báez et al. (2013b), the physical condition of bluefin tuna (*Thunnus thynnus*) caught in this area, is correlate with both NAO and AO.

Blackspot seabream (*Pagellus bogaraveo*) is the most important species targeting by the artisanal fisheries, according to their importance in landed weight (Silva et al. 2002). In this context, the fishery landings and distribution by class of boat are easy to control at small–scale fisheries.

The blackspot seabream is a typical small demersal fish distributed from Eastern Atlantic Ocean to Western Mediterranean Sea, extensively fished from the early 80's by the artisanal fleet home–base in Gibraltar Strait. Fleets of Algeciras and Tarifa fished the blackspot seabream exclusively using a vertical deep water longline called 'voracera' baited with small sardines (*Sardina pilchardus*), while artisanal fleet from Conil used a traditional bottom longline in the western part of the Strait of Gibraltar (for a detailed description of the fishery see Czerwinski et al., 2009; Gil–Herrera, 2010, 2012).

The aim of this study was assessing the responses of exploited marine populations in a context of rapid climate and oceanographic change using the landing of blackspot seabream in the Strait of Gibraltar as study case.

Material and methods

Fisheries data

The study area coincides with the fishing ground, an area within Spanish waters of the Strait of Gibraltar between the Rock of Gibraltar and Cape Trafalgar, and it included the landing harbours of Algeciras, Tarifa and Conil (fig. 1).

Data were collected from two different sources. First, in the period 1983–2011, we used landings per sale, reported in CopeMed II (2012) and Gil–Herrera

(2012) on *Pagellus bogaraveo* in the Gibraltar Strait area, as Landings Per Unit of Effort (LPUE), because each sale is equivalent to the trip per boat (which is typically the fisheries effort). The artisanal Moroccan fleet also fished blackspot seabream in the Strait of Gibraltar. However, we excluded these data because the data available from Moroccan fleet is a short–time series (CopeMed II, 2012).

Second, we used the recorded blackspot seabream landings obtained from the annual fisheries statistics published by the *Junta de Andalucía* (Andalusian Regional Government) (Galisteo et al., 2001a, 2001b, 2002, 2004, 2005; Alonso–Pozas et al., 2007; Galisteo et al., 2007, 2008, 2009a, 2009b, 2011, 2012, 2013) for the period 1985–2012 from Tarifa, the most important landing harbour in the study area (table 1).

Atmospheric data

Monthly values of the NAO index and AO index were taken from the website of the National Oceanic and Atmospheric Administration: http://www.cpc.noaa.gov/products/precip/CWlink/pna/nao_index.html and fttp://www.esrl.noaa.gov/psd/data/correlation/ao.data, respectively.

The atmospheric oscillations present strong inter–annual and intra–annual variability (Hurrell, 1995). However, several studies have shown that changes in NAO/AO trends have a delayed effect on aquatic ecosystems due to ecosystem inertia (Maynou, 2008; Báez et al., 2011). For this reason, we used NAO and NAO in the previous year (NAOpy); and AO and AO in the previous year (AOpy).

Oceanographic data

Ocean temperature and salinity data were obtained using the Simple Ocean Data Assimilation (SODA) package (http://www.atmos.umd.edu/~ocean). SODA uses an ocean model based on Geophysical Fluid Dynamics Laboratory MOM2 physics. Assimilated data include temperature and salinity profiles from the World Ocean Atlas–94 (Levitus & Boyer, 1994), as well as additional hydrography, sea surface temperatures (Reynolds & Smith, 1994), and altimeter sea levels obtained from the Geosat, ERS–1, and TOPEX/Poseidon satellites. Re–analyses of world ocean climate variability are available from 1958 to 2007 at a monthly scale, with a horizontal spatial resolution of 0.5° × 0.5° and a vertical resolution of 40 levels (Carton et al., 2000a, 2000b; Carton & Giese, 2008).

According to previous research, the Mediterranean water mass is produced by the transformation of fresh and warm surface Atlantic water (AW) that enters in the Mediterranean Sea by the Strait of Gibraltar. The surface AW is gradually modified during its displacement eastward in the Mediterranean Sea due to air–sea interactions and mixing processes. A portion of these dense water masses flows back (after seven to 70 years) through the Strait of Gibraltar, mixing with Eastern North Atlantic Central Water (ENACW) to form the Mediterranean Outflow Water (MOW; Bozec et al., 2011). In the eastern Gulf of Cadiz, the MOW is very dense and sinks under water with an Atlantic origin until it reaches an equilibrium level (around 1,100 m). In the western Gulf of Cadiz (8° W), MOW reaches density values similar to those of mid–depth Atlantic layers and splits in two cores separating from the bottom. The upper core is characterized by a maxima temperature (~13°C) and a potential density anomaly between 27.40 and 27.65 kg/m^3, and the lower core is characterized by a maxima salinity (~37.5) and a potential density anomaly between 27.70 and 27.85 kg/m^3. MOW spreads in the North Atlantic westward to the central Atlantic and northward along the coasts of Portugal and the Iberian peninsula. For this reason, the oceanographic analysis was carried out in a region large enough to contain the number of measurement points needed for a suitable oceanographic study of the MW taking into account ocean currents, coastal areas, and water properties. In the present study, the selected area ranged from 8° W to 13.75° W and from 35.25° to 40.25° N (the southern coast of the Iberian peninsula). We used the first 24 vertical levels (which correspond to a water depth of 1,378 m) since the study focuses on the detection of upper Mediterranean water (MW^u hereafter), whose core is located at 800 m. The thickness of the vertical layers increases from 10 m near the surface to 100 m below 300 m. The period under study ranges from 1980 to 2007.

We identified MW^u using temperature, salinity and density values, which should lie within the intervals 10.5–13.5°C, 35.8–36.8, and 27.4–27.65 kg/m^3, respectively. First, the grid points where MW^u was not detected in at least 50% of the samples were discarded from the analysis. Salinity and temperature data for each grid point were averaged to transform them into annual values. All salinity and temperature data corresponding to the intervals mentioned above for a specific year were averaged, regardless of layer, to obtain the mean MW^u salinity and temperature values for that year.

Long–term processes, such as warming–cooling or salinification–freshening, and their effect on the water column stratification were analyzed using annual trends, which were assumed to be linear. All trends were calculated using raw data, without using any filter or running mean. The Spearman rank correlation coefficient was used to analyze the significance of trends due to its robustness to deviations from linearity and its resistance to the influence of outliers (Saunders & Lea, 2008).

Data analysis

In a first step, we analysed the time series for each variable. We searched for common time trends and cyclicity in the time series using spectral analysis, to identify periodicity. Spectral analysis was performed with the software PAST (available from web site: http://folk.uio.no/ohammer/past/) (Hammer et al., 2001; Hammer & Harper, 2006).

We tested the relationship between LPUE of blackspot seabream versus NAOpy and AOpy using linear multiple regressions. We selected the best fit among several significant regressions when different degrees of freedom were involved in accordance with the high-

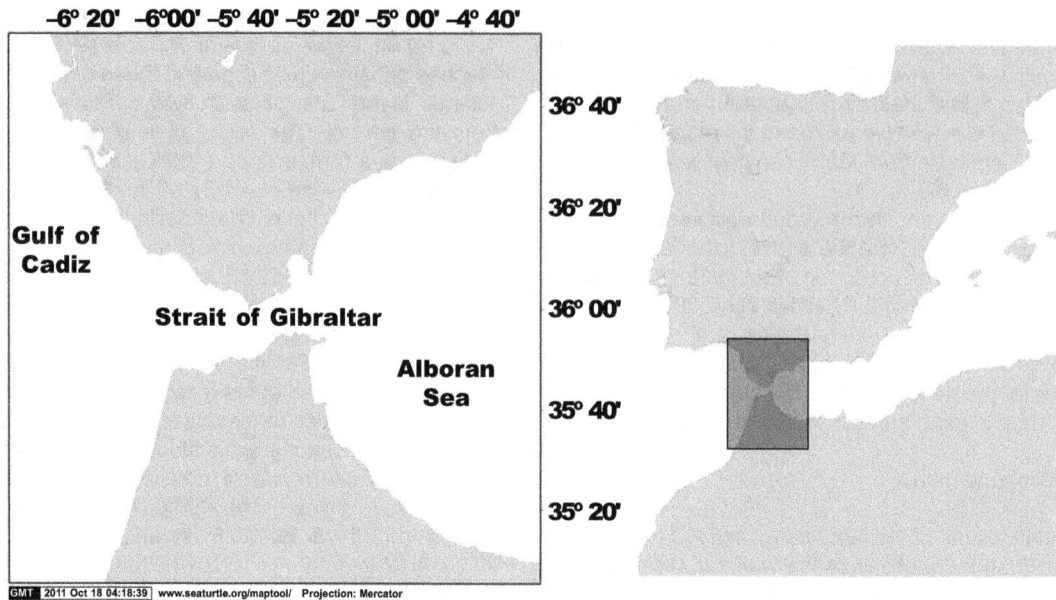

Fig. 1. The study area was centred on the Strait of Gibraltar. The Strait of Gibraltar separates two regions: the Gulf of Cadiz (in the Atlantic Ocean) and the Alboran Sea (within the Mediterranean Sea).

Fig. 1. La zona del estudio tiene en su centro el estrecho de Gibraltar. El estrecho de Gibraltar separa dos regiones: el golfo de Cádiz (en el océano Atlántico) y el mar de Alborán (en el mar Mediterráneo).

est R^2 value. Normality of the data was tested using the Kolmogorov–Smirnov test (Sokal & Rohlf, 1995).

A probabilistic analysis was performed by taking a year at random and calculating the probability that this particular annual landing from Tarifa was higher or lower than the average landing for all the years available pooled together. Báez et al. (2011) used binary logistic regression to model the response of albacore fisheries to changes in the accumulated NAO index. Similarly, using binary logistic regression, we modelled the probability of the value for blackspot seabream landings being higher than the average landings for this species for each specific year. Thus, we assigned a value of 1 or 0, respectively, when the landing in a specific year was higher or lower than the average landing for the 26 years taken together; these were considered to be good and poor landings, respectively. We performed a forward stepwise logistic regression where the independent variables were NAO, NAOpy, AO and AOpy. The goodness–of–fit of the model was assessed using an omnibus test (for model coefficients) and a Hosmer and Lemeshow test, which also follows a Chi–square distribution (Zuur et al., 2007), with the low p–values indicating a lack of fit of the model. We evaluated the discrimination capacity of our model using the area under the receiving operating characteristic (ROC) curve (AUC) (Lobo et al., 2008).

Despite a good fit of the logistic regression model, it is sensitive to the presence/absence ratio (Real et al., 2006). The presence/absence ratio was 0.625 for blackspot seabream. To resolve this difference,

we applied the favourability function (Ff) (Real et al., 2006) based on a logistic regression model, which adjusts the model regardless of the presence/absence ratio. Favourability was easily calculated from the probability obtained from the logistic regression according to the expression:

$$Ff = [P / (1 - P)] / [(n_1 / n_0) + (P / [1 - P])]$$

where P is the probability of the value for blackspot seabream landings per a specific year was higher than the average landings for this species for all years, and n_1 and n_0 are number of years with good or poor blackspot seabream landings, respectively.

The correlation between the different climatic indices and landings can be also analyzed in terms of the accumulated values. Annual values were transformed into anomalies by subtracting the mean value calculated over the whole period 1985–2010. The accumulated variables corresponding to a specific year were then calculated as the sum of the anomalies of the previous years (*e.g.* the accumulated values corresponding to 2000 were calculated as the sum of the anomalies for the period 1985–2000).

Results

The landing of blackspot seabream from Tarifa for 1985–2011 was the only variable with significant periodicity trend (table 2).

Table 1. Blackspot seabream (*Pagellus bogaraveo*) landing per year and corresponding average for the North Atlantic Oscillation (NAOpy) and Arctic Oscillation (AOpy) index in the year before the landing.

Tabla 1. Desembarque por año del besugo de la pinta (Pagellus bogaraveo) y promedio correspondiente para los índices de oscilación del Atlántico Norte (NAO) y de oscilación del Ártico (AO) en el año anterior al desembarque.

Year	Landing	NAOpy	AOpy
1985	209866	0.25	–0.19
1986	249000	–0.18	–0.52
1987	292732	0.5	0.08
1988	318578	–0.12	–0.54
1989	413375	–0.01	0.04
1990	426400	0.7	0.95
1991	421070	0.59	1.02
1992	629668	0.27	0.2
1993	764522	0.58	0.44
1994	854436	0.18	0.08
1995	501569	0.58	0.53
1996	659485	–0.08	–0.27
1997	527186	–0.21	–0.46
1998	282522	–0.16	–0.04
1999	198794	–0.48	–0.27
2000	193408	0.39	0.11
2001	154832	0.21	–0.05
2002	147793.6	–0.18	–0.16
2003	179146.5	0.04	0.07
2004	131692.6	0.1	0.15
2005	165616.8	0.24	–0.19
2006	161772.5	–0.27	–0.38
2007	273035	–0.21	0.14
2008	285481	0.17	0.27
2009	424849.4	–0.38	0.18
2010	227391	–0.24	–0.33

A significant association was found between the LPUE of blackspot seabream from the Gulf of Cadiz and the NAOpy index, according to the following function (fig. 2):

$$LPUE = 66.687 + 15.01 \, NAOpy$$
(adjusted $R^2 = 0.106$; $F = 4.306$; $P = 0.048$)

Table 2. Results of spectral analysis, we show the peaks in observed periodicity (in years), and signification for the time series variables: Pbg–LPUE. Blackspot seabream (*Pagellus bogaraveo*) landing per unit effort (LPUE) from harbours Algeciras, Tarifa and Conil for the period 1983–2011; Pbg. Blackspot seabream (*Pagellus bogaraveo*) landing from Tarifa for the period 1985–2011; NAOpy. Corresponding average for the North Atlantic Oscillation index in the year before at landing; AOpy. Corresponding average for the Arctic Oscillation index in the year before at the landing.

Tabla 2. Resultados del análisis espectral. Mostramos los máximos en la periodicidad observada (en años) y la significación de las variables de series temporales: Pbg–LPUE. Desembarque de besugo de la pinta (Pagellus bogaraveo) por unidad de esfuerzo (LPUE) en los puertos de Algeciras, Tarifa y Conil para el período 1983–2011; Pbg. Desembarque de besugo de la pinta (Pagellus bogaraveo) de Tarifa para el período 1985–2011; NAOpy. Promedio de la oscilación del Atlántico Norte en el año anterior al desembarque; AOpy. Promedio de la oscilación del Ártico en el año anterior al desembarque..

Variables	Periodicity	p (random)
Pbg–LPUE	13.997 years	0.683
Pbg	18.18 years	0.003645
NAOpy	2.59 years	0.7846
AOpy	14.28 years	0.7866

In addition, we obtained a significant model for the probability of obtaining good blackspot seabream landings, according to logit (y) function (fig. 3):

$$y = -0.645 + 3.344 * AOpy$$

The statistical tests for the goodness–of–fit of the model indicated a good fit. An omnibus test for model coefficients obtained $\chi^2 = 6.774$, $p = 0.009$, and the Hosmer AND Lemeshow test obtained $\chi^2 = 8.740$, $p = 0.272$. The AUC of the model was 0.756, which can be considered acceptable discrimination (Hosmer & Lemeshow, 2000). The Nagalkerke test obtained $R^2 = 0.312$.

The favourability function showed that the conditions that favour good blackspot seabream landings for a specific year coincided almost completely with the positive phase of the AOpy (fig. 3).

Accumulated values for the NAO and AO were highly correlated ($R^2 = 0.91$, $p < 0.01$) (fig. 4).

The temperature and salinity trends corresponding to MW[u] over the period 1982–2007 are shown in figures 5A and 5B. Black dots represent the grid points where

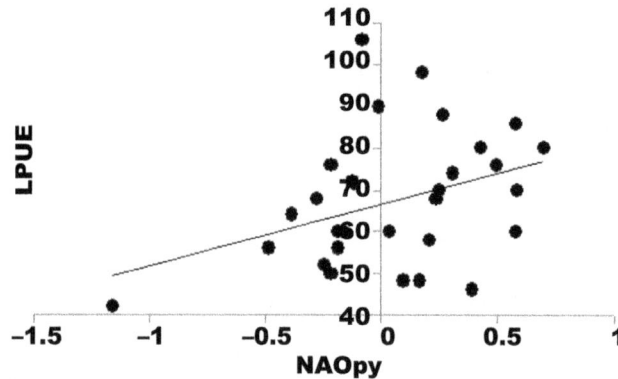

Fig. 2. Linear relationship between the Landing Per Unit Effort of blackspot seabream (LPUE) from the harbours Algeciras, Tarifa and Conil (Gulf of Cadiz) and the NAO previous year (NAOpy) to the landing for the period 1983–2011.

Fig. 2. Relación lineal entre el desembarque por unidad de esfuerzo (LPUE) del besugo de la pinta en los puertos de Algeciras, Tarifa y Conil (golfo de Cádiz) y la oscilación del Atlántico Norte del año anterior (NAOpy) al desembarque para el período 1983–2011.

trends with a significance level greater than 90% were obtained. The blanks areas correspond to points with few measurements of MW^u for the period under study following the protocol described above. The tempera-ture trend (fig. 5A) was positive for the significant area with maximum values close to 0.2°C per decade near to the Portuguese coast. A similar pattern was obser-ved for salinity trends (fig. 5B) with maximum values

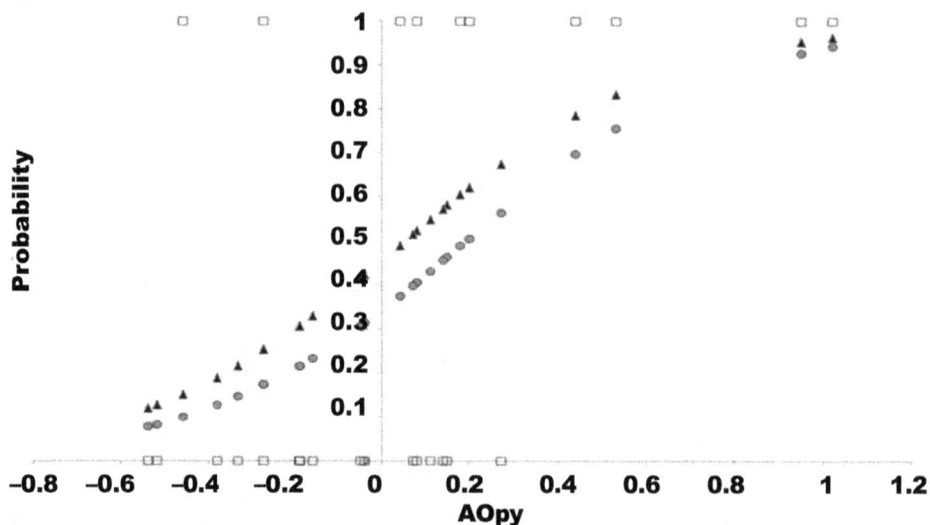

Fig. 3. Probability of obtaining good blackspot seabream landings from Tarifa harbour compared to the average Arctic Oscillation (AO) index for the year prior to landing (AOpy, gray circles), and the adjusted favorability for good blackspot seabream landings (black triangles). We plotted the years with good blackspot seabream landings (top squares) and years with a poor blackspot seabream landings (bottom squares).

Fig. 3. Probabilidad de obtener buenos desembarques de besugo de la pinta en el puerto de Tarifa en comparación con el índice medio de oscilación del Ártico (OA) para el año anterior al desembarque (AOpy, círculos grises) y favorabilidad ajustada de los buenos desembarques de besugo de la pinta (triángulos negros). Elaboramos un gráfico con los años de buenos desembarques de besugo de la pinta (cuadrados superiores) y los malos (cuadrados inferiores).

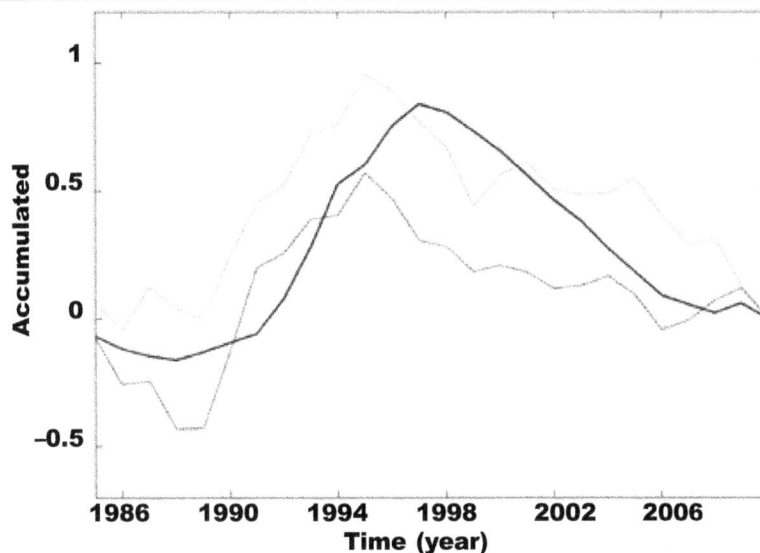

Fig. 4. Time evolution of accumulated values for the NAO (gray line), AO (dashed gray line), and *Pagellus bogaraveo* landings (black line). Signs were normalized so they could be represented in combination.

Fig. 4. Evolución de los valores acumulados para la oscilación del Atlántico Norte (línea gris), la oscilación del Ártico (línea discontinua gris) y los desembarques de Pagellus bogaraveo (línea negra). Se normalizaron los signos para que pudieran representarse en combinación.

close to 0.05 per decade near the Portuguese coast. Warming and salinification were almost negligible at locations far from the coast. Salinity and temperature time series were calculated by averaging the grid points in the area under study where trends with a significance level greater than 90% were obtained. Figure 6 shows the time evolution of *Pagellus bogaraveo* landings and backward averaged MW^u salinity and temperature, where the mean values (S and T) corresponding to a certain year were calculated by averaging the previous 5 years (*e.g.* backward averaged values for 1985 were calculated using values for the period 1980–1984). Both water properties were negatively correlated with landings (salinity: $R = -0.71$, $p < 0.01$; temperature: $R = -0.68$, $p < 0.01$).

Discussion

Few studies have shown that large–scale atmospheric phenomena could affect deep–sea population dynamics (*e.g.* Ruhl & Smith, 2004; Maynou, 2008; and references therein). Maynou (2008) found that the annual strength of red shrimp (*Aristeus antennatus*) landings is affected by variations in NAO (especially in winter) in the previous two or three years our results indicate that the long–term landing of blackspot seabream from the Strait of Gibraltar is associated with the atmospheric oscillations.

The positive NAO results in stronger–than–average westerly winds across northern mid–latitudes, affect-

ing both marine and terrestrial ecosystems, while a positive AO phase is characterized by a strong polar vortex (from the surface to the lower stratosphere). In this situation, storms increase in the North Atlantic and drought prevails in the Mediterranean basin. Strong winds agitate the water, favouring the mixing of deep water and surface water, and thus increasing the supply of nutrients at the surface. When the NAO and AO is in a negative phase, the continental cold air sinks into the Midwestern United States and Western Europe, while storms bring rain to the Mediterranean region (Ambaum et al., 2001).

According to Maynou (2008), 'decreased rainfall during positive NAO years may increase water–mass mixing in the NW Mediterranean, enhancing meso–zooplankton production and food resources to *Aristeus antennatus*, especially in late winter when females are undergoing ovary maturation and require higher energy input. During years of enhanced food resources the reproductive potential of females would increase, and strengthen particular year classes that appear in the landings two to three year later'. Our results suggest the same explanation. We observed a significant negative correlation between blackspot seabream landings and the temperature and salinity values obtained by calculating MW^u. According to Báez et al. (2013a) a positive NAOpy and AOpy increases the amount of snow in the mountains surrounding the Alboran Sea, thus increasing the amount of continental freshwater entering the sea the following year, which in turn reduces surface salinity, and blocks water upwelling.

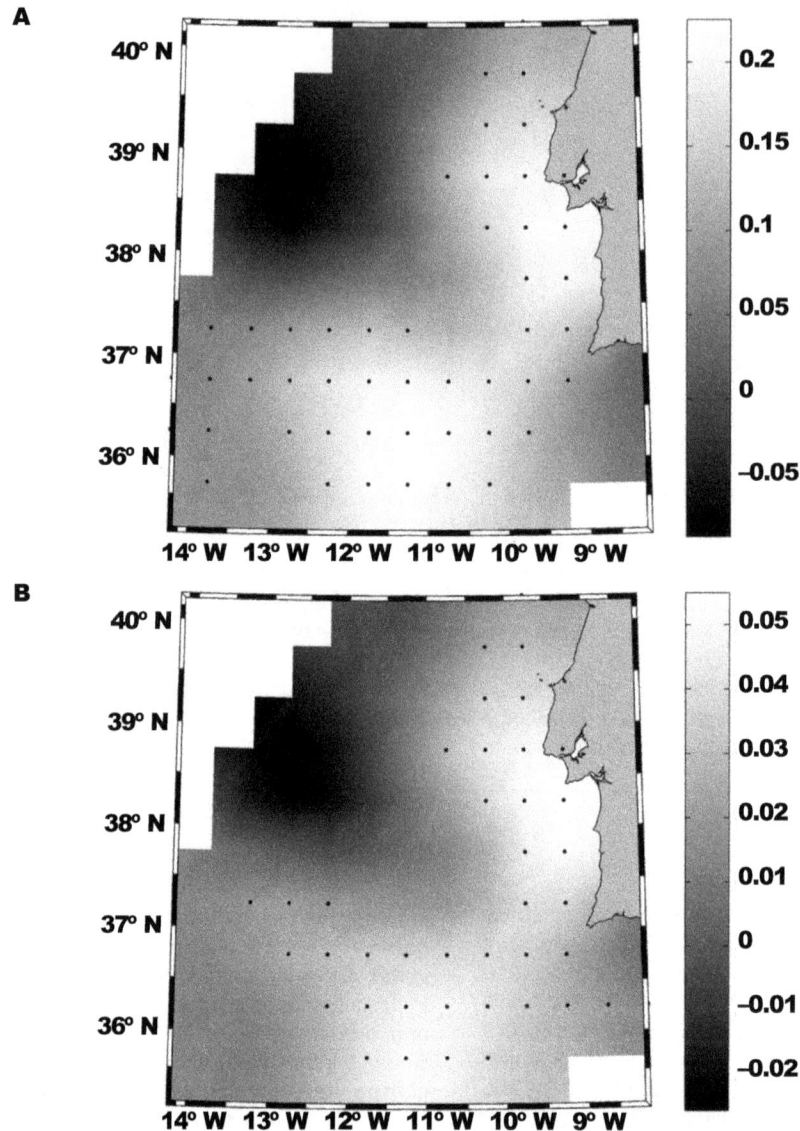

Fig. 5. Temperature (A, in ºC/d) and salinity (B, in psu/d) trends corresponding to upper Mediterranean water (MW^u). Dots mark locations where trends with a significance level greater than 90% were obtained. Blank areas correspond to points with few measurements of MW^u.

Fig. 5. Tendencias de la temperatura (A, en ºC/d) y la salinidad (B, en psu/d) correspondientes a la corriente superior de agua del Mediterráneo (MW^u). Los puntos indican los lugares en los que se obtuvieron tendencias con un grado de significación superior al 90%. Las zonas en blanco corresponden a los puntos con pocas mediciones de la MW^u.

We hypothesize that deep cold waters in the Alboran Sea are prevented from upwelling in the years following positive NAO and AO phases, and appear in the Atlantic as colder MW^u. This chain of events seems to benefit the eco–physiology of blackspot seabream by increasing their biomass. In this context, the dependence link could be due to an increase in survival of larvae related to higher amounts of food. This hypothesis is reinforced by the strongest correlation found for the AO with a lag of two years ($R^2 = 0.95$, $p < 0.01$).

In recent years, a decreasing trend in blackspot seabream landings has been observed. However, this trend has coincided with the end of a long positive NAO and AO cycle between the 1980s and 1990s (Fyfe et al., 1999). Thus, prolonged periods of a

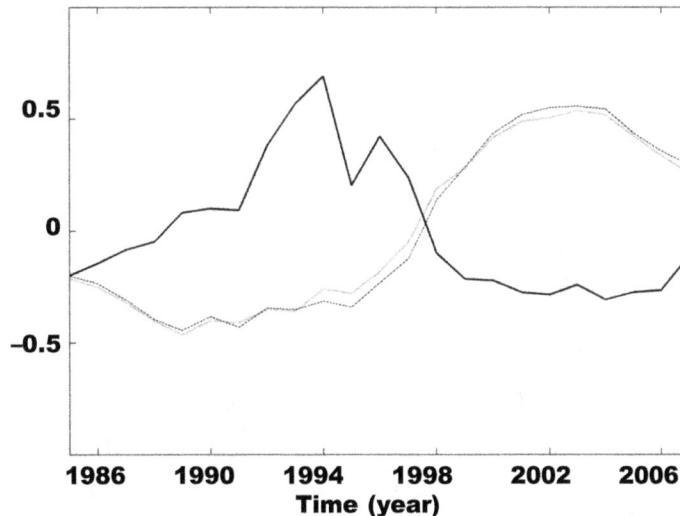

Fig. 6. Time evolution of *Pagellus bogaraveo* landings (black line) and backward averaged MW^u temperatures (gray line) and salinity (dashed gray line). Signs were normalized so they could be represented in combination.

Fig. 6. Evolución de los desembarques de Pagellus bogaraveo *(línea negra) y los promedios de la temperatura (línea gris) y la salinidad (línea gris discontinua) de la MW^u saliente. Se normalizaron los signos para que pudieran representarse en combinación.*

positive AO trend could favour high fishery yields. In contrast, a negative NAO and AO phase drastically reduces production. In the context of global change, this situation could have major implications for fisheries management. Thus, during positive NAO and AO phases, high exploitation levels could be allowed while maintaining the stocks within safety limits. During negative NAO and AO phases, more restrictive management measures should be adopted, such as lower exploitation levels, or temporary fishery closure, to preserve fishery sustainability and population safety.

Changes in the NAO and AO are correlated over long time periods (Feldstein & Franzke, 2006). Given the strong impact of the AO and NAO on the weather and climate of the wealthiest areas of the planet, and their large socioeconomic impact on energy, agriculture, fisheries, industry, traffic and human health throughout the whole of Europe and eastern North America, there has been great interest in quantifying the extent to which the phenomena are predictable and the ability of climate numerical models to simulate them. Bojariu & Gimeno (2003a) provide a good review of the topic. Predictive patterns have been identified in the Atlantic SSTs preceding specific phases of the AO and NAO by up to six months (Rodwell &d Folland, 2003), in Eurasian snow cover by up to one year (Bojariu & Gimeno, 2003b), and in the extent of sea–ice over the Arctic (Deser et al., 2000). Thus, the Atlantic SST, Eurasian snow cover, and Arctic sea–ice are good candidates to explore fisheries in Strait of Gibraltar up to one year in advance.

It is widely accepted that the planet is experiencing a period of rapid global warming (Oreskes, 2004), primarily driven by human activity (Keller, 2007). Although there is increasing concern over the impact of global warming on marine biodiversity and fisheries ecology (Yatsu et al., 2005), it is difficult to predict how the climate could alter marine biodiversity. In this context, climate change simulations with greenhouse gas and aerosol forcing for the period 1900–2100 indicate a positive trend in the AO (Fyfe et al., 1999). On the other hand, the AO responds to natural changes, such as the increase in stratospheric aerosols due to volcanic eruptions (Christiansen, 2007). Thus, via the NAO and AO, global warming could affect the fisheries ecology of blackspot seabream from the Strait of Gibraltar. This possibility could be extrapolated to other northern hemisphere stocks.

Acknowledgements

This work was partially supported by projects from the IEO based in Malaga, GPM–4 programs (IEO) and PNDB (EU–IEO), project CGL 2009–11316/BOS from the Spanish Government and FEDER, and from the Xunta de Galicia under 'Programa de Consolidación e Estruturación de Unidades de Investigación' (Grupos de Referencia Competitiva) funded by FEDER. An anonymous reviewer provided helpful comments on earlier versions of the manuscript. We would also like to thank to Andrew Paterson for style corrections.

References

Alonso–Pozas, C., Arechavaleta, A., Cobo, R., Espinosa, D., Galisteo, A., García, J. J., González, F., Naranjo, S., Nieto, D., Raya, L. & Rebollo, J., 2007. *Producción pesquera andaluza, año 2005*. Consejería de Agricultura y Pesca, Junta de Andalucía, Sevilla.

Ambaum, M. H. P., Hoskins, B. J. & Stephenson, D. B., 2001. Arctic Oscillation or North Atlantic Oscillation? *Journal of Climate*, 14: 3495–3507.

Báez, J. C., Gimeno, L., Gómez–Gesteira, M., Ferri–Yáñez, F. & Real, R., 2013a. Combined effects of the Arctic Oscillation and the North Atlantic Oscillation on Sea Surface Temperature in the Alborán Sea. *PlosOne*, 8(4): e62201. Doi:10.1371/journal.pone.0062201.

Báez, J. C., Macías, D., De Castro, M., Gómez–Gesteira, M.,Gimeno, L. & Real, R., 2013b. Analysis of the effect of atmospheric oscillations on physical condition of pre–reproductive bluefin tuna from the Strait of Gibraltar. *Animal Biodiversity and Conservation*, 36.2: 225–233.

Báez, J. C., Real, R., Camiñas, J. A., Torreblanca, D. & Garcia–Soto, C., 2009. Analysis of swordfish catches and by–catches in artisanal longline fisheries in the Alboran Sea (Western Mediterranean Sea) during the summer season. *Marine Biodiversity Record*, Doi:10.1017/S1755267209990856.

Báez, J. C., Ortiz De Urbina, J. M., Real, R. & Macías, D., 2011. Cumulative effect of the north Atlantic oscillation on age–class abundance of albacore (*Thunnus alalunga*). *Journal Applied Ichthyology*, 27:1356–1359.

Báez, J. C. & Real, R., 2011. The North Atlantic Oscillation affects the landings of Anchovy Engraulis encrasicolus in the Gulf of Cádiz (South of Spain). *Journal Applied Ichthyology*, 27: 1232–1235.

Borja, A. & Santiago, J., 2002. Does the North Atlantic Oscillation control some processes influencing recruitment of temperate tunes? *ICCAT Collective Volume, Scientific Papers,* 54: 964–984.

Bojariu, R. & Gimeno, L., 2003a. Predictability and numerical modelling of the North Atlantic Oscillation. *Earth–Science Reviews*, 63(1–2):145–168.

– 2003b. The role of snow cover fluctuations in multiannual NAO persistence. *Geophysical Research Letters*, 30: 1156.

Bozec, A., Lozier, M. S., Chassignet, E. P. & Halliwell, G. R., 2011. On the variability of the Mediterranean outflow water in the North Atlantic from 1948 to 2006. *Journal of Geophyscal Research*, 116:C09033. Doi: 10.1029/2011JC007191.

Brandt, U. S. & Kronbak, L. G., 2010. On the stability of fishery agreements under exogenous change: An example of agreements under climate change. *Fisheries Research*, 101: 11–19.

Carton, J. A., Chepurin, G. & Cao, X., 2000a. A simple ocean data assimilation analysis of the global upper ocean 1950–95. Part II: Results. *Journal of Physical Oceanography*, 30: 311–326.

Carton, J. A., Chepurin, G., Cao, X. & Giese, B., 2000b. A simple ocean data assimilation analysis of the global upper ocean 1950–95. Part I: Methodology. *Journal of Physical Oceanography*, 30: 294–309.

Carton, J. A. & Giese, B. S., 2008. A reanalysis of ocean climate using Simple Ocean Data Assimilation (SODA). *Monthly Weather Review*, 13: 2999–3017.

Chavez, F. P., Ryan, J., Lluch–Cota, S. E. & Niquen, C. M., 2003. From anchovies to sardines and back: Multidecadal change in the Pacific Ocean. *Science*, 299: 217–221.

Christiansen, B., 2007. The North Atlantic Oscillation or the Arctic Oscillation? Volcanic eruptions as Nature's own experiments. *Geophysical Research Abstract*, 9: 06601, SRef–ID: 1607–7962/gra/EGU2007–A–06601.

CopeMed II, 2012. Report of the Second meeting of the CopeMed II Working Group between Spain and Morocco on blackspot seabream (Pagellus bogaraveo) of the Strait of Gibraltar area. CopeMed II Technical Documents Nº26, Málaga. http://www.faocopemed.org/html/publications.html. [Accessed on 15 May 2013]

Czerwinski, I. A., Gutiérrez–Estrada, J. C., Casimiro–Soriguer–Escofet, M. & Hernando, J. A., 2009. Hook selectivity models assessment for blackspot seabream. Classic and heuristic approaches. *Fisheries Research*, 102: 41–49.

Deser, C., Walsh, J. E. & Timlin, M. S., 2000. Arctic sea ice variability in the context of recent wintertime atmospheric circulation trends. *Journal of Climate*, 13: 618–632.

Douville, H., 2009. Stratospheric polar vortex influence on Northern Hemisphere winter climate variability. *Geophysical Research Letter*, 36: 1–5.

FAO, 2010. The State of World Fisheries and Aquaculture 2010. Food and Agriculture Organization of the United Nations, Rome.

Feldstein, S. B. & Franzke, C., 2006. Are the North Atlantic Oscillation and the Northern Annular Mode distinguishable? *Journal Atmospheric Science*, 63: 2915–2930.

Fromentin, J. M., 2001. Is the recruitment a key biological process in the hypothetical NAO–Atlantic tunas relationships? *ICCAT Collective Volume, Scientific Papers,* 54: 1008–1016.

Fyfe, J. C., Boer, G. J. & Flato, G. M., 1999. The Arctic and Antarctic Oscillations and their Projected Changes Under Global Warming. *Geophysical Research Letter*, 26: 1601–1604.

Galisteo, A., García, C. & Cruz, I., 2001a. *Producción pesquera andaluza, año 2000*. Consejería de Agricultura y Pesca, Junta de Andalucía, Sevilla. (In Spanish.)

Galisteo, A., García, C., Cruz, I. & Zurita, F., 2001b. *Evolución de la producción pesquera andaluza (1985–1999)*. Consejería de Agricultura y Pesca, Junta de Andalucía, Sevilla. (In Spanish.)

Galisteo, A., García, C. & Espinosa, D., 2002. *Producción pesquera andaluza, año 2001*. Consejería de Agricultura y Pesca, Junta de Andalucía, Sevilla. (In Spanish.)

Galisteo, A., González, F., Naranjo, S., Abreu, L., Losa, M. T., Alonso–Pozas, C., Cobo, R. & Espinosa, D., 2011. *Producción Pesquera Andaluza. Año 2010*. Consejería de Agricultura y Pesca, Junta de Andalucía, Sevilla. (In Spanish.)

Galisteo, A., González, F., Naranjo, S., Abreu, L., Losa, M. T., Alonso–Pozas, C., Cobo, R. & Espinosa, D., 2012. *Producción pesquera andaluza. Año 2011*. Junta de Andalucía. Consejería de Agricultura y Pesca. Secretaría General Técnica. Servicio de Publicaciones y Divulgación. Sevilla, Spain.

Galisteo, A., González, F., Naranjo, S., Abreu, L., Losa, M. T., Alonso–Pozas, C., Cobo, R. & Espinosa, D., 2013. *Producción pesquera andaluza. Año 2012*. Junta de Andalucía. Consejería de Agricultura y Pesca. Secretaría General Técnica. Servicio de Publicaciones y Divulgación. Sevilla, Spain.

Galisteo, A., González, F., Naranjo, S., Nieto, D., Espinosa, D., Alonso, C., Arechavaleta, A. & Cobo, R., 2008. *Producción pesquera andaluza, año 2006*. Consejería de Agricultura y Pesca, Junta de Andalucía, Sevilla. (In Spanish.)

– 2009a. *Producción pesquera andaluza, año 2008*. Consejería de Agricultura y Pesca, Junta de Andalucía, Sevilla. (In Spanish.)

Galisteo, A., González, F., Naranjo, S., Nieto, D., Espinosa, D., Alonso, C., Arechavaleta, A., Cobo, R. & Valufo, I., 2009b. *Producción pesquera andaluza, año 2007*. Consejería de Agricultura y Pesca, Junta de Andalucía, Sevilla. (In Spanish.)

Galisteo, A., González, F. & Nieto, D., 2004. *Análisis de la actividad extractiva de la flota andaluza por modalidades de pesca, año 2002*. Consejería de Agricultura y Pesca, Junta de Andalucía, Sevilla. (In Spanish.)

– 2005. *Valor añadido y pesca en Andalucía, año 2003*. Consejería de Agricultura y Pesca, Junta de Andalucía, Sevilla. (In Spanish.)

– 2007. *Valor añadido y pesca en Andalucía, año 2004*. Consejería de Agricultura y Pesca, Junta de Andalucía, Sevilla. (In Spanish.)

Gancedo, U., 2005 Efecto de las variaciones climáticas en la distribución espacio–temporal de *Thunnus thynnus* (Linnaeus, 1758) y *Thunnus alalunga* (Bonnaterre, 1788) en el Océano Atlántico. Gran Canaria: Unpublished PhD thesis, Universidad Palmas de Gran Canaria. (In Spanish.)

Gil Herrera, J., 2010. Spanish information about the red seabream (*Pagellus bogaraveo*) fishery in the Strait of Gibraltar Region. SRWG on shared demersal resources. Ad hoc scientific working group between Morocco and Spain on Pagellus bogaraveo in the Gibraltar Strait area. FAO–CopeMed II Consultant, Málaga. Available from: http://www.faocopemed.org/html/publications.html [Accessed on 15 May 2013]

– 2012. Updated information from the Spanish Blackspot seabream (*Pagellus bogaraveo*) fishery in the Strait of Gibratar area. Working Document presented to the Secound Meeting of the FAO CopeMed II Working Group on Blackspot Seabream (*Pagellus bogaraveo*) of the Strait of Gibraltar area between Spain and Morocco, (Tangiers, Morocco, 19–21 March 2012). GCP/INT/SPA–GCP/INT/006/EC. Tangiers: CopeMed II Occasional

Paper Nº 10.   http://www.faocopemed.org/html/publications.html [Accessed on 15 May 2013]

Graham, C. T. & Harrod, C., 2009. Implications of climate change for the fishes of the British Isles. *Journal of Fish Biology*, 74: 1143–1205.

Hammer, Ø. & Harper, D., 2006. *Paleontological Data Analysis*. Blackwell Publishing, Oxford.

Hammer, Ø., Harper, D. & Ryan, P. D., 2001. PAST: Paleontological Statistics Software Package for Education and Data Analysis. *Palaeontologia Electronica*, 4(1): 9.

Hosmer, D. W. & Lemeshow, L. S., 2000. *Applied logistic regression analysis, 2º edition*. John Wiley and Sons Inc., New York, USA.

Hurrell, J. W., 1995. Decadal trends in the North Atlantic Oscillation: Regional temperatures and precipitation. *Science*, 269: 676–679.

Jones, P. D., Jonsson, T. & Wheeler, D., 1997. Extension to the north Atlantic oscillation using early instrumental pressure observations from Gibraltar and South–Wesr Iceland. *International Journal of Climatology*, 17: 1433–1450.

Keller, C. F., 2007. An update to global warming: the balance of evidence and its policy implications. *The Scientific World Journal*, 7: 381–39.

Levitus, S. & Boyer, T., 1994. *World Ocean Atlas 1994. Vol. 4: Temperature*. Atlas series, NOAA, Washington, D.C.

Lobo, J. M., Jiménez–Valverde, A. & Real, R., 2008. AUC: a misleading measure of the performance of predictive distribution models. *Global Ecology and Biogeography*, 17: 145–151.

Maynou, F., 2008. Influence of the North Atlantic Oscillation on Mediterranean deep–sea shrimp landings. *Climate Research*, 36: 253–257.

Mejuto, J., 2003. Recruit indices of the North Atlantic Swordfish (*Xiphias gladius*) and their possible link to atmospheric and oceanographic indicators during the 1982–2000 periods. *ICCAT Collective Volume, Scientific Papers*, 55: 1506–1515.

Meynecke, J. O., Grubert, M., Arthur, J. M., Boston, R. & Lee, S. Y., 2012. The influence of the La Nina–El Nino cycle on giant mud crab (*Scylla serrata*) catches in Northern Australia. *Estuaries Coast Shelf Science*, 100: 93–101.

Oreskes, N., 2004. The scientific consensus on climate change. *Science*, 306: 1686.

Overland, J. E., Alheit, J., Bakun, A., Hurrell, J. W., Mackas, D. L. & Miller, A. J., 2010. Climate controls on marine ecosystems and fish populations. Journal *of Marine System*, 79: 305–315.

Pitcher, T. J., 2005. Back–to–the–future: a fresh policy initiative for fisheries and a restoration ecology for ocean ecosystems. *Philosophical Transactions of the Royal Society B: Biological Sciences*, 360: 107–121.

Pranovi, F., Caccin, A., Franzoi, P., Malavasi, S., Zucchetta, M. & Torricelli, P., 2013. Vulnerability of artisanal fisheries to climate change in the Venice Lagoon. *Journal of Fish Biology*. Doi:10.1111/jfb.12124.

Real, R., Barbosa, A. M. & Vargas, J. M., 2006. Obtaining environmental favourability functions from logistic regression. *Environmental and Ecological Statistics*, 13: 237–245.

Reynolds, R. W. & Smith, T. M., 1994. Improved global sea surface temperature analysis using optimum interpolation. *Journal of Climate*, 7: 929–948.

Rodwell, M. J. & Folland, C. K., 2003. Atlantic air–sea interaction and seasonal predictability. Quarterly *Journal of the Royal Meteorological Society*, 128: 1413–1443.

Ruhl, H. A. & Smith, K. L. Jr, 2004. Shifts in deep–sea community structure linked to climate and food supply. *Science*, 305: 513–515.

Saunders, M. A. & Lea, A. S., 2008. Large contribution of sea surface warming to recent increase in Atlantic hurricane activity. *Nature*, 451: 557–560.

Silva, L., Gil, J. & Sobrino, I., 2002. Definition of fleet components in the Spanish artisanal fishery of the Gulf of Cádiz (SW Spain ICES division IXa). *Fisheries Research*, 59(2): 117–128.

Sokal, R. R. & Rohlf, F. J., 1995. *Biometry*. 3rd edition. W. H. Freeman and Co, New York.

Straile, D. & Stenseth, N. C., 2007. The North Atlantic Oscillation and ecology: links between historical time–series, and lessons regarding future climate warming. *Climate Research*, 34: 259–262.

Thompson, D. W. J. & Wallace, J. M., 1998. The Arctic Oscillation signature in the wintertime geopotential height and temperature fields. *Geophysical Research Letters*, 25: 1297–1300.

Thompson, D. W. J., Wallace, J. M. & Hegerl, G. C., 2000. Annular modes in the extratropical circulation: Part II: Trends. *Journal of Climate*, 13: 1018–1036.

Thurstan, R. H., Brockington, S. & Roberts, C. M., 2010. The effects of 118 years of industrial fishing on UK bottom trawl fisheries. *Nature Communication*. Doi: 10.1038/ncomms1013.

Walker, G. T. & Bliss, E. W., 1932. World weather V. *Memoirs of the Royal Meteorological Society*, 4: 53–84.

Worm, B., Barbier, E. B., Beaumont, N., Duffy, J. E., Folke, C., Halpern, B. S., Jackson, J. B. C., Lotze, H. K., Micheli, F., Palumbi, S. R., Sala, E., Selkoe, K. A., Stachowicz, J. J. & Watson, R., 2006. Impacts of biodiversity loss on ocean ecosystem services. *Science*, 314: 787–790.

Worm, B., Hilborn, R., Baum, J. K., Branch, T. A., Collie, J. S., Costello, C., Fogarty, M. J., Fulton, E. A., Hutchings, J. A., Jennings, S., Jensen, O. P., Lotze, H. K., Mace, P. A., McClanahan, T. R., Minto, C., Palumbi, S. R., Parma, A. M., Ricard, D., Rosenberg, A. A., Watson, R. & Zeller, D., 2009. Rebuilding global fisheries. *Science*, 325: 578–585.

Worm, B. & Myers, R. A., 2004. Managing fisheries in a changing climate – No need to wait for more information: industrialized fishing is already wiping out stocks. *Nature*, 429: 15.

Yatsu, A., Watanabe, T., Ishida, M., Sugisaki, H. & Jacobson, L. D., 2005. Environmental effects on recruitment and productivity of Japanese sardine *Sardinops melanostictus* and chubmackerel *Scomber japonicus* with recommendations for management. *Fisheries Oceanography*, 14: 263–278.

Zuur, A. K., Ieno, E. N. & Smith, G. M., 2007. *Analysing Ecological Data*. Springer, New York.

14

Minimally invasive blood sampling method for genetic studies on *Gopherus* tortoises

L. M. García–Feria, C. A. Ureña–Aranda & A. Espinosa de los Monteros

García–Feria, L. M., Ureña–Aranda, C. A. & Espinosa de los Monteros, A., 2015. Minimally invasive blood sampling method for genetic studies on *Gopherus* tortoises. *Animal Biodiversity and Conservation*, 38.1: 31–35.

Abstract

Minimally invasive blood sampling method for genetic studies on Gopherus *tortoises.*— Obtaining good quality tissue samples is the first hurdle in any molecular study. This is especially true for studies involving management and conservation of wild fauna. In the case of tortoises, the most common sources of DNA are blood samples. However, only a minimal amount of blood is required for PCR assays. Samples are obtained mainly from the brachial and jugular vein after restraining the animal chemically, or from conscious individuals by severe handling methods and clamping. Herein, we present a minimally invasive technique that has proven effective for extracting small quantities of blood, suitable for genetic analyses. Furthermore, the samples obtained yielded better DNA amplification than other cell sources, such as cloacal epithelium cells. After two years of use on wild tortoises, this technique has shown to be harmless. We suggest that sampling a small amount of blood could also be useful for other types of analyses, such as physiologic and medical monitoring.

Key words: Blood extraction, DNA source, Tortoises

Resumen

Método de extracción de sangre mínimamente invasivo para estudios genéticos en tortugas terrestres del género Gopherus.— La obtención de muestras de tejido de buena calidad es la primera dificultad en cualquier estudio molecular. Esto es especialmente cierto en los estudios de gestión y conservación de la fauna silvestre. En el caso de las tortugas terrestres, la fuente más habitual de ADN son las muestras de sangre obtenidas principalmente de las venas braquial y yugular por contención química, o de individuos conscientes mediante métodos de manipulación y sujeción que pueden causar estrés en el animal. Se requiere una cantidad mínima de sangre para los ensayos del PCR. A continuación, presentamos una técnica mínimamente invasiva que ha resultado eficaz para extraer pequeñas cantidades de sangre apropiadas para realizar análisis genéticos. Además, las muestras obtenidas producen una amplificación de ADN mejor que otras fuentes celulares, como las células epiteliales cloacales. Después de dos años de aplicación en tortugas terrestres silvestres, esta técnica ha demostrado ser inofensiva. Sugerimos que el muestreo de pequeñas cantidades de sangre con esta técnica podría ser útil para otro tipo de análisis, como el seguimiento fisiológico y médico.

Palabras clave: Extracción de sangre, Fuente de ADN, Tortugas terrestres

Luis M. García–Feria, Red de Biología y Conservación de Vertebrados, Inst. de Ecología A. C., carretera antigua a Coatepec 351, El Haya, CP 91070, Xalapa, Veracruz, México.– Cinthya A. Ureña–Aranda & Alejandro Espinosa de los Monteros, Lab. de Sistemática Filogenética, Red de Biología Evolutiva, Inst. de Ecología A.C., carretera antigua a Coatepec 351, El Haya, CP 91070, Xalapa, Veracruz, México.

Corresponding author: L. M. Gracía–Feria. E–mail: luis.garcia@inecol.mx

Introduction

Many behavioral, ecological, physiological, and medicinal studies for conservation purposes require the use of blood samples. To reduce physical risk during animal handling, a minimal invasive method has been tested to obtain blood samples for these as well as for other kinds of studies (*e.g.,* genetic, systematic, toxicological and stable isotope analyses). In many cases, the use of molecular markers is currently one of the prime tools in wildlife research, management, and conservation (DeWoody, 2005, DeYoung & Honeycutt, 2005). Gathering good quality samples has sometimes been a problem under field conditions. In the past, collectors used to sacrifice individuals for museum collections and sometimes for obtaining accessory material such as tissue samples. The rapid development of PCR methods and kits for DNA extraction has made it possible to obtain suitable genetic material from minuscule samples (*e.g.,* one hair). In many instances; however, collecting samples requires restraining the animals by physical or chemical means (*i.e.,* invasive techniques), resulting in prolonged stress and even injuring the subject. New non–invasive methods have been developed for specific taxa groups (García–Feria, 2008); nonetheless, for many species or for field conditions, these techniques are not an option. Isolation of DNA from stool samples is among the common non–invasive methods used in many wildlife studies (Dalén et al., 2004; Lukacs & Burnham, 2005), with the DNA source being the epithelial cells. However, these cells are usually scarce, and the feces may contain PCR inhibitory substances. Besides, there is a high risk of contamination from alien DNA (Taberlet et al., 1999; Broquet et al., 2007).

For reptiles, particularly for live turtles and tortoises, DNA has been extracted by means of oral scrapes and cloacal swabs (Nagy & Medica, 1986; Mautino & Page, 1993; Van der Kuyl et al., 2005; Wendland et al., 2009). Even so, whole blood is the best source, and several blood extraction techniques have been developed (Gandal, 1958; Avery & Vitt, 1984; Gottdenker & Jacobson, 1995; Knotková et al., 2002; López–Olvera et al., 2003; Rohilla & Tiwari, 2008). The sample is usually extracted either from the main veins (*e.g.,* jugular, brachial, femoral, iliac vein) (Mader, 2005), the subcarapacial venous plexus (Hernández–Divers, et al., 2002), or the occipital sinuses, or by cardiac puncture (Mautino & Page, 1993; Fowler, 1995). Nonetheless, the anatomy and behavior of tortoises (Testudinidae) makes blood extraction rather difficult by any of these methods. The thick scales on the skin and the characteristic retraction of the limbs and the head within the shell block access to the veins. Sometimes the use of forceps and anesthetics for safe handling of the animals is necessary (Fowler, 1995). Additionally, lymphatic vessels running beside the main body veins may be damaged (Wendland et al., 2009). These methods are therefore excessive for PCR purposes. Turtles, like other reptiles, have nucleated erythrocytes (Knotková et al., 2002), so small quantities of blood are sufficient to obtain good quality genomic DNA.

Material and methods

When disturbed, *Gopherus flavomarginatus* (Bolson tortoise) can strongly retract its head and limbs, blocking access to the vessels used to draw blood. While in the retracted position, the exposed soft parts are covered with dense and thick scales (fig. 1). However, between the fingers and the dorsal area of the forelegs there is a characteristic thin line of almost naked skin (fig. 2). This is the recommended area for obtaining small amounts of blood. We have used the following method for over two years in a study that assesses the genetic variations of 76 wild individuals of the Bolson tortoise (Ureña–Aranda & Espinosa de los Monteros, 2012). According to Germano (1994), maturity is attained once the carapace reaches at least 28 cm; therefore all the handled specimens can be considered adults. Before the present field work, the blood sampling method was tested on two species of captive tortoises, *G. berlandieri* (*n* = 1) and *G. agassizii* (*n* = 5), also sampled by means of cloacal swabs. Blood was sampled as follows. First, we cleaned the dorsal side of the hand with a cotton swab soaked in 75% ethanol to avoid any possible infection or contamination of organic material. Then, a puncture was performed using a sterile hypodermic needle (27G x 13 mm; Becton, Dickinson & Company, Franklin Lakes, NJ) in the bare line of skin located at the distal edge of the hand just before the fingers. The needle should be introduced between the second and third fingers in an angle of 45° (approximately) toward the third finger. There may be no bleeding if the needle is introduced elsewhere or at a different angle. Without practice, no more than three puncture attempts were required to obtain the blood sample. Immediately after removing the needle, the blood can be collected in a borosilicate glass capillary tube that does not contain heparin. Finally, we placed a cotton swab with 75% ethanol on the puncture for a few seconds, applying little pressure as to stop any extra bleeding. The cloacal swabs were taken after cleaning the cloacal area with a cotton swab soaked in 75% ethanol. We then introduced and softly spun a rayon swab (Medical Wire and Equipment, 100–100 MW, Biomerieux) in the cloaca to obtain the epithelium cells.

Results

In the field, we usually collected up to 30 µl of whole blood and transferred this to 500 µl vials containing 100 µl lysis buffer (Longmire et al., 1997); the animals were released immediately after manipulation. The samples preserved in this buffer do not require refrigeration, which is an advantage for field conditions. However, if the genetic study involves protein analyses, blood samples must be stored by different means (*e.g.,* liquid nitrogen). Once in the laboratory, we extracted DNA from the cloacal swabs, and dried tissue from shells samples and from the small aliquots of the blood–buffer mixture (10–20 µl) using Chelex 100© resin (Walsh et al., 1991). We obtained greater yields of high molecular weight DNA from the blood

Fig. 1. Bolson tortoise in retracted position. Note the inaccessibility to the blood vessels commonly used to draw blood.

Fig. 1. Tortuga del Bolsón en posición retraída. Nótese la imposibilidad de acceder a los vasos sanguíneos que habitualmente se utilizan para la extracción de sangre.

samples than from other tested tissue sources (*i.e.,* cloacal swabs, and dry tissue from shells; fig. 3). DNA amplification was conducted in Peltier–effect thermocyclers (ABI GeneAmp PCR system 2400) using the following parameters: one initial cycle at 95° for 120 s, followed by 32 cycles of 95° for 20 s, 47° for 20 s, 74° for 60 s, with one final cycle at 72° for 180 s.

This minimally invasive blood sampling method used on *Gopherus* species has been extremely useful. Performing the whole procedure takes no more than two minutes, even for those people who have been trained only once, and have little or no experience in animal handling. Using this technique, stress for the animal is kept to a minimum, and there is practically no risk of injuring the tortoise.

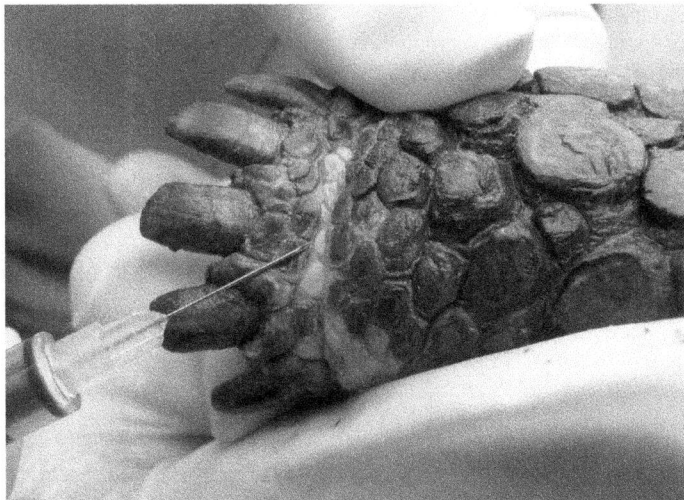

Fig. 2. Suggested puncturing site for the minimally invasive blood sampling method.

Fig. 2. Lugar sugerido para la punción como método de extracción de sangre mínimamente invasivo.

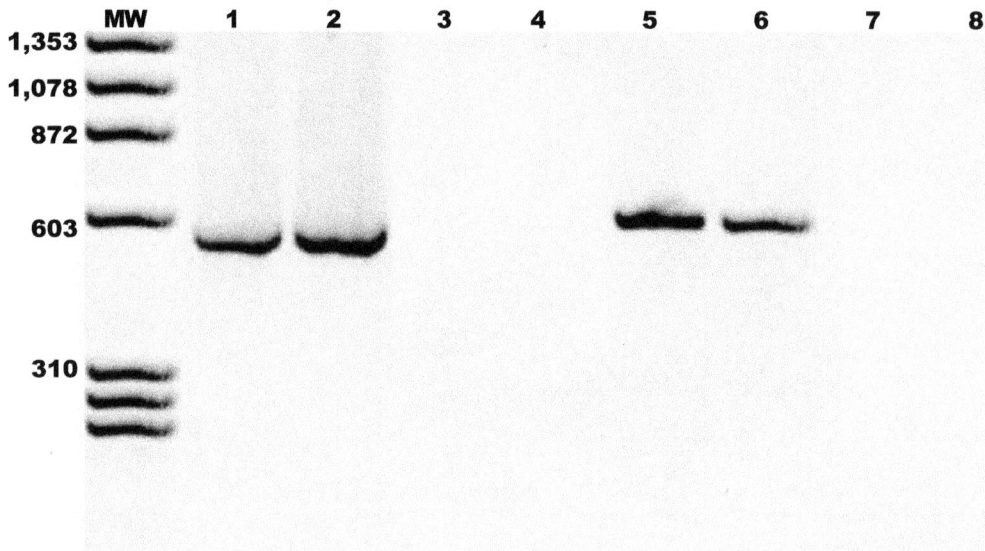

Fig. 3. Ethidium bromide stained 2% agarose gel showing PCR products from two regions of the mitochondrial D–loop gene. Amplifications from several specimens were attempted from forelimb puncture (lines 1, 2, 5 and 6), cloacal swab (lines 3 and 7), and carcass tissue (lines 4 and 8): MW indicates the molecular weight bands in base pairs.

Fig. 3. Productos de PCR de dos regiones del gen mitocondrial D–loop mostrados en un gel de agarosa teñido con bromuro de etidio al 2%. Se intentó realizar amplificaciones de diferentes muestras obtenidas a partir de la punción de la pata delantera (líneas 1, 2, 5 y 6), hisopos cloacales (líneas 3 y 7) y tejidos de carcasas (líneas 4 y 8): MW indica las bandas del peso molecular en pares de bases.

A remarkable characteristic of this species is its burrowing behavior. Several nerves and ligaments are located in the hand area, and some concerns may result from puncturing around this area. We have used this method for two years (*i.e.,* between 2009 and 2011), to collect blood samples from 76 individuals of the Bolson tortoise. This species is highly philopatric, and the adults do not have natural predators. In recent visits to the study area, we have been able to verify the health status of every manipulated individual, and none of them showed any apparent problem. Therefore, we are confident that this technique of foot puncture is harmless compared to other invasive blood sampling methods (*e.g.,* Gandal, 1958; Avery & Vitt, 1984; Fowler, 1995). Several analyses for physiological and medical surveys require larger volumes of blood than those extracted with the foot puncture method (from 0.5 ml to microtainer until 1.8 ml or more). However, the amount of blood that is obtained with the suggested method (≈ 30–40 µl, approximately two–thirds to three–quarters of a capillary tube; Kerr, 2002) is adequate for implementing laboratory and clinical analyses other than PCR. This minimally invasive method can be applied to obtain blood for a morphological characterization of peripheral blood cells from blood smears (Knotková et al., 2002), packed cell volume (PCV) measurement by microhaematocrit, refractometry to assess the total protein level, specific gravity and refractive index of serum, glucose level by blood glucose strips or pocket glucose meter (Kerr, 2002).We therefore recommend the use of this minimally invasive method before attempting more aggressive blood extraction techniques for genetic analysis or for any other survey that requires only small amounts of blood.

Acknowledgements

The Bolson tortoise individuals were captured with collecting permits from the Secretaría de Medio Ambiente y Recursos Naturales (No. SGPA/ DGVS/04690/09). We thank Gustavo Aguirre–León for his advice on the biology of the Bolson tortoise, the Centro Ecológico del Estado de Sonora, CEDES, for permission to conduct captive tortoise sampling. Rolando González Trapaga and Francisco Herrera for providing invaluable assistance during field work, and Marco A. L. Zuffi and three anonymous reviewers for their pertinent comments.

References

Avery, H. W. & Vitt, L. J., 1984. How to get blood from a turtle. *Copeia*, 1984: 209–210.

Broquet, T., Ménard, N. & Petit, E., 2007. Noninvasive population genetics: A review of sample source, diet, fragments length and microsatellite motif effects on amplification success and genotyping error rates. *Conservation Genetics*, 8: 249–260.

Dalén, L., Götherström, A. & Angerbjörn, A., 2004. Identifying species from pieces of feces. *Conservation Genetics*, 5: 109–111.

DeYoung, R. W. & Honeycutt, R. L., 2005. The molecular toolbox: genetic techniques in wildlife ecology and management. *Journal of Wildlife Management*, 69: 1362–1384.

DeWoody, J. A., 2005. Molecular approaches to the study of parentage, relatedness, and fitness: practical applications for wild animals. *Journal of Wildlife Management*, 69: 1400–1418.

Fowler, M. E., 1995. *Restraint and handling of wild and domestic animals*, 2nd Ed. Iowa State University Press, USA.

Gandal, C. P., 1958. A practical method to obtaining blood from anesthetized turtles by means of cardiac puncture. *Zoologica*, 43: 93–94.

García–Feria, L. M., 2008. Remote sampling of hair for genetic analysis of wild mammals. *Revista de Ecología Latinoamericana*, 13: 13–15.

Germano, D. J., 1994. Comparative life histories of North American tortoises. In*: Biology of North American Tortoises*: 174–185 (R. B. Bury & D. J. Germano, Eds.). Fish and Wildlife Research 13, Technical Report Series, U. S. Department of the Interior, National Biological Survey, Washington DC.

Gottdenker, N. L. & Jacobson, E. R., 1995. Effect of venipuncture sites on hematologic and clinical biochemical values in desert tortoises (*Gopherus agassizii*). *American Journal of Veterinary Research*, 56: 19–21.

Hernández–Divers, S. M., Hernandez–Divers, S. J. & Wyneken, J., 2002. Angiographic, anatomic, and clinical technique descriptions of a subcarapacial venipuncture site for chelonians. *Journal of Herpetological Medicine and Surgery*, 12: 32–37.

Kerr, M. G., 2002. *Veterinary laboratory medicine. Clinical biochemistry and haematology.* Blackwell Science Ltd., UK.

Knotková, Z., Doubek, J., Knotek, Z. & Hájková, P., 2002. Blood cell morphology and plasma biochemistry in Russian tortoises (*Agrionemys horsfieldi*).

Acta Veterinaria BRNO, 71: 191–198.

Longmire, J. L., Maltbie, M. & Baker, R. J., 1997. Use of 'lysis buffer' in DNA isolation and its implications for museum collections. *Occasional Papers, Museum of Texas Tech University,* 163: 1–3.

López–Olvera, J. R., Montane, J., Marco, I., Martinez–Silvestre, A., Soler, J. & Lavin, S., 2003. Effect of venipuncture site on hematologic and serum biochemical parameters in marginated tortoise (*Testudo marginata*). *Journal of Wildlife Diseases*, 39: 830–836.

Lukacs, P. M. & Burnham, K. P., 2005. Review of capture–recapture methods applicable to noninvasive genetic sampling. *Molecular Ecology*, 4: 3909–3919.

Mader, D. R., 2005. *Reptile medicine and surgery*, 2nd ed. Elsevier Saunders, St Louis (MO).

Mautino, M. & Page, D., 1993. Biology and medicine of turtles and tortoises. *Veterinary Clinics of North American: Small Animal Practice,* 23: 1251–1270.

Nagy, K. & Medica, P. A., 1986. Physiological ecology of desert tortoises in southern Nevada. *Herpetologica*, 42: 73–92.

Rohilla, M. S. & Tiwari, P. K., 2008. Simple method of sampling from Indian freshwater turtles for genetic studies. *Acta Herpetologica*, 3: 65–69.

Taberlet, P., Waits, L. P. & Luikart, G., 1999. Noninvasive genetic sampling: look before you leap. *Trends in Ecology amd Evolution*, 14: 323–327.

Ureña–Aranda, C. A. & Espinosa de los Monteros, A., 2012. The genetic crisis of the Mexican Bolson Tortoise (*Gopherus flavomarginatus*: Testudinidae). *Amphibia–Reptilia*, 33: 45–53.

Van der Kuyl, A. C., Ballasina, D. L. P. & Zorgdrager, F., 2005. Mitochondrial haplotype diversity in the tortoise specie *Testudo graeca* North Africa and the Middle East. *BMC Evolutionary Biology*, 5: 29–36.

Walsh, P. S., Metzger, D. A. & Higuchi, R., 1991. Chelex 100 as a medium for simple extraction of DNA for PCR–based typing from forensic material. *Biotechniques*, 10: 506–513.

Wendland, L., Balbach, H., Brown, M., Berish, J. D., Littell, R. & Clarck, M., 2009. *Handbook of Gopher Tortoise (Gopherus polyphemus).* U. S. Army Corps Engineer Research and Development Center, Washington DC.

15

Critical steps to ensure the successful reintroduction of the Eurasian red squirrel

B. P. Vieira, C. Fonseca & R. G. Rocha

Vieira, B. P., Fonseca, C. & Rocha, R. G., 2015. Critical steps to ensure the successful reintroduction of the Eurasian red squirrel. *Animal Biodiversity and Conservation*, 38.1: 49–58.

Abstract

Critical steps to ensure the successful reintroduction of the Eurasian red squirrel.— Wildlife reintroduction strategies aim to establish viable long–term populations, promote conservation awareness and provide economic benefits for local communities. In Portugal, the Eurasian red squirrel (*Sciurus vulgaris*) became extinct in the 16th century and was reintroduced in urban parks in the 1990s, mainly for aesthetic and leisure purposes. We evaluated the success of this reintroduction in two urban parks and here describe the critical steps. We assessed habitat use, population density and abundance, and management steps carried out during reintroduction projects. Reintroductions have been successful to some extent given squirrels are present 20 years after release. However, populations in both parks are declining due to the lack of active management and poor quality habitat. Successful reintroduction of Eurasian red squirrel in areas without competition of alien tree squirrels involves three critical main stages. The pre–project stage includes studies on habitat quality, genetic proximity between donors and closest wild population, and health of donor stocks. In the release stage, the number of individuals released will depend on resource variability, and the hard release technique is an effective and economically viable method. Post–release activities should evaluate adaptation, mitigate mortality, monitor the need for supplementary feeding, provide veterinary support, and promote public awareness and education.

Key words: Conservation, Management, Release, Rodentia, *Sciurus vulgaris*, Urban park

Resumen

Pasos fundamentales para garantizar la eficacia de la reintroducción de la ardilla roja.— El objetivo de las estrategias de reintroducción de fauna silvestre es establecer poblaciones viables a largo plazo, fomentar la concienciación con respecto a la conservación y aportar beneficios económicos para las comunidades locales. La ardilla roja (*Sciurus vulgaris*), que estaba extinta en Portugal desde el s. XVI, fue reintroducida en varios parques urbanos en la década de los años 90, principalmente con fines estéticos y recreativos. Evaluamos la eficacia de esta reintroducción en dos parques urbanos y describimos los pasos fundamentales de la misma. Se evaluaron la utilización del hábitat, la densidad y abundancia de la población y las medidas de gestión adoptadas durante los proyectos de reintroducción. Las reintroducciones fueron relativamente eficaces dado a que las ardillas seguían presentes 20 años después de la liberación. No obstante, las poblaciones en ambos parques están disminuyendo debido a la falta de una gestión activa y a la mala calidad del hábitat. La reintroducción eficaz de la ardilla roja en zonas donde no hay ardillas arborícolas exóticas conlleva tres etapas fundamentales. La etapa previa al proyecto comprende estudios sobre la calidad del hábitat; la proximidad genética entre los donantes y la población silvestre más cercana, y la salud de las poblaciones donantes. En la etapa de liberación, el número de individuos liberados dependerá de la variabilidad de los recursos disponibles; asimismo, se ha observado que la técnica de liberación dura es un método eficaz y viable desde el punto de vista económico. Las actividades posteriores a la liberación deberían analizar la adaptación, mitigar la mortalidad, hacer un seguimiento de la necesidad de aportar alimentación complementaria, prestar apoyo veterinario y fomentar la sensibilización pública y la educación.

Palabras clave: Conservación, Gestión, Liberación, Rodentia, *Sciurus vulgaris*, Parque urbano

Bianca P. Vieira, Post–graduate Research Program, Inst. of Biodiversity, Animal Health and Comparative Medicine, Univ. of Glasgow, G12 8QQ, Glasgow, U. K.– Carlos Fonseca & Rita G. Rocha, Depto. de Biologia & CESAM, Univ. de Aveiro, Campus Santiago, 3810–193, Aveiro, Portugal.

Corresponding author: Rita G. Rocha. E–mail: rgrocha@ua.pt

Introduction

Animal translocation is an ancient process used by humans to relocate species from one place to another (Griffith et al., 1989; Hodder & Bullock, 1997; Armstrong & Seddon, 2007; Seddon et al., 2007; Ewen et al., 2012). Griffith et al. (1989) defined animal translocation as the intentional release to establish, re–establish or increase the population of a given species. Reintroduction is currently one of the most popular translocation strategies used in the management of species (Armstrong & Seddon, 2007; Seddon et al., 2007; Ewen et al., 2012). Wildlife reintroductions are conducted to establish viable populations, enhance long–term survival of a given species, settle long–term economic benefits for local communities, and to promote conservation awareness (IUCN, 1998). Reintroductions should be carefully planned by a multidisciplinary team, and follow a three–step protocol, focusing on the pre–project activities, release stages and post–released activities (IUCN, 1998). Such projects require complex planning, implementing and monitoring species and habitats according to their biology, socio–economic impact on local communities, and legal requirements (Caughley & Gunn, 1996; IUCN, 1998; Armstrong & Seddon, 2007; Seddon et al., 2007; Ewen et al., 2012; Harrington et al., 2013). Parameters of success change in each project but should follow the principles of long–term survival of species while providing benefits for the local community and fostering conservation awareness (IUCN, 1998, 2012). The potential positive impact of reintroductions depends on temporal, spatial, and taxonomic factors (Ewen et al., 2012), and if reintroductions are not properly carried out they can damage both donor and receptor populations as well as ecosystems (Hodder & Bullock, 1997; Armstrong & Seddon, 2007; Seddon et al., 2007; Ewen et al., 2012). Therefore, publication and dissemination of successful and unsuccessful cases contribute to improve current reintroduction protocols (Armstrong & Seddon, 2007; Seddon et al., 2007; Ewen et al., 2012; IUCN, 2012).

Mammals, together with birds, are the most frequently chosen groups for releases with conservation purposes (Griffith et al., 1989; Seddon et al., 2005, 2007). Although most of reintroductions focus on ungulates and carnivores (Seddon et al., 2005), rodents such as the edible dormouse *Glis glis* in Poland (Jurczyszyn, 2006) and the European ground squirrel *Spermophilus citellus* in Central Europe (Matĕjů et al., 2010) have also been released in the last 20 years. The most commonly reported reintroductions among rodents are those concerning the Eurasian red squirrel *Sciurus vulgaris* reintroductions, with a considerable number of programmes being implemented in Europe over the last 30 years (Swinnen, 1988; Fornasari et al., 1997; Wauters et al., 1997a, 1997b; Poole & Lawton, 2009).

Although the Eurasian red squirrel is a widespread Palearctic species (Lurz et al., 2005; Shar et al., 2008; Bosch & Lurz, 2012), some of its populations, particularly in the United Kingdom and Italy, are threatened or extinct due to habitat loss, hunting, disease and competition with alien tree squirrels

Sciurus carolinensis, Callosciurus erythraeus and *C. finlaysonii* (Gurnell, 1987; Wood et al., 2007; Bosch & Lurz, 2012; Bertolino & Lurz, 2013). In Portugal, the Eurasian red squirrel became extinct in the 16th century due to significant habitat loss, and it only reappeared in extreme northern areas around the 1980s (Mathias & Gurnell, 1998; Ferreira et al., 2001). One decade later, isolated reintroductions occurred in some urban parks, but no monitoring has been conducted since then to understand population dynamics and their status or to evaluate management and success. In order to determine whether Eurasian red squirrel reintroductions carried out in Jardim Botânico da Universidade de Coimbra and in Parque Biológico de Gaia were successful or not, we estimated population viability through density, abundance, and habitat use in released sites. We also evaluated stepwise reintroduction in both urban parks, based on the IUCN guidelines (IUCN, 1998, 2012) to highlight critical steps and suggest actions ensure the long–term persistence and viability of these Eurasian red squirrel populations.

Material and methods

Study area

The Parque Biológico de Gaia (PBG), which was created in 1983, initially covered 2 ha but has been extended to include 35 ha (Oliveira, 2013). It is situated in Vila Nova de Gaia, northern Portugal (41° 05' N and 8° 33' W; fig. 1A) and it is managed by a municipal company (Oliveira, 2013). This urban park is composed of open areas, a wildlife rehabilitation center and monospecific forests of black alder *Alnus glutinosa*, oak *Quercus robur* or cork oak *Q. suber* (fig. 1A). It also has enclosures distributed throughout open areas in the park containing wildlife that could not be rehabilitated or that provide examples of native and exotic fauna. The aim is to promote environmental education, as the case of the Eurasian red squirrel.

The Jardim Botânico da Universidade de Coimbra (JBUC) was created in 1772 as part of the Museu de História Natural da Universidade de Coimbra (UC, 2013). It covers 13 ha in Coimbra city, central Portugal (40° 12' N and 8° 25' W; fig. 1B). This park comprises mainly gardens and forests of alien flora (fig. 1B).

Eurasian red squirrel survey

From 7th October to 11th November 2013, 15 walking transects of 100 m were established in each study site. All transects were surveyed in the morning and afternoon for seven days to avoid biases from squirrel behavior (Gurnell et al., 2001). Transects were performed one after other to avoid double counting of individuals moving from one transect to another. Transects were selected to include all habitat types and to cover most of the area at each of the two urban parks, but with at least 20 m distance from each

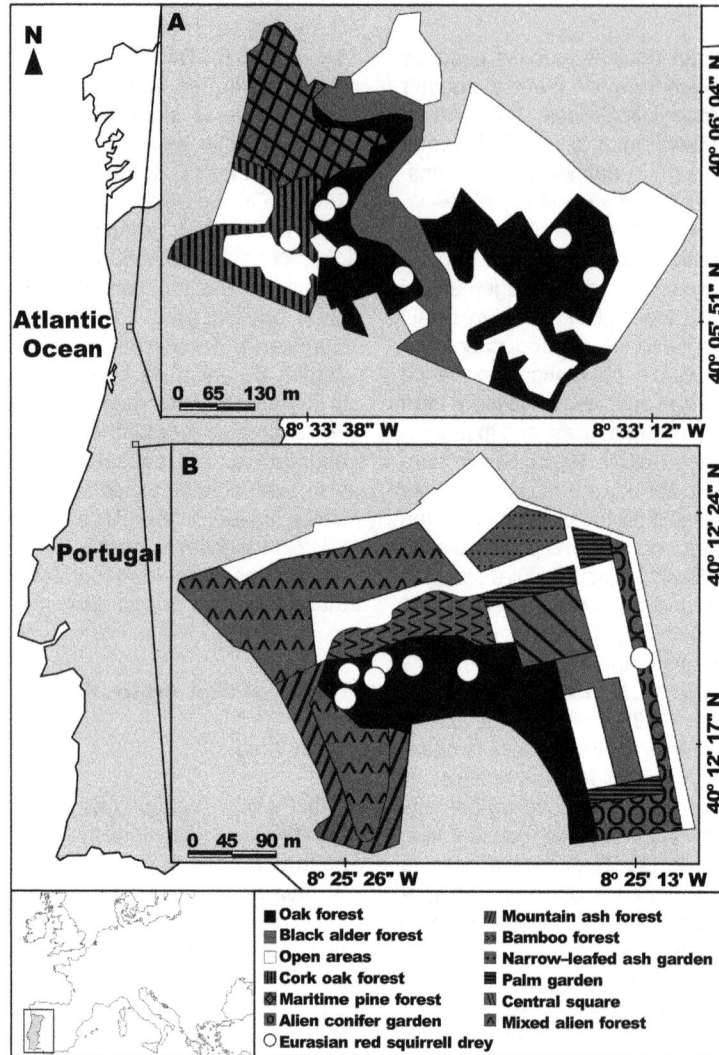

Fig. 1. Study area in northern and central Portugal with inset showing vegetation type and distribution of dreys (white circles) in the Parque Biológico de Gaia at Vila Nova de Gaia (A), and in the Jardim Botânico da Universidade de Coimbra at Coimbra (B).

Fig. 1. Zona de estudio situada en el norte y el centro de Portugal con recuadros que muestran el tipo de vegetación y la distribución de los nidos de ardilla (círculos blancos) en el Parque Biológico de Gaian en Vila Nova de Gaia (A) y en el Jardín Botánico de la Universidad de Coimbra, en Coimbra (B).

other also to avoid double counting (fig. 1; Gurnell et al., 2001). We counted squirrels using the distance sampling method with direct observation using binoculars 8–16 x 40 (Gurnell et al., 2001), given that both parks had great visibility with small and clear forests. Squirrel surveys were conducted between 8:00 and 16:00 in autumn when higher numbers of squirrels can be found (Tonkin, 1983; Wauters et al., 1992; Bosch & Lurz, 2012). The distance from the observer to the squirrel was measured using a telemeter, and compass bearings were taken to determine the angle between the animal and the transect line (Buckland et al., 1993; Gurnell et al., 2001). We measured

the distance of squirrels once and did not consider individuals again after moving to a different position.

Population density and abundance were estimated using Distance Sampling 6.0 software (Thomas et al., 2010). Estimates were stratified and based on Conventional Distance Sampling. Half–normal, hazard and negative exponential rate models for the detection function were fixed against the records using a cosine function (Thomas et al., 2010). Models assumed certainty of detection and measurements (Thomas et al., 2010). The selection of the best model and adjustment term were based on the lowest Akaike information criterion (AIC).

Table 1. Best–fitting models according to Akaike information criterion (AIC) and degree of freedom (df) values to estimate the population density of Eurasian red squirrels at the Parque Biológico de Gaia (PBG) and at the Jardim Botânico da Universidade de Coimbra (JBUC), Portugal, in autumn 2013.

Tabla 1. Los mejores modelos según el criterio de información de Akaike (AIC) y los valores del grado de libertad (df) para estimar la densidad de la población de ardilla roja en el Parque Biológico de Gaia (PBG) y en el Jardín Botánico de la Universidad de Coimbra (JBUC), en Portugal, en otoño de 2013.

	PBG			JBUC		
Model	Negative exponential	Half–normal	Hazard	Negative exponential	Half–normal	Hazard
AIC	378.09	378.28	376.65	17.51	17.56	19.56
df	60	59	59	3	3	2

Habitat use

Eurasian red squirrels prefer mature native forests that can provide them with an abundant supply of food (Bosch & Lurz, 2012). We assessed vegetation type, location of dreys (*i.e.* squirrel nests) and food availability to understand habitat use in both urban parks. The survey was conducted in the PBG in October 2013 and in the JBUC in November 2013. Vegetation type (fig. 1) was mapped with a geographic information system in ArcView GIS 9.2 software (ESRI, 2008). The geographical limits of forests and gardens having the same composition and dominance were confirmed in the field. The number of trees to determine dominant species was verified in 10 x 10 m quadrats randomly within the study sites. Due to different area sizes, 80 quadrats were located in PBG and 60 in JBUC.

Dreys were mapped to determine preferences in relation to vegetation type (fig. 1). Drey counts were obtained by direct observation in a 3 km transect at each site. Transects to count dreys were larger than transects to count individuals because dreys were fixed and double counting was unlikely. We determined the position of dreys, tree species chosen and drey height (Cagnin et al., 2000; Kopij, 2009). Old or abandoned dreys were excluded from counts (Wauters & Dhondt, 1988; Cagnin et al., 2000; Kopij, 2009). The significance of the distribution of dreys in relation to height was measured using a one–way ANOVA in Bioestat 5.0 software (Ayres et al., 2007). Tukey's *post hoc* test (*F*) was applied to determine the significance of any differences (Zar, 1999).

Food availability focused on three aspects: number of feeders, relative abundance and richness of edible mushrooms, and energetic content of natural seeds (cones, acorns, hackberries, and nuts). Feeders with supplementary food were counted directly. Relative abundance of Basidiomycota was estimated by counting fungal bodies or remains with characteristic squirrel bites in stipe and cap on the ground in the same 10 x 10 m quadrats where the vegetation type was measured. Only mushrooms eaten by squirrels during the surveys or reported in the literature were considered as a component of Eurasian red squirrel diet (Fogel & Trappe, 1978; Bertolino et al., 2004). Fungi identification and nomenclature follows Crous et al. (2004).

In each 10 x 10 m quadrat, the number of trees with fruits of each species was counted. Only tree species already reported in the literature (Lurz et al., 2005; Bosch & Lurz, 2012) or those seen being consumed during fieldwork were considered as a component of the Eurasian red squirrel diet. A quadrat of 5 x 5 m was placed below every tree bearing fruit inside the 10 x 10 m quadrat to count fallen cones, acorns, hackberries or nuts. The remains of fruits consumed by Eurasian red squirrels were also recorded and identified by characteristic squirrel bites. Only natural sources of seeds were evaluated given that the composition of seeds offered in feeders varied widely. Seed counts provided an estimate of seed availability (calculated as 10^3 seeds/ha, Bosch & Lurz, 2012). We used data on seed production and calorific content obtained from the literature (Grodziński & Sawicka–Kapusta, 1970; Demir et al., 2002; Wauters et al., 2002; Bosch & Lurz, 2012; Stock et al., 2013) to measure the mean energy value (10^3 kJ/ha^{-1}) and standard deviation (± SD) related to seed counts per habitat.

Reintroduction management

Reintroductions were considered successful as viable populations were established, long–term benefit for local communities were achieved, and improvements in conservation awareness were made, in accordance with IUCN guidelines (IUCN, 1998, 2012). Qualitatively data on management were assessed by unstructured interviews with park managers and employees, and by consulting official documents. We investigated release histories according to motivation, year of release, reintroduction technique (*e.g.* soft release in which animals are first acclimatized with new habitat in enclosures before release, or hard release in which individuals are directly released into the new environment; see Ewen et al., 2012), supplementary feeding, veterinary support, choice of donor population, number of individuals released, reinforcements, population expansion or decrease, presence/absence of dispersal into surrounding areas (distances greater than 500 m and less than 35 km from release sites), and the presence/absence of squirrels killed on roads.

Results

Eurasian red squirrel population

During the surveys, we observed 61 individuals in the PBG and four in the JBUC. The best relative fit model and adjustment term for the population in the PBG was a hazard–rate cosine based on the lowest AIC score (table 1). In contrast, the best fit for the JBUC population was a negative exponential cosine model based on the lowest AIC score (table1). Estimated abundance and density were higher in PBG (N = 47 squirrels, D = 1.33 squirrel/ha) than in JBUC (N = 2 squirrels, D = 0.17 squirrel/ha). The detection probability in the PBG was 44% whereas in the JBUC it was 51.5%. The encounter rate was 56% and 48.5% for the PBG and JBUC, respectively.

Habitat use

Seven squirrel dreys were found placed in the oak and cork oak forest at PBG (fig. 1B). Squirrels placed a significant portion of dreys in the height of 13 m in forests dominated by *Q. robur* and *Castanea sativa* (F = 8.35, df = 11, P < 0.01). Seven dreys were found around 14 m high in the oak forest in the JBUC (F = 15.74, df = 14, P < 0.01). One drey was found on *Pseudotsuga menziesii* in the alien conifer garden (fig. 1B).

During autumn 2013, only three tree species were fruiting in the PBG: *Q. robur*, *C. sativa* and *P. pinaster* (fig. 1A, table 2). The black alder forest had higher seed productivity (143.8 ± 163.6 x 10^3 seeds/ha) and energetic content (9,524 x 10^3 kJ/ha^{-1}) due to the high concentration of fruiting *C. sativa* (table 2). We counted 471 fungal bodies from 28 species in the PBG. Only 16% of mushrooms were edible to the Eurasian red squirrel (table 3). *Russula* spp. showed significant relative abundance of edible fungi in the PBG, with *R. cyanoxantha* and *R. decipiens* together accounting for 59.4% (table 3). Ongoing supplementary feeding in the PBG consisted of five feeders daily supplied with birdseed to attract birds, but these were also used by squirrels. Squirrels were observed eating mainly sunflower seeds.

Only the oak and mixed forests had fruiting trees in the JBUC during surveys (table 2), namely *Pinus pinea*, *Quercus robur*, and *Celtis australis*. Fruits of this last tree were seen being eaten by squirrels during fieldwork. The oak forest had higher seed productivity (974.0 ± 534.9 x 10^3 seeds/ha) and energetic content (20,779 x 10^3 kJ/ha^{-1}) due to the high productivity of seeds per cone of *P. pinea* (table 2). In the JBUC, we counted 33 fungal bodies of seven species and 70% of them were edible to the Eurasian red squirrel (table 3). As in PBG, the genus *Russula* was also

Table 2. Estimation of seed production in each habitat at the Parque Biológico de Gaia (PBG) and at the Jardim Botânico da Universidade de Coimbra (JBUC), Portugal, in autumn 2013: S. Seed (10^3 seeds/ha); Sec. Seed energetic content (10^3 kJ/ha^{-1}). Habitats: Of. Oak forest; Baf. Black alder forest; Oa. Open area; Cof. Cork oak forest; Mpf. Maritime pine forest; Ecg. Exotic conifer garden; Maf. Mountain ash forest; Bf. Bamboo forest; Nag. Narrow–leafed ash garden; Pg. Palm garden; Cs. Central square; Mef. Mixed exotic forest.

Tabla 2. Estimación de la producción de semillas en cada hábitat en el Parque Biológico de Gaia (PBG) y en el Jardín Botánico de la Universidad de Coimbra (JBUC), Portugal, en otoño de 2013: S. Semillas (10^3 seeds/ha); Sec. Contenido energético de las semillas (10^3 kJ/ha^{-1}). Hábitats: Of. Robledal; Baf. Alisal; Oa. Zona despejada; Cof. Alcornocal; Mpf. Pinar de pino negral; Ecg. Plantación de coníferas exóticas; Maf. Bosque de eucalipto regnans; Bf. Bosque de bambú; Nag. Plantación de fresno de hoja pequeña; Pg. Palmeral; Cs. Cuadrado central; Mef. Bosque mixto exótico.

Urban parks	Measured parameters	
Habitat	S	Sec
PBG		
Of	3.6 ± 2.2	96.4
Baf	143.8 ± 163.6	9,524
Oa	–	–
Cof	–	–
Mpf	74.6 ± 35.2	1,859
JBUC		
Of	974.0 ± 534.9	20,779
Oa	–	–
Ecg	–	–
Maf	–	–
Bf	–	–
Nag	–	–
Pg	–	–
Cg	–	–
Mef	56.3 ± 19.5	3,821

an important food source in JBUC, with a relative abundance of 30.4% for *R. foetens* in the diet, which together with *Amanita gemmata* represented 91.2% of available edible mushrooms in the JBUC (table 3). Supplementary feeding was not recorded in this urban park during the study.

Table 3. Relative abundance of mushrooms (Basidiomycota) recorded in the diet of Eurasian red squirrel (*Sciurus vulgaris*) and found at the Parque Biológico de Gaia (PBG) and at the Jardim Botânico da Universidade de Coimbra (JBUC), Portugal, in autumn 2013: * Consumption seen during fieldwork.

*Tabla 3. Abundancia relativa de hongos (Basidiomycota) observada en la alimentación de la ardilla roja (Sciurus vulgaris) y encontrada en el Parque Biológico de Gaia (PBG) y en el Jardín Botánico de la Universidad de Coimbra (JBUC), en Portugal, en otoño de 2013: * Consumo visto durante el trabajo de campo.*

Urban park

Taxon	Fungal bodies	Relative abundance (%)	Source
PBG			
Boletus aestivalis	1	1.3	Fogel & Trappe (1978)*
Cantharellus cibarius	12	16.2	Fogel & Trappe (1978)
Pholiota alnicola	9	12.1	Fogel & Trappe (1978)
Amanita rubescens	1	1.3	Fogel & Trappe (1978)
Russula cyanoxantha	30	40.5	*
Russula decipiens	14	18.9	*
Xerocomus chrysenteron	7	9.4	Fogel & Trappe (1978)*
Total	**74**	**100**	
JBUC			
Agaricus campestris	2	8.7	Fogel & Trappe (1978)
Amanita gemmata	14	60.8	Fogel & Trappe (1978)
Russula foetens	7	30.4	*
Total	**23**	**100**	

Reintroduction management

Table 4 summarizes reintroductions attendance to IUCN stepwise. Both reintroductions in PBG and JBUC aimed at enhancing parks aesthetics and enable people to become familiar with this species (table 4). Park managers used the Eurasian red squirrel historical population observations of Antunes (1985) as proving of the species historical range in Portugal. Both urban parks acquired squirrels from commercial creators with veterinary control and support which lowered possibilities of diseases or parasites.

The PBG released 12 squirrels In 1997, and a further 40 couples between 1998 and 2001 using a hard release approach. The animals were from Azé (France). The squirrels in the PBG have continuous veterinary support, because a wildlife rehabilitation center is located therein (table 4), and continuous feeding is provided through bird feeders. Two main failures were detected in the reintroduction project in PBG: the absence of genetic comparison between donors and the closest wild population, and a lack of long–term, technical monitoring.

Twelve squirrels from Madrid (Spain) were hard released at JBUC in 1994. Four squirrel feeders in the forest were active only during the first year (table 4). As in the PBG, the reintroduction project at this park did not consider genetic comparison between donors and the closest wild population. This project ended one year after releases and no post–project management measures and/or long–term monitoring were conducted (table 4).

Discussion

To date, reintroductions of Eurasian red squirrels in Portugal have been successful to some extent given that squirrels are still present in the urban parks almost 20 years later. However, the populations of squirrels are decreasing in both urban parks. Studies found densities from 0.03 to 1.80 squirrels/ha in mixed woodlands (Wauters & Dhondt, 1988; Cagnin et al., 2000; Magris & Gurnell, 2002; Vilar, 1997), figures that are similar to our estimate (0.17 squirrels/ha in JBUC and 1.33 squirrels/ha in PBG). Considering urban parks of limited area and resources, the density in the PBG is similar to densities found in Belgium (Wauters et al., 1997a) and Spain (Vilar, 1997, while the density in the JBUC is lower. The difference in density of the reintroduced populations is mainly related to post–release management since the PBG had squirrel population reinforcements but the JBUC did not. Habitat quality also regulates species

Table 4. Conditions of Eurasian red squirrel (*Sciurus vulgaris*) reintroductions in the Parque Biológico de Gaia (PBG) and in the Jardim Botânico da Universidade de Coimbra (JBUC), Portugal.

Tabla 4. Condiciones de las reintroducciones de ardilla roja (Sciurus vulgaris) en el Parque Biológico de Gaia (PBG) y en el Jardín Botánico de la Universidad de Coimbra (JBUC), en Portugal.

Phase			
Aspect		PBG	JBUC
Pre–project			
	Main motivation	Aesthetic, leisure and environmental education	Aesthetic, leisure and environmental education
	Origin of donor population	Azé (France)	Madrid (Spain)
	Captive or wild squirrels	Captive	Captive
	Subspecies of donor population	*Sciurus vulgaris fuscoater*	*Sciurus vulgaris infuscatus*
	Study of historical range of extinct populations	Antunes (1985)	Antunes (1985)
	Study of genetic individual variability of donor population	No	No
	Governmental permits	Not required	Not required
	Veterinary certification of health and absence of parasites	Yes	Yes
	Other certifications	Origin and transportation	Origin and transportation
Release			
	Year of release	November 1997	June 1994
	Number of squirrels released	12 (6♀♀ and 6♂♂)	12 (6♀♀ and 6♂♂)
	Method (soft or hard)	Hard release	Hard release
	Supplementary feeding	Five feeders for birds, but used by squirrels	Four squirrel feeders
	Kind of supplementary feeding	Birdseed with sunflower seed sunflower seed	Walnuts, hazelnuts, and others
Post–project			
	Population monitoring	No	No
	Veterinary support	Yes	No
	Continuity of supplementary feeding	Yes, to date	Stopped after one year
	Manager's general feeling about squirrels abundance	Decrease and need for reinforcement	Population explosion in next three years
	Population reinforcement	Yes (three)	No
	Subspecies used to reinforcement	*Sciurus vulgaris fuscoater*	–
	Number of individuals (origin, and year of population reinforcement)	10 couples (Epe, Netherlands, X 1998) 15 couples (Epe, Netherlands, VII 2001) 15 couples (Azé, France, VIII 2001	– – –
	Manager's general feeling about squirrel abundance after reinforcement	Population explosion in next five years	–
	Squirrels seen in nearby areas (> 500 m and < 35 km from release site)	Yes	Yes
	Name of five localities where squirrels were seen	Sermonde, Vila Chã, Serra da Agrela, Serra da Freita, and Marco de Canaveses	Mata do Buçaco, Serra da Lousã, Alfarelos, Serra do Sicó, and Soure
	First year of squirrels seen in these nearby localities	2010	2001
	Squirrels killed on nearby roads	Yes	Yes

abundance and density and is of great importance for the success or failure of reintroductions (Ewen et al., 2012). Squirrel dreys in both urban parks were predominantly placed in native oak forests and near food sources, reinforcing the need for high quality habitat and food diversity for the maintenance of these populations. The studied parks had few fruiting trees compared with other studies (Bosch & Lurz, 2012). Forests in the JBUC had a higher energetic content than those in the PBG but the diversity of native food items was poorer. In contrast, forests in the PBG had less energetic content, but they presented richer and more abundant additional food items, such as edible mushrooms. Additionally, the PBG had continuous supplementary food, mainly through bird feeders also used by squirrels, whereas the JBUC only had feeders in the year following the reintroduction.

In terms of species identity for conservation purposes, genetic proximity was only adequately considered in the JBUC where the subspecies *Sciurus vulgaris infuscatus* was reintroduced, while in the PBG the subspecies *S. v. fuscoater* was released. Although both subspecies occur in the Iberian Peninsula, only *S. v. infuscatus* occurs naturally in Portugal (Mathias & Gurnell, 1998; Lurz et al., 2005; Bosch & Lurz, 2012). Further studies on Eurasian red squirrel distribution, taxonomy and genetic diversity in the Iberian Peninsula should consider the influence of *S. v. fuscoater* presence in Portugal, as has been done to other subspecies in the United Kingdom (see Hale & Lurz, 2003; Hale et al., 2004).

Post–project monitoring was not explicitly considered in either park. It is there not fully understood whether dispersal to vicinity (table 3) was natural or due to stress of limited resources. Deficiency in post–release actions, such as monitoring health and abundance, is responsible for the long–term decrease in Eurasian red squirrel populations in both urban parks but active adaptive management could improve the current situation (Ewen et al., 2012; Runge, 2013). Future actions should consider improving habitat quality by means of specific feeders for Eurasian red squirrels, and replacement of alien trees for native oak forest. Monitoring population health, adaptation and demographic variation will endorse the long–term success of the reintroductions. In addition, managers should ensure active human community involvement so that effective education would not only foster knowledge of species but also concern for its needs (IUCN, 2012).

Critical steps for successful reintroduction of Eurasian red squirrels in areas without competition of alien tree squirrels should follow three stages, consisting of pre–project activities, release stages and post–release activities (IUCN, 1998, 2012). Pre–project activities should include studies on (1) habitat quality, (2) genetic proximity between donors and the closest wild population, and (3) the health of donor stocks.

In the release stage, (1) the number of individuals released should consider 35 to 85 individuals to achieve a long–term viable population in an area of high resource variability, and 55 to 175 individuals in areas of low resource variability (Wood et al., 2007) and (2) hard release technique proved to be a good

and cheaper method to Eurasian red squirrel reintroductions (Swinnen 1988; Fornasari et al., 1997). Finally, post–release activities should (1) evaluate population adaptation, (2) mitigate mortality, (3) monitor the need for supplementary feeding, (4) provide veterinary support, and (5) promote continuous public awareness and education (IUCN, 1998, 2012).

Reintroductions for aesthetic and leisure purposes are not usually concerned about strictly following conservation protocols unless required by law. However, these reintroductions for aesthetic and leisure purposes have significant effects on wildlife management and conservation (Hodder & Bullock, 1997). Therefore, we strongly suggest that reintroductions with aims other than conservation should also have standardized international guidelines, regulations and monitoring.

Acknowledgments

We thank Maria A. Neves for helping with mushroom identification, and Paulo Trincão (Jardim Botânico da Universidade de Coimbra) and Nuno G. Oliveira (Parque Biológico de Gaia) for permits, institutional support, and data on reintroductions. We also thank the reviewers for improving this manuscript. Bianca P. Vieira had support from the Brazilian National Council for Scientific and Technological Development (CNPq) under a Science without Borders fellowship (nº 221.575/2012–0). The project was partially supported by European Funds through Operational Program for Competitiveness Factors and by National Funds through the Portuguese Science Foundation (PEst–C/ MAR/LA0017/2013).

References

Antunes, M. T., 1985. *Sciurus vulgaris* no Cabeço da Arruda, Muge: presença e extinção em Portugal. *Arqueologia*, 12: 1–16.

Armstrong, D. P. & Seddon, P. J., 2007. Directions in reintroduction biology. *Trends in Ecology and Evolution*, 23: 20–25.

Ayres, M., Ayres, M. Jr., Ayres, D. L. & Santos, A. A. S., 2007. *BioEstat 5.0: aplicações estatísticas nas áreas das ciências biológicas e médicas*. Sociedade Civil Mamirauá, Belém.

Bertolino, S. & Lurz, P. W., 2013. *Callosciurus* squirrels: worldwide introductions, ecological impacts and recommendations to prevent the establishment of new invasive populations. *Mammal Review*, 43: 22–33.

Bertolino, S., Vizzini, A., Wauters, L. A. & Tosi, G., 2004. Consumption of hypogeous and epigeous fungi by the red squirrel (*Sciurus vulgaris*) in subalpine conifer forests. *Forest ecology and management*, 202: 227–233.

Bosch, S. & Lurz, P. W. W., 2012. *The Eurasian red squirrel*. Westarp Wissenschaften, Hohenwarsleben.

Buckland, S. T., Anderson, D. A., Burnham, K. P. &

Laake, J. L., 1993. *Distance sampling: estimating abundance of biological populations.* Chapman & Hall, London.

Cagnin, M., Aloise, G., Fiore, F., Oriolo, V. & Wauters, L. A., 2000. Habitat use and population density of the red squirrel, *Sciurus vulgaris meridionalis*, in the Sila Grande mountain range (Calabria, South Italy). *Italian Journal of Zoology*, 67: 81–87.

Caughley, G. & Gunn, A., 1996. *Conservation biology in theory and practice.* Blackwell Science, Cambridge and Massachusetts.

Crous, P. W., Gams, W., Stalpers, J. A., Robert, V. & Stegehuis, G., 2004. MycoBank: an online initiative to launch mycology into the 21st century. *Studies in Mycology*, 50: 19–22.

Demir, F., Doğan, H., Özcan, M. & Haciseferoğullari, H., 2002. Nutritional and physical properties of hackberry (*Celtis australis* L.). *Journal of Food Engineering*, 54: 241–247.

Ewen, J. G., Armstrong, D. P., Parker, K. A. & Seddon, P. J., 2012. *Reintroduction biology: integrating science and management.* Wiley–Blackwell, New Jersey.

ESRI, 2008. *ArcView: Release 9.2.* Environmental Systems Research Institute, Redlands.

Ferreira, A. F., Guerreiro, M., Álvares, F. & Petrucci–Fonseca, F., 2001. Distribución y aspectos ecológicos de *Sciurus vulgaris* en Portugal. *Galemys*, 13: 155–170.

Fogel, R. & Trappe, J. M., 1978. Fungus consumption (mycophagy) by small animals. *Northwest Science*, 52: 1–31.

Fornasari, L., Casale, P. & Wauters, L., 1997. Red squirrel conservation: the assessment of a reintroduction experiment. *Italian Journal of Zoology*, 64: 163–167.

Griffith, B., Scott, J. M., Carpenter, J. W. & Reed, C., 1989. Translocation as a species conservation tool: status and strategy. *Science*, 245: 477–480.

Grodziński, W. & Sawicka–Kapusta, K., 1970. Energy values of tree–seeds eaten by small mammals. *Oikos*, 21: 52–58.

Gurnell, J., 1987. *The natural history of squirrels.* Facts on File Publications, Oxford.

Gurnell, J., Lurz, P. W. W. & Pepper, H., 2001. Practical techniques for surveying and monitoring squirrels. *Forestry Commission Practice Note*, 11: 1–12.

Hale, M. L. & Lurz, P. W. W., 2003. Morphological changes in a British mammal as a result of introductions and changes in landscape management: the red squirrel (*Sciurus vulgaris*). *Journal of Zoology*, 260: 159–167.

Hale, M. L., Lurz, P. W. W. & Wolff, K., 2004. Patterns of genetic diversity in the red squirrel (*Sciurus vulgaris* L.): Footprints of biogeographic history and artificial introductions. *Conservation Genetics*, 5: 167–179.

Harrington, L. A., Moehrenschlager, A., Gelling, M., Atkinson, R. P., Hughes, J. & Macdonald, D. W., 2013. Conflicting and complementary ethics of animal welfare considerations in reintroductions. *Conservation Biology*, 27: 486–500.

Hodder, K. H. & Bullock, J. M., 1997. Translocations of native species in the UK: implications for biodi-versity. *Journal of Applied Ecology*, 34: 547–565.

IUCN, 1998. *Guidelines for reintroductions.* IUCN/SSC Re–introduction Specialist Group, Gland and Cambridge.

– 2012. *Guidelines for reintroductions and other conservation translocations.* IUCN/SSC Species Survival Commission, Gland and Cambridge.

Jurczyszyn, M., 2006. The use of space by translocated edible dormice, *Glis glis* (L.), at the site of their original capture and the site of their release: radio–tracking method applied in a reintroduction experiment. *Polish Journal of Ecology*, 54: 345–350.

Kopij, G., 2009. Habitat and drey sites of the red squirrel *Sciurus vulgaris* Linnaeus 1758 in suburban parks of Wroclaw, SW Poland. *Acta Zoologica Cracoviensia Series A: Vertabrata*, 52: 107–114.

Lurz, P. W. W., Gurnell, J. & Magris, L., 2005. *Sciurus vulgaris. Mammal Species*, 769: 1–10.

Magris, L. & Gurnell, J., 2002. Population ecology of the red squirrel (*Sciurus vulgaris*) in a fragmented woodland ecosystem on the Island of Jersey, Channel Islands. *Journal of Zoology*, 256: 99–112.

Matějů, J., Říčanová, Š., Ambros, M., Kala, B., Hapl, E. & Matějů, K., 2010. Reintroductions of the European ground squirrel (*Spermophilus citellus*) in Central Europe (Rodentia: Sciuridae). *Lynx*, 41: 175–191.

Mathias, M. L. & Gurnell, J., 1998. Status and conservation of the red squirrel (*Sciurus vulgaris*) in Portugal. *Hystrix*, 10: 13–19.

Oliveira, N. G., 2013. *Parque Biológico de Gaia: 1983/2013.* Parque Biológico de Gaia, Vila Nova de Gaia.

Poole, A. & Lawton, C., 2009. The translocation and post release settlement of red squirrels *Sciurus vulgaris* to a previously uninhabited woodland. *Biodiversity Conservation*, 18: 3205–3218.

Runge, M. C., 2013. Active adaptive management for reintroduction of an animal population. *Journal of Wildlife Management*, 77: 1135–1144.

Seddon, P. J., Armstrong, D. P. & Maloney, R. F., 2007. Developing the science of reintroduction biology. *Conservation Biology*, 21: 303–312.

Seddon, P. J., Soorae, P. S. & Launay, F., 2005. Taxonomic bias in reintroduction projects. *Animal Conservation*, 8: 51–58.

Shar, S., Lkhagvasuren, D., Bertolino, S., Henttonen, H., Kryštufek, B. & Meinig, H., 2008. *Sciurus vulgaris.* IUCN Red List of Threatened Species, v.2013.1. http://www.iucnredlist.org. (Accessed on 6 November 2013).

Stock, W. D., Finn, H., Parker, J. & Dods, K., 2013. Pine as fast food: foraging ecology of an endangered cockatoo in a forestry landscape. *Plos One*, 8: e61145.

Swinnen, C., 1988. Reintroduction of the red squirrel (*Sciurus vulgaris* L.) in an isolated park habitat. *Parasitica*, 44: 89–91.

Thomas, L., Buckland, S. T., Rexstad, E. A., Laake, J. L., Strindberg, S., Hedley, S. L., Bishop, J. R. B., Marques, T. A. & Burnham, K. P., 2010. Distance software: design and analysis of distance sampling surveys for estimating population size. *Journal of Applied Ecology*, 47: 5–14.

Tonkin, J. M., 1983. Activity patterns of the red squirrel (*Sciurus vulgaris*). *Mammal Review*, 13: 99–111.

Vilar, J. P., 1997. Ecoetologia i biologia de l'esquirol (*Sciurus vulgaris*, Linnaeus, 1758) en dos hàbitats de predictibilitat alimentària contínua que difereixen en l'abundància d'aliment. Ph. D. Thesis, University of Barcelona.

Wauters, L. A., Casale, P. & Fornasari, L., 1997b. Post-release behaviour, home range establishment and settlement success of reintroduced red squirrels. *Italian Journal of Zoology*, 64: 169–175.

Wauters, L. A. & Dhondt, A. A., 1988. The use of red squirrel (*Sciurus vulgaris*) dreys to estimate population density. *Journal of Zoology*, 214: 179–187.

Wauters, L. A., Somers, L. & Dhondt, A. A., 1997a. Settlement behaviour and population dynamics of reintroduced red squirrels *Sciurus vulgaris* in a park in Antwerp, Belgium. *Biology Conservation*, 82: 101–107.

Wauters, L. A., Swinnen, C. & Dhondt, A. A., 1992. Activity budget and foraging behaviour of red squirrels (*Sciurus vulgaris*) in coniferous and deciduous habitats. *Journal of Zoology*, 227: 71–86.

Wauters, L. A., Tosi, G. & Gurnell, J., 2002. Interspecific competition in tree squirrels: do introduced grey squirrels (*Sciurus carolinensis*) deplete tree seeds hoarded by red squirrels (*S. vulgaris*)? *Behavioral Ecology and Sociobiology*, 51: 360–367.

Wood, D. J., Koprowski, J. L. & Lurz, P. W. W., 2007. Tree squirrel introduction: a theoretical approach with population viability analysis. *Journal of Mammalogy*, 88: 1271–1279.

Zar, J. H., 1999. *Biostatistical analysis, 4th ed.* Prentice–Hall Inc., New Jersey.

16

Habitat use pattern and conservation status of smooth–coated otters *Lutrogale perspicillata* in the Upper Ganges Basin, India

M. S. Khan, N. K. Dimri, A. Nawab, O. Ilyas & P. Gautam

Khan, M. S., Dimri, N. K., Nawab, A., Ilyas, O. & Gautam, P., 2014. Habitat use pattern and conservation status of smooth–coated otters *Lutrogale perspicillata* in the Upper Ganges Basin, India. *Animal Biodiversity and Conservation*, 37.1: 69–76.

Abstract

Habitat use pattern and conservation status of smooth–coated otters Lutrogale perspicillata *in the Upper Ganges Basin, India.*— Smooth–coated otters inhabit several major river systems in southern Asia, and their environmental requirements link them to food and water security issues as the region is so densely populated by humans. The lack of baseline data on their distribution and ecology is another major constraint that the species is facing in India. The present study was stimulated by the rapid decline in the otter's population in the country and focuses on estimating the conservation status, habitat use pattern, and associated threats in the upper Ganges River Basin (N India). Our findings contribute towards a better understanding of the complex ecological interactions and the design of effective conservation measures. Coupled with the habitat preferences, the study also provides new locations in the species distribution. This paper highlights the gap areas in the conservation of the species and suggests areas that should be prioritized for management.

Key words: Otter, Ganges Basin, Conservation status, Habitat use.

Resumen

Modelo de uso del hábitat y estado de conservación de las nutrias lisas Lutrogale perspicillata *en la zona alta de la cuenca del Ganges, India.*— Las nutrias lisas habitan en varios sistemas fluviales importantes del Asia meridional y sus necesidades medioambientales las vinculan con problemas de seguridad alimentaria e hídrica, debido a la elevada densidad de humanos. La falta de datos de referencia sobre su distribución y ecología es otra limitación notable que la especie está afrontando en la India. El presente estudio se vio impulsado por el rápido descenso de la población de nutrias en el país y se centra en estimar el estado de conservación, el modelo de uso del hábitat y las amenazas asociadas en la zona alta de la cuenca del río Ganges (Asia septentrional). Nuestros resultados contribuyen a comprender mejor las complejas interacciones ecológicas y a elaborar medidas de conservación eficaces. Junto con las preferencias de hábitat, en el estudio también se informa sobre nuevas ubicaciones en la distribución de la especie. Asimismo se ponen de relieve las deficiencias existentes en la conservación de la especie y se sugieren las zonas cuya ordenación debería ser prioritaria.

Palabras clave: Nutria, Cuenca del Ganges, Estado de conservación, Uso del hábitat.

Mohd. Shahnawaz Khan & Asghar Nawab, WWF India, 172 B Lodi Estate, New Delhi, 110 003 (India).– Nand Kishor Dimri, WII, 18, Chandrabani, Dehradun, Uttarakhand, (India).– Orus Ilyas, AMU, Aligarh, Uttar Pradesh 202 002 (India).– Parikshit Gautam, FES, NDDB House PB. 4906 Safdarjung Enclave, New Delhi, 110 029 (India).

Introduction

Natural floodplains are biologically the most productive and diversified ecosystems on earth (Mitsch & Gosselink, 2000) but due to their very slow recovery they are also the most threatened (Vitouesk et al., 1997; Ravenga et al., 2000). The Ganga River Basin is among the world's largest productive floodplain ecosystems with enormous ecological, cultural and economical value (Ambastha et al., 2007). It has an extraordinary variety in altitude, climate, land use and biodiversity (O'Keeffe et al., 2012) The entire span of the Ganga River Basin in India can be divided into three stretches *i.e.* the upper reach from the origin to Narora, the middle reach from Narora to Ballia, and the lower reach from Ballia to its delta.

The upper Ganga River Basin is a dynamic, bio–spatial complex eco–region. The natural landscape has been severely fragmented by anthropogenic factors and most of the wildlife endowments are restricted either to the Shivalik hills and their adjacent Bhabar–Terai tract or to protected areas (Rodgers & Panwar, 1988). These pockets in the upper Ganga River Basin provide refuge to some threatened populations of endangered aquatic and semi–aquatic mammalian species like the Ganges river dolphin *Platanista gangetica* and the smooth–coated otter *Lutrogale perspicillata*, respectively.

The amphibious life styles of otters allow them to disperse over wide areas of riverine landscape, and as a result, they influence the ecological processes of the river floodplain in a direct and expansive manner. Smooth–coated otters play a vital role in balancing the freshwater ecosystems as a top carnivorous species (Sivasothi, 1995; Acharya & Lamsal, 2010), and they may therefore significantly influence the overall spatio–temporal dynamics of the eco–region over a long period of time (Naiman et al., 2000). There is little information available on the status of otter populations in India, although there seems to have been a rapid decline due to loss of habitat and intensive trapping (Hussain, 1999; Nawab, 2007, 2009; Nawab & Gautam, 2008). Presently, the population is severely fragmented throughout its distribution range and isolated populations are restricted mostly to protected areas (Hussain, 1999; Nawab, 2007, 2009). Although otter occurrence in the upper Ganga River Basin has been previously reported from the National Chambal Wildlife Sanctuary (Hussain, 1993), Corbett Tiger Reserve (Nawab, 2007), Dudhwa Tiger Reserve and Katerniaghat Wildlife Sanctuary (Hussain, 2002), the present study appends new geographical locations in the distribution range of smooth–coated otter, *i.e.* (i) Alaknanda–Ganga Basin in Uttarakhand and (ii) Hastinapur Wildlife Sanctuary in Uttar Pradesh. The present study was triggered by the rapid decline in the otter's population in the country and it focuses mainly on assessing the otter's conservation status, its habitat use pattern, and associated threats in the upper Ganga River Basin (N India). This will improve the understanding of the complex ecological interactions and will help to design effective conservation measures for this species (Stanford et al., 1996). The purpose of this paper is to highlight the gap areas in the conservation of the species and to suggest areas for management in the upper Ganga River Basin.

Material and methods

Study sites

The Ganga River Basin is the largest river basin in India, constituting 26% of the country's land mass and supporting about 43% of its population (448.3 million as per the 2001 census) (Ambastha et al., 2007). Rainfall and melt water from snow and glaciers are the main sources of water in the River Ganga (O'Keeffe et al., 2012). The present study was carried out at two selected sites, one in Uttar Pradesh and the other in Uttarakhand, states of India where the species has not been studied previously.

Site I. Alaknanda–Ganga Basin (from Rudraprayag to Rishikesh)

The River Alaknanda originates from the confluence of the Sathopanth and Bhagirathi Kharak Glacier and forms a unified stream of the upper Ganga River by merging with the River Bhagirathi at Devprayag. The Alaknanda–Ganga Basin (fig. 1) is characterized by rugged topography with major landforms comprising moderate to steep precipitous sloping mountainous terrain, narrow and broad valleys and highly dissected ridges with the formation of deep gorges (Anbalagan et al., 2008). Despite its unprotected status, the basin holds a good variety of wildlife, including endangered freshwater fauna like Golden Mahasheer *Tor putitora*. The general vegetation in the area is dominated by *Pinus roxburghii, Anogeissus latifolius, Acacia catechu, Holoptelea integrifolia, Syzgium cumini* and *Aegle marmelos*. The drainage system of the basin has been extensively regulated for hydroelectric production.

Site II. Hastinapur Wildlife Sanctuary

Hastinapur Wildlife Sanctuary spreads over an area of 2,073 km² along the banks of the River Ganges in western Uttar Pradesh (fig. 1). The Sanctuary was established in 1986 to conserve the fast vanishing, unique Ganga River grassland–wetland complex, locally known as Khadar. It is unique in the sense that it presents a variety of landforms and habitat types that include wetland, marshes, dry sandy beds and gently sloping ravines.

River Ganga and its old bed, locally called Boodhi Ganga, forms the drainage system of the Sanctuary. River Ganga enters the Sanctuary area at Bijnor and leaves it at Garmukteshwer after flowing for 125 km. During summers, Boodhi Ganga becomes fragmented into a series of small swampy patches with nil or very insignificant water current. Because of this discontinuous belt of highly marshy land, there is profuse growth of vegetation like *Phragmites* species, *Arundinella* species and *Typha* species.

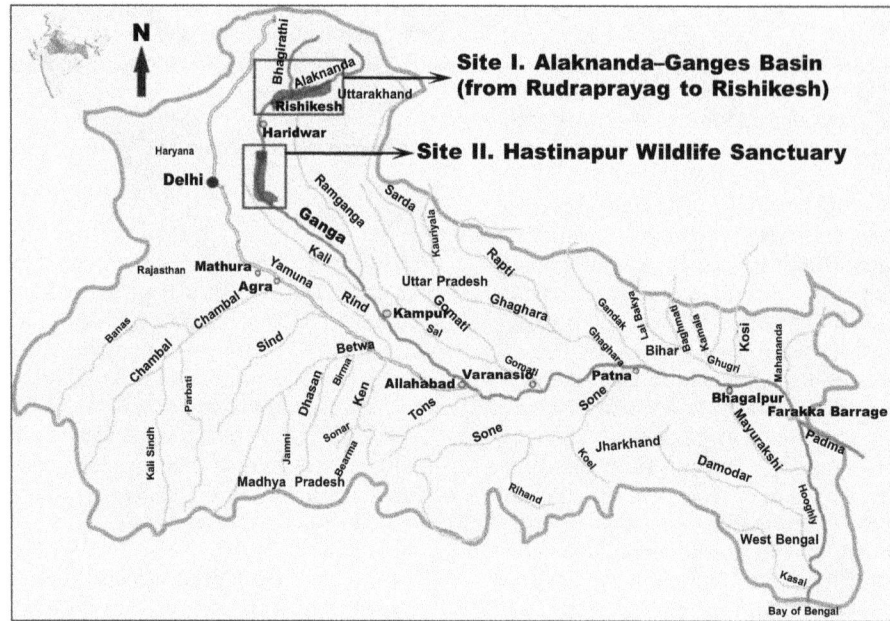

Source: The original map is downloaded from www.sandrp.in on 20 January 2014

Fig. 1. Location of study sites in the Ganges River Basin.

Fig. 1. Ubicación de las localidades de estudio en la cuenca del río Ganges.

Data collection

During the summer in 2010 we surveyed 35 kilometers of the River Alaknanda–Ganges (sampling sections, n = 7) and 145 km stretch of River Ganga (main stream) and its old bed Boodhi Ganga (sampling sections, n = 29). The selected river stretches were divided into 5 km sections using a Survey of India's 1:50,000 topographic maps (Macdonald & Mason, 1983; Kruuk et al., 1994; Hussain & Choudhury, 1997; Nawab, 2007). Data on the habitat parameters and indirect evidences of otter occurrence such as tracks, spraints, den sites or scent marks were recorded from each section. Searches were made in 15 m wide strips along the edge of the river with the help of two trained researchers, by walking along both banks. In each study section, any location where spraints, tracks, den sites and other signs of otter presence were found was defined as a 'used plot' with dimensions 100 × 15 m; additionally, for each used plot, two available plots, one each at 500 m downstream as well as upstream, were considered. In case of spraint sites, a new site was registered only when spraints were separated by more than 5 m (Melquist & Hornocker, 1983; Newman & Griffin, 1994; Medina, 1996; Nawab, 2007).

At each section habitat parameters and human activities which are considered potentially threatening to otters were also recorded (Prenda & Granado–Lorencio, 1996; Prenda et al., 2001; Anoop & Hussain, 2004) (table 1). Species habitat selection was analyzed at plot scale.

Data analysis

The present study was based on the premise that otters live at low densities and are shy and often nocturnal or crepuscular, and hence difficult to track and to make direct estimates of population size and density. The distribution and frequency of occurrence of spraints and tracks were considered as the index of habitat use by the otters. The preference of habitat covariates was established following Bonferroni confidence intervals in combination with Chi–square goodness of fit test (Neu et al., 1974; Byers et al., 1984).

Bonferroni confidence interval equation:

$$\bar{P}_i - Z_{\alpha/2k}\sqrt{\bar{P}_i(1 - \bar{P}_i)/n} \leqslant P_i \leqslant \bar{P}_i + Z_{\alpha/2k}\sqrt{\bar{P}_i(1 - \bar{P}_i)/n}$$

where P_i is the proportion of indirect evidences in the ith habitat category, n is the sample size, k is the number of categories of habitat studied, α is confidence interval while Z is the tabular value of standard curve.

Chi–Square equation:

$$\chi^2 = \frac{\sum (O_i - E_i)^2}{E_i}$$

where O_i is the observed number of indirect evidence in the ith habitat category and E_i is expected number of indirect evidence in the ith habitat category.

An independent sample t–test was performed to know the significance of difference between the used and available habitat covariates following Neu et al. (1974), Byers

Table 1. Ecological parameters and human activities affecting the occurrence of smooth–coated otter, recorded during the study.

Tabla 1. Parámetros ecológicos y actividades humanas registrados durante el estudio que afectan a la presencia de la nutria lisa.

Variable	Data type	Description and measurement details
Width of river (m)	Continuous	Distance between shorelines visually estimated
Average depth of river (m)	Continuous	The depth of the river was measured at both banks and middle of the river and mean depth was calculated
Shoreline substrate type (%)	Categorical	Approximate percentage of total area (100 m × 15 m) of the plot covered by rock/boulder, sand, mud, clay or alluvial deposit was visually estimated
Water current (m/s)	Continuous	The surface water velocity was calculated via floating ball method.
River bank slope (degree)	Continuous	Measured via Clinometers
Shoreline vegetation cover (%)	Categorical	Approximate percentage of total area (100 m × 15 m) of the plot covered by tree, shrub, herb or grass was visually estimated
Escape distance (m)	Continuous	Nearest distance from water's edge to shoreline vegetation which provides cover for otter measured by measuring tape
Disturbance (present/absent)	Binary	Presence of disturbing activities/evidences was recorded at every plot

et al. (1984) and Zar (1984). Statistical package SPSS 7.3 (Norusis, 1994) was used for computing purposes.

Results

Site I. Alaknanda–Ganga Basin (from Rudraprayag to Rishikesh)

The thirty–five kilometer stretch of the River Alaknanda–Ganga was divided into seven sampling sections of five kilometers. Otter occurrence was recorded only from two of these sections (*i.e.* 28.57% occupancy), at village Malysu and Papdasu (district Rudraprayag). Informal interviews with locals suggested occurrence of otters in the study area was common in the 1990s, but due to human disturbance, the habitat quality had declined and consequently the numbers of otters in the area had decreased.

Sandy substrate was preferred over other available substrates by the species in the area (table 2). Of the 16 habitat parameters, the means of shoreline vegetation cover ($P < 0.05$), percentage of clay substrate ($P < 0.001$) and bank slope ($P < 0.001$) were used significantly different from their availability (table 3).

Site II. Hastinapur Wildlife Sanctuary

The total 145 km stretch of River Ganga (main stream) and its old bed Boodhi Ganga was surveyed. The findings of the survey append the new locality record in the distribution range of smooth coated otter in north India. From a total of 29 sampling sections, only 6.89% ($n = 2$) were found occupied by otters. Interviews with locals revealed that the occurrence of otters in the sanctuary was common a decade before. However, excessive changes in land–use pattern and human disturbance led to a vast decline in habitat quality and hence the otter population also decreased.

The result of Bonferroni confidence intervals indicates that smooth–coated otter prefer the most remote muddy parts of the river and avoid alluvial, sand and areas with clay as dominant substrate (table 2) as they are found adjacent to cultivated fields and easily accessible. Of the 15 parameters, the respective means of used and available plots of ten parameters were found significantly different at $P < 0.001$ level, while the differences between the mean of used and available plots for % sand was found significant at $P < 0.05$ level (table 3).

Table 2. Preference of shelter sites by the smooth–coated otter along site I and II: S. Substrate type; Pio. Proportion of total sampling plots; O_i. Number of used plots; E_i. Expected number of used plots; Pi. Proportion of indirect evidences at each sampling plot; χ^2. Chi–square distribution; Bonferroni. Bonferroni confidence interval proportions; C. Conclusion (+ Used more than available; – Used less than available)

Tabla 2. Preferencia de la nutria lisa por los lugares de cobijo en las localidades I y II: S. Tipo de sustrato; Pio. Proporción en el total de parcelas de muestreo; Oi. Número de parcelas utilizadas; E_i. Número esperado de parcelas utilizadas; Pi. Proporción de pruebas indirectas en cada parcela de muestreo; χ^2. Distribución de la χ^2; Bonferroni. Intervalo de confianza de Bonferroni para las proporciones; C. Conclusión (+ Más utilizado de lo esperado; – Menos utilizado de lo esperado)

S	Pio	O_i	E_i	Pi	χ^2	Bonferroni	C
Site I							
Sand	0.29 (N = 14)	4	2.00	0.57	2.00	0.395 ≤ Pi ≤ 0.748	+
Clay	0.02 (N = 1)	0	0.14	0.00	0.14	0.000 ≤ Pi ≤ 0.000	–
Boulder	0.65 (N = 32)	3	4.57	0.43	0.54	0.252 ≤ Pi ≤ 0.605	–
Alluvial	0.04 (N = 2)	0	0.29	0.00	0.29	0.000 ≤ Pi ≤ 0.000	–
Site II							
Sand	0.16 (N = 71)	0	4.02	0.00	4.02	0.000 ≤ Pi ≤ 0.000	–
Mud	0.49 (N = 218)	18	12.33	0.72	2.16	0.667 ≤ Pi ≤ 0.773	+
Clay	0.29 (N = 126)	7	7.13	0.28	0.00	0.227 ≤ Pi ≤ 0.333	–
Alluvial	0.06 (N = 27)	0	1.53	0.00	1.53	0.000 ≤ Pi ≤ 0.000	–

Discussion

Mainly due to habitat loss and over–exploitation, the population of smooth–coated otters is declining throughout their range of distribution and the trend of population decline is expected to continue (Hussain et al., 2008). A deficiency of baseline data on the ecology of the species is another constraint for its conservation. Information on habitat selection by otters is further sketchier as compared to other aspects of their ecology (Hussain, 1996). In Europe and North America, many studies on *Lutra lutra* and *Lutra canadensis* have led to an increasing understanding of otter habitat preferences in temperate regions (Melisch et al., 1996), whereas in the case of the smooth–coated otter, availability of food, freshwater and shelter for resting, grooming and breeding are the important factors known to govern the process of habitat selection by otters (Mason & Macdonald, 1986; Kruuk, 1995; Anoop & Hussain, 2004; Nawab, 2009).

In site I (Alaknanda–Ganga Basin), otters showed preference for sandy stretches in all the seasons, as these stretches provide sites for dens and grooming (Hussain, 1993); while in site II (Hastinapur Wildlife Sanctuary), the species preferred to use the muddy stretches of Boodhi Ganga which is almost inaccessible to humans and thus less disturbed. This ability of the species to adapt to diverse aquatic habitats accounts for its broad geographic distribution (Pocock, 1941).

Otter occurrence was associated with shallow and calmer regions (with low water velocity) along the Gan-

ga River Basin in site I, as these conditions increase the rate of prey capture per efforts. Ease in capturing prey was interpreted to be the most important factor in selecting the habitat by the species, as also suggested by other studies (Kruuk, 1995; Anoop, 2001; Nawab, 2007; Acharya & Lamsal, 2010).

Hastinapur Wildlife Sanctuary is one of the most populated and disturbed protected areas in Uttar Pradesh. As most of its land is cultivated, the area is highly accessible to humans, imposing an adverse effect on the inhabiting wildlife. Therefore, despite being a protected area, only 6.89% (n = 2) of otter occupancy was recorded in the area, far below the 28.57% (n = 2) recorded for otter occupancy at site I. Moreover, most of the animals like otters restricted themselves to the remaining inaccessible parts of the sanctuary, such as the swampy patches of the Boodhi Ganga River. Habitat features of Boodhi Ganga, such as deep waters forming pools, prey availability, presence of shoreline vegetation and gentle bank slopes, endorse the occurrence of otters. Other authors have also found a positive correlation between otter signs and the percentage of vegetation cover (Macdonald & Mason,1983; Melisch et al., 1996; Anoop & Hussain, 2004; Nawab, 2007). Gentle bank slopes are favored by otters as they reduce energy expenditure while foraging or grooming (Kruuk, 1995).

Otters are facing extreme threats by human–induced habitat destruction. The expansion of agriculture has led to the destruction of huge areas of natural habitats, including forests, grasslands and wetlands, in nearly all regions of the world (Ottino & Giller, 2004).

Table 3. Habitat variables influencing otter distribution along site I and II. (SE. Standard error)

Talba 3. Variables del hábitat que influyen en la distribución de la nutria en las localidades I y II.

Variables	Available plots		Used plots			
	Mean	SE	Mean	SE	t	Sig.
Site I						
River bank characteristics						
% Alluvial	10.12	1.69	7.14	1.84	-0.70	0.486
% Boulder	55.86	4.22	55.71	13.60	-0.01	0.990
% Clay	5.29	1.53	0.00	0.00	-3.45	**0.001**
% Grass cover	19.88	2.52	22.14	2.86	0.36	0.723
% Herb cover	17.98	2.15	22.14	3.91	0.75	0.454
% Mud	1.55	0.65	1.43	1.43	-0.07	0.944
% Sand	27.19	4.20	35.71	13.07	0.74	0.463
% Shrub cover	30.95	4.03	46.43	8.29	1.48	0.146
% Total veg. cover	28.33	3.02	47.14	6.44	2.39	**0.021**
% Tree cover	9.76	1.69	9.29	2.02	-0.18	0.859
Escape distance	7.07	0.82	5.29	2.01	-0.83	0.412
Slope	50.76	3.91	14.29	2.02	-8.29	**< 0.001**
River characteristics						
Average depth	4.91	0.44	3.06	0.74	-1.64	0.108
Average width	28.58	2.96	26.14	6.24	-0.32	0.753
Water current	1.28	0.08	0.93	0.23	-1.68	0.101
pH	7.81	0.02	7.83	0.02	0.75	0.461
Site II						
River bank characteristics						
% Alluvial	9.70	1.09	0.00	0.00	8.89	**< 0.001**
% Clay	32.81	1.96	25.80	7.34	0.86	0.393
% Grass cover	86.47	1.20	94.00	1.35	-4.16	**< 0.001**
% Herb cover	6.16	0.46	4.40	0.97	1.64	0.109
% Mud	41.49	2.05	74.20	7.34	-4.30	**< 0.001**
% Sand	16.01	1.59	0.00	0.00	2.46	**0.014**
% Shrub cover	0.94	0.14	1.40	0.68	-0.77	0.444
% Total veg. cover	31.74	0.61	65.60	2.13	-13.28	**< 0.001**
% Tree cover	0.35	0.09	0.60	0.33	-0.69	0.491
Escape distance	48.68	8.67	2.78	0.49	5.29	**< 0.001**
Slope	14.17	0.	9.00	0.82	5.74	**< 0.001**
River characteristics						
Average depth	0.86	0.04	0.48	0.03	8.02	**< 0.001**
Average width	145.61	7.94	19.72	4.96	3.88	**< 0.001**
Water current	0.65	0.04	0.02	0.00	14.33	**< 0.001**
pH	8.55	0.02	7.88	0.07	9.92	**< 0.001**

The expansion and development of urbanization and riverfront infrastructural developments, such as the construction of dams, has broken the continuum of natural habitats into small fragments (Nawab, 2007) and these patches of suitable habitat may be too small to support a breeding pair or a functional social group. It is of note that area sensitive species (Lambeck, 1997) like otter, that have a low dispersal capacity, are unable to re–colonize such patches following extinction (Collinge, 1996).

Recommendations

Site I. Alaknanda–Ganga Basin (from Rudraprayag to Rishikesh)

Maximum evidence of otter occurrence was concentrated around the villages Malysu and Papdasu in the Rudraprayag district. These areas therefore merit special attention in terms of habitat management and protection. As evident from this study, otters are confined to small areas and the population seems to be vulnerable to anthropogenic and other stochastic disturbances. Detailed research on the population ecology of the species is necessary to implement better management practices to conserve the species in the region. Education and awareness programmes should be launched, focusing special emphasis on fishing and immigrant communities known to be involved in otter killings for meat and skin.

Although otters are often in direct conflict with fishermen who view them as competitors for fish and kill them (Foster–Turley, 1992), in the Alaknanda–Ganga Basin, a tolerable association of otters and human presence was observed. From local sources we heard that otters damage nets and steal fish from the fishermen's catch, but the conflict remains negligible; locals also appreciate the aesthetic and ecological importance of otters, accepting it within their environment and making co–existence possible.

Site II. Hastinapur Wildlife Sanctuary

Until the mid–twentieth century, extensive tracts of grassland–wetland complex (locally known as Khadar) harbored rich biodiversity all along the River Ganga. After India gained independence in 1947, Khadar received a large influx of Pakistani emigrants and in the following decades (*i.e.* 1980s) Punjabi emigrants also settled in the area, converting the Khadar into agricultural farms (Agarwal, 2009).

Presently, the Hastinapur Wildlife Sanctuary is subjected to human disturbance, mainly due to large scale commercial exploitation of grasses (*Phragmites*), livestock grazing and illegal cultivation (Khan et al., 2003). Many swamps have been drained and converted into crop fields, or are in the process of such activity, like Boodhi Ganga. Modernised farming, *i.e.* unabated use of chemical fertilizers and pesticides in these agriculture fields, is deteriorating water quality (Agarwal, 2009). Indiscriminate fishing by use of gillnet, hooks and poison poses a major threat to aquatic fauna (Khan, 2010).

There is a need for locals, especially fishermen and farmers, to become aware of the importance of aquatic ecosystems both for the conservation of wildlife and for their own sustenance. Local communities should be helped to obtain better educational opportunities.

Otters are confined to small swampy patches of Boodhi Ganga and the population is vulnerable to anthropogenic and other stochastic disturbances in the sanctuary. The solution for their long–term survival in the sanctuary lies not only in taking stringent protection measures but also in developing and implementing long–term monitoring programs for otters along Boodhi Ganga in and around the Sanctuary. The illegal encroachment and clearing of Boodhi Ganga that is currently in progress and encouraged by some migrant farmers severely affects the survival of the area's wild inhabitants. The government needs to apply strict measures and stringently implement the law to prevent such illegal activities.

Acknowledgements

The data were collected during an M.Sc. internship of the first and second author, in the project Otter Conservation under the sponsorship of Living Ganga Programme of WWF India. We are thankful to Mr. Ravi Singh and Dr. Sejal Worah (WWF India) for constant encouragement and support. The help rendered by staff of WWF India's Hastinapur field office is highly appreciated. We are also grateful to Dr. Anjana Pant (WWF India), Dr. Satish Kumar, Dr. Faiza Abbasi and Zarreen Syed (AMU, Aligarh) for their valuable suggestions on the manuscript.

References

Acharya, P. M. & Lamsal, P., 2010. A Survey for Smooth coated Otter *Lutrogale perspicillata* on the River Narayani, Chitwan National Park, Nepal. *Hystrix Italian Journal of Mammology.* 21(2): 203–207. DOI: 10.4404/Hystrix–21.2–4464

Agarwal, S., 2009. Angiosperm species diversity and ecological assessment of Hastinapur Wildlife Sanctuary, India. Ph. D. Thesis, Department of Botany, Aligarh Muslim University, Aligarh, India.

Ambastha, K., Hussain, S. A. & Badola, R., 2007. Social and Economic Considerations in Conserving Wetlands of Indo–Gangetic Plains: A Case Study of Kabartal Wetland, India. *Environmentalist*, 27: 261–273.

Anbalagan, R., Kohli, A. & Chakraborty, D., 2008. *Stability Analysis of Harmony Landslide in Garhwal Himalaya, Uttarakhand State, India.* Department of Earth Sciences, Indian Institute of Technology, Roorkee, India.

Anoop, K. R., 2001. Factors affecting habitat selection and feeding habits of Smooth–coated Otter *Lutra perspicillata* in Periyar Tiger Reserve, Kerala. M. Sc. Dissertation, Saurashtra University, Rajkot, India.

Anoop, K. R. & Hussain, S. A., 2004. Factors affecting habitat selection by Smooth–coated otter *Lutra perspicillata* in Kerala, India. *Journal of Zoology*, 263: 417–423.

Byers, C. R., Randall, C., Stienhorst, R. K. & Krausman, P. R., 1984. Clarification of a technique for analysis of utilization availability data. *Journal of Wildlife Management*, 48: 1050–1053.

Collinge, S. K., 1996. Ecological consequences of habitat fragmentation: implication of landscape architecture and planning. *Landscape Urban Planning,* 36: 59–77.

Foster–Turley, P., 1992. Conservation ecology of sympatric Asian Otters *Aonyx cinerea* and *Lutra perspicillata*. Ph. D. Thesis, University of Florida.

Hussain, S. A., 1993. Aspect of the ecology of Smooth–coated Otter in National Chambal Sanctuary.

Ph. D. Thesis, Centre for Wildlife and Ornithology, Aligarh Muslim University, India.

– 1996. Group Size, Group Structure and Breeding in Smooth–Coated Otter Lutra *perspicillata* (Geoffroy) (Carnivora, Mustelidae) in National Chambal Sanctuary, India. *Mammalia*, 60: 289–297.

– 1999. Otter conservation in India. Envis Bulletin – Wildlife and Protected Areas, 2(2): 92–97.

– 2002. Conservation status of otters in the Tarai and Lower Himalayas of Uttar Pradesh, India. In: Otter Conservation – An Example for a Sustainable use of Wetlands (R. Dulfer, J. Conroy, J. Nel & A. Gutleb, eds.). *IUCN Otter Specialist Group Bulletin*, 19: 131–142. Trebon, Czech Republic.

Hussain, S. A. & Choudhury, B. C., 1997. Status and distribution of Smooth–coated Otter Lutra *perspicillata* in National Chambal Sanctuary. *Biological Conservation*, 80: 199– 206.

Hussain, S. A., De Silva, P. K. & Mostafa Feeroz, M., 2008. *Lutrogale perspicillata*. In: IUCN 2013. IUCN Red List of Threatened Species. Version 2013.2. <www.iucnredlist.org> downloaded on 22 January 2014.

Khan, M. S., 2010. Conservation status and habitat use pattern of Otters in Hastinapur Wildlife Sanctuary, Uttar Pradesh; India. M. Sc. Dissertation, Department of Wildlife Sciences, Aligarh Muslim University, India.

Khan, J. A., Khan, A. & Khan, A. A., 2003. *Structure and composition of barasingha habitat in Hastinapur Wildlife Sanctuary*. Technical Report. Wildlife Society of India, Aligarh Muslim University, Aligarh: 5–7.

Kruuk, H., 1995. *Wild Otters – Predation and populations*. Oxford University Press.

Kruuk, H., Kanchanasaka, B. O'Sullivan, S. & Wanghongsa, S., 1994. Niche separation in three sympatric Otters Lutra *perspicillta*, L. *Lutra* and Aonyx *cinerea*. *Biological Conservation*, 69: 115–120.

Lambeck, R. J., 1997. Focal species: a multi species umbrella for nature conservation. *Conservation Biology*, 11(4): 849–856.

Macdonald, S. M. & Mason, C. F., 1983. Some factors influencing the distribution of Otters Lutra *lutra*. *Mammlian Review*, 13(1): 1–10.

Mason, C. F. & Macdonald, S. M., 1986. *Otters: ecology and conservation*. Cambridge University Press, Cambridge, London.

Medina, G., 1996. Conservation and status of *Lutra provocax* in Chile, Pacific. *Conservation Biology*, 2: 414–419.

Melisch, R., Asmoro, P. B., Kusumawardhani, L. & Lubis, I. R., 1996. *The Otters of west java: A survey of their distribution and habitat use and a strategy toward a species conservation programme*. PHPA/ Wetlands International–Indonesia Programme.

Melquist, W. E. & Hornocker, M. G., 1983. Ecology of Otters in West Central Idaho. *Wildlife Monograph*, 83: 60.

Mitsch, W. J. & Gosselink, J. G., 2000. *Wetlands* 3rd (edu.). John Wiley and Sons Inc., New York.

Naiman, R. J., Bilby, R. E. & Bisson, P. A., 2000. Riparian ecology and management in the Pacific coastal rain forest. *BioScience*, 50: 996–1011.

Nawab, A., 2007. Ecology of Otters in Corbett Tiger Reserve, Uttarakhand; India. Ph. D. Thesis, Forest Research Institute, Dehradun, India.

– 2009. Aspects of the ecology of Smooth–coated Otter *Lutrogale perspicillata* Geoffroy St. Hilaire, 1826: A Review. *Journal of Bombay Natural History Society*, 106(1): 5–10.

Nawab, A. & Gautam, P., 2008. Living on the edge: Otters in developing India. In Wetlands – The Heart of Asia. *Proceedings of the Asian Wetland Symposium*, Hanoi, Vietnam.

Neu, C. W., Byers, C. R. & Peek, J. M., 1974. A technique for analysis of utilization – availability data. *Journal of Wildlife Management*, 38: 541–545.

Newman, D. G. & Griffin, R., 1994. Wetland use by river otters in Massachusetts. *Journal of Wildlife Management*, 58: 18–23.

Norusis, M. J., 1994. SPSS/PC+ statistic 7.3. for 1BMPC/XT/AT and PS/2 SPSS. International Br, Netherlands.

O'Keeffe, J., Kaushal, N., Smakhtin, V. & Bharati, L. 2012. *Assessment of Environmental Flows for the Upper Ganga Basin*. Summary Report. WWF India.

Ottino, P. & Giller, P., 2004. Distribution, density, diet and habitat use of the Otter in relation to land use in the Araglin valley, Southern Ireland. Biology and Environment: *Proceedings of the Royal Irish Academy*, 104B(1): 1–17.

Pocock, R. I., 1941. *The Fauna of British India including Ceylon and Burma*. Vol. II. Taylor and Francis, London.

Prenda, J. & Granado–Lorencio, C., 1996. The relative influence of riparian habitat structure and fish availability on otter Lutra *lutra* sprainting activity in a small Mediterranean catchment. *Biological Conservation*, 76: 9–15.

Prenda, J., López–Nieves, P. & Bravo, R., 2001. Conservation of otter (*Lutra lutra*) in a Mediterranean area: the importance of habitat quality and temporal variation in water availability. *Aquatic Conservation: Marine and Freshwater Ecosystem*, 11: 343–355.

Ravenga, C., Brunner, J., Henninger, N., Kassem, K. & Payne, R., 2000. *Pilot analysis of global ecosystems. Freshwater Systems*. World Resources Institute, Washington D. C.

Rodgers, W. A. & Panwar, H. S., 1988. *Planning wildlife protected area network in India*. Vol. 2 Project FO: IND/82/003. FAO, Dehra Dun.

Sivasothi, N., 1995. The status of Otters in Singapore and Malaysia, and the diet of Smooth coated Otter *Lutrogale perspicillata* in Penang, West Malaysia. M. Sc. Thesis, National University of Singapore, Singapore.

Stanford, J. A., Ward, J. V., Liss, W. J., Frissell, C. A., Williams, R. N., Lichatowich, J. A. & Coutant, C. C., 1996. A general protocol for restoration of regulated rivers. *Regulated Rivers: Research and Management*, 12: 391–413.

Vitouesk P. M., Mooney, H. A., Lubchenco, J. & Melillo, J. M., 1997. Human domination of earth's ecosystem. *Science*, 227: 494–499.

Zar, J. H., 1984. *Biostatistical analysis*. IInd. Edn. Prentice–Hall Inc., New Jersey.

17

Diversity of large and medium mammals in Juchitan, Isthmus of Tehuantepec, Oaxaca, Mexico

M. Cortés–Marcial, Y. M. Martínez Ayón &
M. Briones–Salas

Cortés–Marcial, M., Martínez Ayón, Y. M. & Briones–Salas, M., 2014. Diversity of large and medium mammals in Juchitan, Isthmus of Tehuantepec, Oaxaca, Mexico. *Animal Biodiversity and Conservation*, 37.1: 1–12.

Abstract

Diversity of large and medium mammals in Juchitan, Isthmus of Tehuantepec, Oaxaca, Mexico.— The Isthmus of Tehuantepec in Oaxaca, Mexico, is one of the country's most important regions from a zoogeographical perspective due to the large number of endemic Neotropical species found there. Between September 2007 and August 2008, we sampled medium–sized and large mammals in the Juchitan municipality and compared their diversity in two areas with distinct levels of anthropogenic impact, defined according to estimates of human activities, livestock density and habitat degradation, We obtained 167 records of 18 species, with a 79% representation according to species accumulation models in both areas. The highest species richness and alpha diversity were recorded in the preserved area, whereas the disturbed area exhibited half the diversity found in the preserved area. A high interchange of species was also observed between zones. The two species with the largest number of records were *Urocyon cinereoargenteus* (n = 52) and *Didelphis virginiana* (n = 42). In both areas, the highest relative abundance occurred during the rainy season. Habitat degradation and human activities seem to affect the diversity of mammal species in the region.

Key words: Biodiversity, Conservation, Disturbance, Isthmus of Tehuantepec, Tropical deciduous forest.

Resumen

La diversidad de los mamíferos de talla grande y mediana en Juchitán, istmo de Tehuantepec, Oaxaca, México.— El istmo de Tehuantepec en Oaxaca, México, es una de las regiones más importantes del país desde el punto de vista zoogeográfico, ya que alberga una gran cantidad de especies endémicas neotropicales. Entre septiembre de 2007 y agosto de 2008, se realizó un muestreo de mamíferos de talla mediana y grande en el municipio de Juchitán, y comparamos su diversidad en dos zonas con distintos niveles de impacto antropogénico definido de acuerdo con las estimaciones de las actividades humanas, la densidad de ganado y la degradación del hábitat. Se obtuvieron 167 registros de 18 especies, con una representatividad del 79% según el modelo de acumulación de especies en ambas zonas. La mayor riqueza de especies y de diversidad alfa se registraron en la zona conservada, mientras que la zona perturbada presenta la mitad de la diversidad encontrada en la zona conservada. Se observó un fuerte intercambio de especies entre ambas zonas. Dos especies, *Urocyon cinereoargenteus* (n = 52) y *Didelphis virginiana* (n = 42), tuvieron el mayor número de registros. En ambas zonas, la mayor abundancia relativa se observó durante la época de lluvias. La degradación del hábitat y las actividades humanas al parecer afectan a la diversidad de especies de mamíferos en la región.

Palabras clave: Biodiversidad, Conservación, Perturbación, Istmo de Tehuantepec, Bosque deciduo tropical.

Malinalli Cortés–Marcial, Yazmín del Mar Martínez Ayón & Miguel Briones–Salas, Lab. de Vertebrados Terrestres (Mastozoología), Centro Interdisciplinario de Investigación para el Desarrollo Integral Regional (CIIDIR–Unidad Oaxaca), Inst. Politécnico Nacional, Oaxaca, México.

Corresponding author: M. Cortés–Marcial. E–mail: mali_cor@yahoo.com.mx

Introduction

One of the issues of greatest interest in ecology is the relationship between habitat structure and the structure of animal communities. Habitat disturbance and habitat fragmentation influence both the original plant communities and the heterogeneity and complexity of the entire ecosystem. This, in turn, influences the availability of resources, and affects the birth and death rates of several species, thus affecting vertebrate diversity (August, 1983; Soule et al., 1992; Collins et al., 1995; Murcia, 1995; Zarza, 2001). Large and medium–sized mammals are particularly sensitive to habitat changes, and they are common victims of poaching and illegal trading (Michalski & Peres, 2005; Laurance et al., 2006). The functional significance of these species lies in their ecological roles, such as seed dispersal and predation on numerous plant species. These functional roles may change the structure and composition of the ecosystem. Moreover, these species influence the community structure and complexity on the trophic levels in which they are involved, due to their regulatory role as preys and predators (Roemer et al., 2009). The loss of these organisms could have devastating effects because they contribute in many ways to the functioning of the natural ecosystem (Alonso et al., 2001; Bolaños & Naranjo, 2001). Given the importance of these species, studies identifying and predicting the environmental changes that may affect their diversity are essential, and in such studies, relative abundance and species diversity are usually used as indicators (Carrillo et al., 2000).

The Isthmus of Tehuantepec (Mexico) is one of most diverse regions within this country (Briones–Salas & Sánchez–Cordero, 2004; González et al., 2004). Furthermore, this area has a particular importance from a zoogeographical perspective because it lies in the zone where the Nearctic and Neotropical regions overlap. This important corridor between the Atlantic and costal Pacific plains represents a significant barrier for highland mammal species, and also favors a high degree of endemicity (Peterson et al., 1999; García–Trejo & Navarro, 2004; Barragan et al., 2010). However, this diversity may be declining dramatically, due to hunting and habitat modification derived from crops and livestock. Therefore, the aim of this study was to identify the differences in diversity, in terms of abundance and heterogeneity, of medium–sized and large mammals in two areas with differing degrees of anthropogenic disturbance. If anthropogenic environmental changes affect mammal communities, we hypothesized that the area with greater human disturbance would exhibit a lower diversity of medium and large mammals.

Methods

The study area is located in the coastal plain of Tehuantepec, northeast of the city of Juchitan, Oaxaca, Mexico, at 200 m a.s.l., within the coordinates 94° 55' to 94° 50' W, and 16° 38' to 16° 30' N (fig. 1). The climate is sub–humid and warm. There is a marked dry season from December to May, and a rainy season from June to November, with an average annual rainfall of 932.2 mm. The annual average temperature is 27.6°C (Garcia, 1988). The first sampling area was located on the hill of Tolistoque, northeast of Juchitan (16° 35' 5.91" N, 94° 52' 20.63" W) within an area —protected by the regional indigenous communities— known as Ojo de Agua Tolistoque Protected Communal Area (Ortega et al., 2010). The vegetation is tropical deciduous forest. The second sampling area was south of the Protected Communal Area northeast of Juchitan (16° 32' 12.95" N, 94° 50' 53.95" W), in an area of secondary vegetation. This area is dedicated to farming activities, with gallery forest areas around irrigation canals, and tropical deciduous forest remnants (fig. 1).

We applied an indirect sampling method. Such methods are sometimes the only option available to study the distribution and abundance of inaccessible vertebrates such as medium–sized and large mammals (Sutherland, 1996). These methods also have some advantages over direct methods as they are easier to implement and independent of the time of day, which is important when target species are nocturnal, cryptic and difficult to capture or recapture because their traces remain for long periods of time (Bilenca et al., 1999; Simoneti & Huareco, 1999; Aranda, 2000; Carrillo et al., 2000; Ojasti, 2000).

In both areas the level of disturbance was evaluated according to the index proposed by Peters & Martorell (2000) and Martorell & Peters (2005). In order to measure the contribution of different agents, we recorded 14 metrics at each site by means of two 50 m long transects at each site (table 1). Disturbance was measured on a scale of 0–100, where zero is the least disturbance. The values were calculated as follows:

$$Disturbance = 3.41\ Goat - 1.37\ Catt + \\ + 27.62\ Brow + 49.20\ Ltra - 1.03\ Comp + \\ + 41.01\ Fuel + 0.12\ Tran + 24.17\ Prox + \\ + 8.98\ Core + 8.98\ Luse - 0.49\ Fire + \\ + 26.94\ Eros + 17.97\ Isla + 26.97\ Toms + 0.2$$

The medium and large mammals were classified using the system of Robinson & Redford (1986), who divided mammals into four categories based on a logarithmic scale of average weight: small < 100 g; medium > 100~ < 1,000 g; large > 1,000 g < 10,000 g; very large > l0,000 g. To search for traces of medium and large mammals, monthly samples were taken from September 2007 to August 2008. During each period, four transects (two in each zone) of 4.5 km each were sampled, resulting in a total sampling of 108 km walked in each zone.

We used a Mexican mammal field guide (Aranda, 2000) to identify tracks and feces, and compared these with the reference material on traces of mammals of Oaxaca, of the Collection of Mammalogy (OAX.MA.026.0497) at the Centro Interdisciplinario de Investigación para el Desarrollo Integral Regional (CIIDIR–Oaxaca), National Polytechnic Institute (IPN).

Ten camera traps (Cuddeback Expert ®) were also used for the last six sampling periods to confirm the presence of the species (five in each zone). These were placed at approximately 1.5 km from each other.

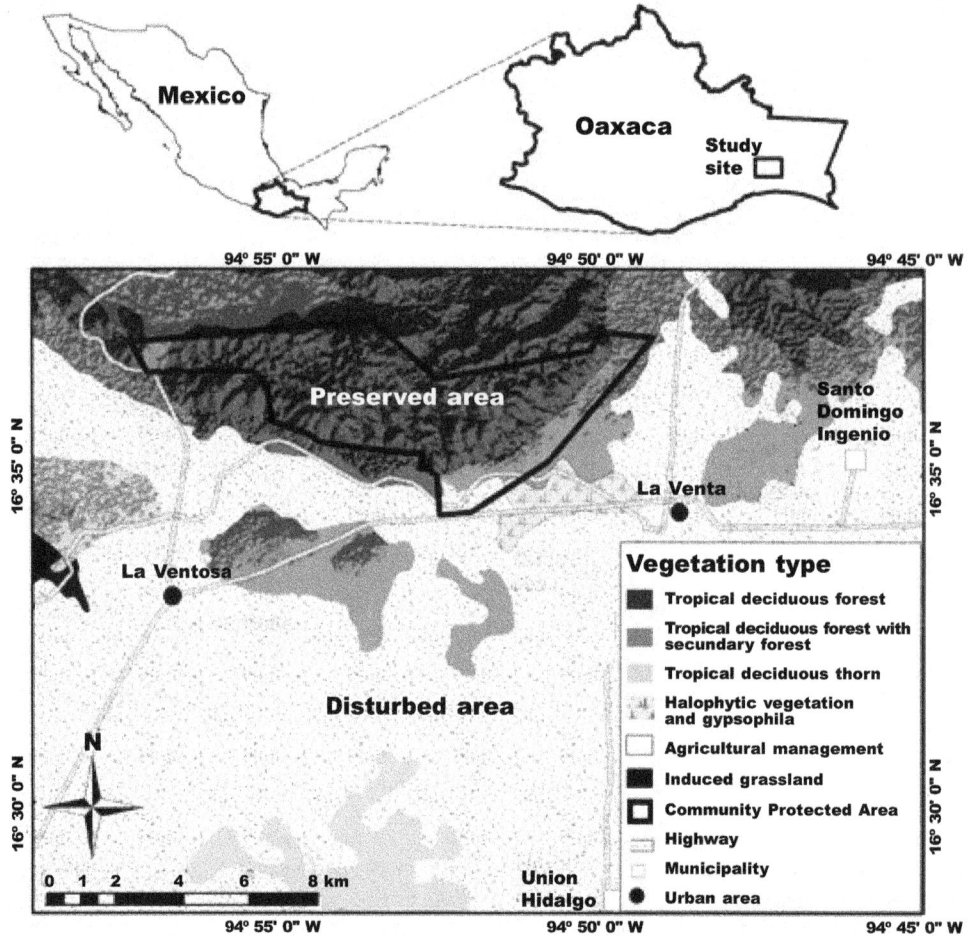

Fig. 1. Geographic location and vegetation types around the study area in the Isthmus of Tehuantepec, Oaxaca, Mexico.

Fig. 1. Ubicación geográfica y tipos de vegetación en el entorno de la zona de estudio en el istmo de Tehuantepec, Oaxaca, México.

Each camera trap was installed approximately 40–50 cm above ground level, depending on the topography and slope of the sampling area. The camera circuit was programmed to remain active for 24 hours, and the camera locations were geo–referenced with a GPS (Garmin Etrex®). Cameras were checked monthly. Photographic records were prepared according to Botello et al. (2007) and deposited in the Collection of Mammalogy (OAX. MA.026.0497) of CIIDIR–OAX.

Data analysis

Species inventories were evaluated using Clench's asymptotic models of species accumulation with the program Species Accumulation, for which the data were previously randomized 100 times with the EstimateS program, version 8.0 (Colwell, 2000). We also calculated the sampling effort required to include 95% of the species in the inventories.

The relative species abundance index for each area and season (dry and rainy) was calculated as the total number of signs found per species, divided by the distance sampled (Carrillo et al., 2000). A Mann–Whitney U test was applied to determine whether there were significant differences in relative abundance between areas and seasons (Zar, 1999).

The species diversity of each area and season was determined according to the Shannon–Wiener entropy index (H'). Dominance (D) was estimated with the Berger–Parker index (Whittaker, 1972), which is an indirect method to measure species diversity: The lower the dominance, the higher the species diversity, and vice versa. Pielou's evenness index (J') was determined as the proportion of diversity observed in relation to the maximum diversity expected (Magurran, 1988). To compare the Shannon index between areas, we applied the Student's t test modified by Hutchenson (Magurran, 1988).

Table 1. Metrics of the disturbance index of livestock density variables, human activities variables and land degradation variables.

Tabla 1. Valores del índice de perturbación de las variables relativas a la densidad de ganado, las actividades humanas y la degradación del suelo.

Variable	Acronym	Description
Livestock density		
Goat droppings frequency	Goat	Computed from presence of goat dung in ten randomly chosen 1 m squares along the transect; frequency was defined as the fraction of squares with positive records.
Cattle droppings frequency	Catt	Bovine and equine dung, computed as for Goat.
Browsing	Brow	All shrubs and trees that were rooted within the transect were thoroughly examined for signs of browsing. The ratio of browsed to total plants was calculated as an index of browsing intensity.
Livestock trail density	Ltra	Livestock uses well–defined trails to move while browsing. The number of these per meter along the transect was recorded.
Soil compaction	Comp	The constant trampling of livestock along tracks causes soil compaction, which affects water infiltration. A cylinder of 10.4 cm of diameter was driven 4 cm into the ground in a randomly chosen trail. 250 ml of water were then poured into the cylinder, and the time needed for complete infiltration was recorded. This procedure was repeated on a spot with no evidence of trampling. The degree of soil compaction was calculated as the ratio of the time recorded on the trail and in the untrampled terrain.
Human activities		
Fuelwood extraction	Fuel	Peasants cut branches for fuel. This metric was measured as Brow, but taking machete cuts into account.
Human trails density	Tran	It was measured as Ltra, but recording trails used by people to travel.
Settlement proximity	Prox	Proximity was defined as the multiplicative inverse of the distance to the closest towns in km.
Contiguity to activity cores	Core	A core was defined as a place where human activities normally take place, such as houses, cornfields, mines and chapels. Contiguity was recorded at each transect if a core was less than 200 m away. The fraction of transects contiguous to a core was used as a metric.
Land use	Luse	In several studies the percent of land cover devoted to agriculture, cultivated or induced pastures, or urban areas is used as a measure of disturbance. Here, the fraction of the study area used for these purposes was visually estimated.
Evidence of fires	Fire	Most of these are initiated by people, either to clear an area, promote pasture growth for livestock, or accidentally. The presence or absence of evidence at a study site was recorded as one or zero.

Table 1. (Cont.)

Variable	Acronym	Description
Land degradation		
Erosion	Eros	Overgrazing and human activities increase erosion. We only considered spots where the soil showed tracks of strong and frequent removal of material by water (such as ravines) as unequivocal evidence of erosion. Twenty points were selected randomly along the transect for its estimation, and the fraction of eroded spots was recorded.
Presence of soil islands	Isla	When severe erosion takes place, soil is only held where large shrubs are rooted. As a result, a landscape of small mounds can be observed. The presence or absence of these "islands" was recorded either as one or zero.
Totally modified surfaces	Toms	Land may be so severely modified that measuring most of the previous metrics makes no sense, as it can happen on a paved road, a house, or on artificial water–ways. When the transect crossed such surfaces, their cover was measured by means of the line intercept method.

Furthermore, to analyze diversity more effectively, we calculated the effective number of species (true diversity) to know how much diversity was lost or gained between areas and between seasons. We used the exponential Shannon–Wiener index, in which all the species in the community are weighted in exact proportion to their abundance (Jost, 2006; Moreno et al., 2011).

Beta diversity (change in species composition) between areas was evaluated using the Whittaker index (Wilson & Schmida, 1984; Magurran, 1988), which in this case can have values between 1 and 2, and the degree of similarity between habitats was evaluated according to the Jaccard similarity index (Magurran, 1988).

Results

The least disturbed area was located on the Tolis-toque hill, hereafter called the 'preserved area'. The area located southeast of La Venta was named the 'disturbed area', and it showed greater disturbance due to its proximity to centers of activity, changes in land use, and islands (table 2).

Clench's species accumulation model was the best choice for the data, although asymptote was not reached in the study area. The model predicted 23 species (a = 6,806 and b = 0.297), meaning that our mammal inventory was 79% complete. According to this model, a total of 63 months would be required to record 95% of the medium and large mammal species living in the study site.

We obtained 167 records, of which 61% were traces and 28% were feces. Of all the records, 79 (47.30%) were found in the preserved area and 88 (52.70%) in the disturbed area (table 3). The records belonged to 18 species, 18 genera, 12 families and six orders of medium and large mammals (table 4). Through the use of camera traps, 82 photographs of mammals were obtained, confirming the presence of ten of the species recorded by indirect methods.

In terms of relative abundance, *Urocyon cinereoargenteus* was the species with the highest abundance in the preserved area (0.23/km), while in the distur-bed area the most abundant species were *Didelphis virginiana* (0.29/km) and *U. cinereoargenteus* (0.25/km) (table 4). According to the Mann–Whitney test, significant differences were found between the relative abundance in the two study areas (N_1 = 79, N_2 = 88, U = 63.5, p = 0.032). *U. cinereoargenteus* and *D. virginiana* were the most abundant species during the two seasons. In both areas, the highest relative abundance of species was observed in the rainy season. However, the seasonal variation in relative abundance was not statistically significant (U = 72, p = 0.76, and U = 23.5, p = 0.72, in preserved and disturbed areas, respectively).

The preserved area exhibited the highest diversity (H' = 2.33) and evenness (J' = 0.82), and the lowest dominance (D = 30.86) (table 4). Significant differen-ces were observed in the Shannon–Wiener index between the diversity of the preserved and disturbed areas (t = 4.9, d.f. = 160). The highest diversity was recorded during the rainy season in the preserved area

Table 2. Values obtained with the disturbance index in two areas with different levels of perturbation near the Isthmus of Tehuantepec, Oaxaca. (For the abbreviations of variables see table 1.)

Tabla 2. Valores obtenidos con el índice de perturbación en dos zonas con diferentes grados de perturbación en el istmo de Tehuantepec, Oaxaca. (Para las abreviaturas de las variables ver tabla 1.)

	Area	
	Preserved	Disturbed
Livestock density		
Goat	0.000	0.000
Catt	0.400	0.400
Fire	0.000	0.000
Brow	0.027	0.050
Ltra	0.020	0.020
Comp	0.185	0.260
Human activities		
Fuel	0.126	0.046
Fire	0.000	0.000
Tran	0.051	0.040
Prox	0.191	0.301
Core	0.000	1.000
Luse	0.000	1.000
Land degradation		
Eros	0.125	0.700
Isla	0.000	1.000
Toms	0.000	0.050
Total disturbance	14.339	67.074

(H' = 2.30). This area also showed lower dominance (D = 18.18) and higher evenness (J' = 0.92). No significant differences were found in the Shannon index between seasons for the preserved and disturbed areas (t = 1.40, g.l. = 80.56 and t = 1.68, g.l. = 73.60, respectively).

According to the measure of true diversity, the diversity of medium and large mammals in the preserved area was double that of the disturbed area. During the dry season, the diversity of mammal species was lower than during the rainy season in both the preserved (24%) and disturbed areas (28%).

Our data revealed a high turnover of species between zones (βw = 1.48). Of the 17 species recorded in this study, eight were found in both areas, while nine species were exclusively found in the preserved area. *Spilogale gracilis* was recorded only in the disturbed area. Finally, the two areas showed a similarity of 47% in species composition according to the Jaccard similarity index.

Discussion

The indirect method was an efficient way to study mammal diversity in this study. Using this method we recorded 18 species of medium–sized and large mammals, whereas the camera traps only recorded the presence of ten species. However, this sampling was not standardized, as camera trapping was only used during the last six months of sampling. Consequently, we recommend the use of complementary methods to record a greater number of species. Indirect methods could however underestimate species richness and abundance as they focused mainly on recording terrestrial species and can overlook tree–dwellers (Aranda, 2000). Combining various techniques also reduces the influence of environmental and methodological factors, providing a more reliable estimate of diversity and abundance in a particular study site (Botello et al., 2008). Zarco (2007) recorded the same number of species as in this study using camera traps in the same vegetation type. This technique facilitates the determination of species' activity patterns, but it is expensive to implement compared to indirect methods.

The species richness found in the area is equivalent to 34.62, 45.00, 57.89 and 66.67% of the total species, genera, families and orders of medium and large mammals present in Oaxaca. These values are higher than the 17 species reported by Santos–Moreno & Ruiz–Velásquez (2011) in the region of Isthmus of Tehuantepec in similar vegetation type, while Monroy–Vilchis et al. (2011) recorded 19 species with camera traps in an area where the main vegetation type, was tropical deciduous forest. These results show that the study area maintains a diverse community of medium and large mammals, despite the effects of disturbance (habitat deterioration and a high presence of human activities) in the south of the Protected Communal Area.

The species richness found at the site, however seems low compared to the study by Cervantes & Yepez (1995) around Salina Cruz, in the coastal plain of Tehuantepec, Oaxaca. In their study, the authors recorded 30 species of medium–sized and large mammals. This difference may be due to the fact that Cervantes & Yepez (1995) conducted their study in tropical deciduous forest, mangrove forest, thorn scrub and dune vegetation, so a greater number of species occupying different ecological niches and ecosystems was recorded. This was seen in the case of *Lontra longicaudis*, for example, which is located only in aquatic environments.

The number of species found in our study was similar to that reported by Lavariega et al. (2012) in the municipality of Santiago Camotlán. However, their study was conducted in cloud forest, oak forest, evergreen forest, crop fields and coffee plantations. Species typical of highly conserved sites, such as *Panthera onca* and *Tamandua mexicana*, are reported in some of these habitats. They are also recorded in association with cattle in disturbed areas, but in a lower proportion (Treves & Karanth, 2003).

According to the Clench model, the species inventory is not fully represented, and it is likely that more species are still to be found in the area. We

Table 3. Number of records of medium and large mammals recorded in La Venta, Juchitan. Record types: F. Footprint; f. Feces; Sr. Skeletal remains; S. Sighting.

Tabla 3. Número de registros de especies de mamíferos de talla mediana y grande registrados en La Venta, Juchitán. Tipos de registros: F. Huella; f. Excrementos; Sr. Restos óseos; S. Avistamiento.

	Preserved area					Disturbed area				
	F	f	Sr	S	Total	F	f	Sr	S	Total
Canis latrans	1	4	–	–	5	–	–	–	–	–
Coendou mexicanus	–	–	3	–	3	–	–	–	–	–
Conepatus leuconotus	3	–	–	–	3	–	–	–		–
Dasypus novemcinctus	1	–	2	–	3	12	–	1	–	13
Didelphis virginiana	9	–	1	–	10	32	–	–	–	32
Herpailurus yagouaroundi	1	–	–	–	1	1	–	–	–	1
Leopardus pardalis	2	–	–	–	2	–	–	–	–	–
Mustela frenata	1	–	–	–	1	–	–	–	–	–
Nasua narica	1	–	–	–	1	1	–	–	–	1
Odocoileus virginianus	9	1	1	–	11	–	–	–	–	–
Pecari tajacu	–	–	3	–	3	–	–	–	–	–
Philander opossum	1	–	–	–	1	1	–	–	–	1
Procyon lotor	1	–	–	–	1	7	–	–	–	7
Puma concolor	2	1	–	–	3	–	–	–	–	–
Spilogale putorius	–	–	–	–	–	3	–	–	–	3
Sciurus aureogaster	–	–	–	3	3	–	–	–	–	–
Sylvilagus floridanus	2	–	–	1	3	2	1	–	–	3
Urocyon cinereoargenteus	6	–	18	1	25	8	18	–	1	27

recorded the presence of *Ateles geoffroyi* at the north of the Tolistoque hill on April 2007 (16° 35' 52.97" N / 94° 52' 35.56" W), although its presence had not been reported by Ortiz–Martinez et al. (2008) in a study on the distribution of *Alouatta palliata* and *A. geoffroyi*. We did not include this latest species in our analysis given that we saw it only once, several months before the present study, in the north of the preserved area. By including *A. geoffroyi*, our inventory would reach 83% of completeness, and we would be missing only three species.

One factor that could affect estimates of the relative abundance of species is the difference in the detectability of their traces, which is related to the size of the species (Litvaitis et al., 1994), their habits, their inclination while walking, and the type of substrate. It is therefore more likely to find tracks of *D. virginiana* because their weight facilitates track impressions and makes them easier to detect. On the contrary, the genus *Sciurus* may be more abundant than deer *Odocoileus* sp., but their habits are primarily arboreal, making track observations more difficult. It is noteworthy that the rainy season facilitated the record of tracks, mainly in areas of flooding, and at this season we recorded the greatest abundance of species.

The relative abundance of *Dasypus novemcinctus* was lower than that reported by Navarro (2005) in secondary forest and oak forest, as this author reported densities of 0.2 individuals/km at each vegetation type. Likewise, Perez–Irineo & Santos–Moreno (2012) reported an even higher relative abundance for the same species (0.07 individuals/km) in a deciduous forest in northeastern Oaxaca. In our study, particularly the disturbed area is affected by strong human intervention, which may explain the low observed abundance of this species. Hunting may also contribute to decrease the abundance and increase the secretive and evasive behavior of some species. It is well known that medium and large sized mammal species are the most affected by hunting. In our study area local inhabitants and people from the surroundings were observed hunting. The most hunted species for meat consumption are armadillos *D. novemcinctus*, squirrels *Sciurus aureogaster* and rabbits *Silvilagus floridanus*.

The high abundance of *U. cinereoargenteus* and *D. virginiana* corresponds with the findings of Orjuela & Jimenez (2004) and Luna (2005), who report that the fox has the highest relative abundance values. These high values of abundance may be related to

Table 4. List of species of medium and large mammals recorded in La Venta, Juchitan, following the taxonomic arrangement proposed by Ramirez et al. (2005) and including the number of records (n) and relative abundance (Relab) in each of the areas and seasons. Index of diversity α and β. Status conservation NOM 059 (* Threatened, ** Endangered). (For the abbreviations of record types, Rec, see table 3.)

Tabla 4. Lista de las especies de mamíferos de talla mediana y grande registradas en La Venta, Juchitán, siguiendo la taxonomía propuesta por Ramírez et al. (2005) e incluyendo el número de registros (n) y la abundancia relativa (Relab) en cada zona y temporada. Índices de diversidad α y β. Estado de conservación NOM 059 (* Amenazada, ** En peligro). (Para las abreviaturas de los tipos de registro, Rec, véase la tabla 3.)

| | | Preserved area | | | | | | Disturbed area | | | | | |
| | | Rainy | | Dry | | Total | | Rainy | | Dry | | Total | |
Taxonomic list	Rec	n	Relab	n	Relab	n	Relab	n	Relab	n	Relab	n	Relab
O. Didelphimorphia / F. Didelphidae													
Didelphis virginiana	F, P	3	0.0556	7	0.1296	10	0.0926	21	0.3889	11	0.2037	32	0.2963
Philander oposum	F	1	0.0185	0	0.0000	1	0.0093	1	0.0185	0	0.0000	1	0.0093
O. Cingulata / F. Dasypodidae													
Dasypus novemcinctus	F, Sr, P	3	0.0556	0	0.0000	3	0.0278	11	0.2037	2	0.0370	13	0.1204
O. Canivora / F. Canidae													
Canis latrans	F, f, P	3	0.0556	2	0.0370	5	0.0463	–		–		0	0.0000
Urocyon cinereoargenteus													
	F, f, S, P	6	0.1111	19	0.3519	25	0.2315	11	0.2037	16	0.2963	27	0.2500
O. Canivora / F. Felidae													
*Herpailurus yagouaroundi**	F	0	0.0000	1	0.0185	1	0.0093	1	0.0185	0	0.0000	1	0.0093
Puma concolor	F	1	0.0185	2	0.0741	3	0.0463	–		–		0	0.0000
*Leopardus pardalis***	F, P	0	0.0000	2	0.03704	2	0.0185					0	
O. Canivora / F. Mustelidae													
Mustela frenata	F	0	0.0000	1	0.0185	1	0.0093	–		–		0	0.0000
O. Carnivora / F. Mephitidae													
Conepatus leuconotus	F, P	1	0.0185	2	0.0370	3	0.0278	–		–		0	0.0000
Spilogale putorius	F, P					0	0.0000	1	0.0185	2	0.0370	3	0.0278
O. Carnivora / F. Procyonidae													
Nasua narica	F	0	0.0000	1	0.0185	1	0.0093	1	0.0185	0	0.0000	1	0.0093
Procyon lotor	F, P	0	0.0000	1	0.0185	1	0.0093	5	0.0926	2	0.0370	7	0.0648
O. Artiodactyla / F. Tayassuidae													
Pecari tajacu	Sr	3	0.0556	0	0.0000	3	0.0278	–		–		0	0.0000
O. Artiodactyla / F. Cervidae													
Odocoileus virginianus	F, f, Sr, P	6	0.1111	5	0.0926	11	0.1019	–		–		0	0.0000
O. Rodentia / F. Sciuridae													
Sciurus aureogaster	F, S	1	0.0185	2	0.0370	3	0.0278	–		–		0	0.0000
O. Rodentia / F. Erethizontidae													
*Coendou mexicanus**	Sr	3	0.0556	0	0.0000	3	0.0278	–		–		0	0.0000
O. Lagomorpha / F. Leporidae													
Sylvilagus floridanus	F, S, P	2	0.0370	1	0.0185	3	0.0278	2	0.0370	1	0.0185	3	0.0278
Total records		33		46		79		54		34		88	
Total species						17						9	

Table 4. (Cont.)

Diversity α	Preserved area			Disturbed area		
	Rainy	Dry	Total	Rainy	Dry	Total
Shannon–Wiener	2.304	2.023	2.331	1.653	1.324	1.597
Evenness (J)	0.927	0.789	0.823	0.752	0.739	0.727
Dominance	0.181	0.395	0.308	0.388	0.470	0.363
Effectivenes diversity	10.010	7.563	10.289	5.225	3.757	4.939
% Diversity loss/areas	52.00					
% Loss/seasons	24.449			28.099		

Diversity β		
Whittaker	1.48	
Jaccard	47%	

the characteristics of the species; as omnivores, they are more likely to find food. Consequently, its presence is favored on disturbed areas, or in crops such as sorghum, one of the crops found in the region. We found evidence of sorghum consumption by foxes.

The diversity values recorded for both the preserved and the disturbed areas are lower than those reported by Cueva et al. (2010) (H' = 2.4). However, their study area represents a very well preserved area with a greater extension, since it belongs to a biological reserve of about 730 ha in the reserve community Santa Lucía (Ecuador). Contrary, the mammal diversity in our study area is higher than that reported by Perez–Irineo & Santos–Moreno (2012) in a deciduous forest in Oaxaca (H' = 0.89). Therefore, our results are significant because this index is usually between 1.5 and 3.5 (Magurran, 1988). It also has been observed that H' decreases as disturbance increases, varying from 0.98 to 2.16 according to the degree of the environmental disturbance. The results obtained in this study show that the preserved area is the most diverse, since in this area we found the lowest dominance and the highest evenness.

The total values of diversity indexes in both areas of study show that the populations of medium–sized and large mammals respond to anthropogenic factors, which is reflected in a decrease in their diversity. The preserved area offers the best conditions under which species can develop their activities: find shelter, search for food, and reproduce. The greatest diversity in the preserved area may be due to the greater vegetation richness and greater canopy height, which increases the potential niches and provides more food resources, shelter, protection and escape opportunities to mammals (Gallina et al., 2007). In this area we also found species such as Puma concolor, Leopardus pardalis and Pecari tajacu, which can be considered indicators of well–preserved environments (Cruz–Lara et al., 2004).

The disturbed area may present lower diversity due to several processes found in the area. Human activities such as deforestation, the opening of roads, and noise pollution, affect the habitat directly and indirectly, and modify wildlife activity (Herrera–Flores et al., 2002). Nevertheless, this area still maintains moderate diversity because of the fast–growing vegetation used as habitat and a food source for mammals (Soto & Herrera–Flores, 2003). The presence of water bodies near the site also attracts some mammal species that can find food, water and shelter in the surrounding vegetation (Guzmán–Lenis & Camargo–Sanabria, 2004).

Species diversity can change or remain stable in response to disturbances in the forest. Certain groups of animals, such as foxes, can increase their abundance. Thus, some species may increase their dominance, while the community species richness remains constant in the area. A change like this may decrease the diversity in the area. According to Rocha & Dalponte (2006), the absence of deer and puma in the disturbed area may be because the site does not meet the needs of a predator at on the top of the food chain, such as P. concolor, and does not provide a suitable habitat for the occurrence of O. virginianus.

The values of beta diversity and similarity suggest a high species turnover. The medium and large mammals found in both areas are considered different communities according to the proposal of Sanchez & Lopez (1988), who propose that for two communities to be similar they should have a similarity of above 66.6%. The high species turnover may be mainly due to the fragmentation of local populations throughout the environment, derived from disturbances such as the presence of the Panamerican Highway 185, which separates the two areas and thus creates a barrier that limits the movement of organisms between areas. Also, the isolation of populations may cause local extinctions due to lack of genetic exchange with other individuals from different populations (Arroyave et al., 2006). In this way the presence of human activity can have also an adverse effect on the dispersion pattern of animals.

Acknowledgements

We thank the authorities of La Venta, Juchitan for their support and the facilities provided. M. Cortes thanks CONACyT for the scholarship granted during the period August 2007 and June 2009. The study was supported by the Secretaría de Investigación y Posgrado of the IPN (SIP 20100263). The authors would also thank C. Moreno for her valuable suggestions and comments. M. Briones was funded by Estímulos al Desempeño de la Investigación, Comisión de Operación y Fomento a las Actividades Académicas of IPN and Sistema Nacional de Investigadores. We thank M. Lavariega for the map design.

References

Alonso, A., Dallameier, F. & Campbell, P., 2001. *Urubamba: The biodiversity of a Peruvian rainforest.* Smithsonian Institution, Washington, D.C.

Aranda, M., 2000. *Huellas y otros rastros de los mamíferos grandes y medianos de México.* Primera edición. Ed. Instituto de Ecología, AC, Veracruz–México.

Arroyave, M. P., Gómez, C., Gutiérrez, M. E., Múnera, D. P., Zapata, P. A., Vergara, I. C., Andrade, I. M. & Ramos, K. C., 2006. Impactos de las carreteras sobre la fauna silvestre y sus principales medidas de manejo. *Revista Escuela de Ingeniería Antioquia (EIA)*, 5: 45–57.

August, P., 1983. The role of habitat complexity and heterogeneity in structuring tropical mammal communities. *Ecology*, 64: 1495–1507.

Barragán, F., Lorenzo, C., Morón, A., Briones–Salas, M. & López, S., 2010. Bat and rodent diversity in a fragmented landscape on the Isthmus of Tehuantepec, Oaxaca, Mexico. *Tropical Conservation Science*, 3(1): 1–16.

Bilenca, D., Balla, P., Álvarez, M. & Zalueta, G., 1999. Evaluación de dos técnicas para determinar la actividad y abundancia de mamíferos en el bosque chaqueño, Argentina. *Revista Ecológica Latino Americana*, 6(1): 13–18.

Bolaños, C. & Naranjo, J. E., 2001. Abundancia, densidad y distribución de las poblaciones de ungulados en la cuenca del río Lacatún, Chiapas, México. *Revista Mexicana de Mastozoología*, 5: 45–57.

Botello, F., Monroy, G., Illoldi–Rangel, P., Trujillo–Bolio, I. & Sánchez–Cordero, V., 2007. Sistematización de imágenes obtenidas por fototrampeo: una propuesta de ficha. *Revista Mexicana de Biodiversidad*, 78: 207–210.

Botello, F., Sánchez–Cordero, V. & González, G., 2008. Diversidad de carnívoros en Santa Catarina Ixtepeji, Sierra Madre de Oaxaca, México. In: *Avances en el estudio de los mamíferos de México.* Publiciones

Especiales, Vol. II: 335–354 (C. Lorenzo, E. Espinoza & J. Ortega, Eds.). Asociación Mexicana de Mastozoología A.C., México, D.F.

Briones–Salas, M. A. & Sánchez–Cordero, V., 2004. Mamíferos. In: *Biodiversidad de Oaxaca*: 423–447 (A. J. García Mendoza, M. J. Ordóñez & M. Briones–Salas, Eds.). Instituto de Biología, UNAM–Fondo Oaxaqueño para la Conservación de la Naturaleza–World Wildlife Found, México.

Carrillo, E., Wong, G. & Cuarón, A. D., 2000. Monitoring mammal populations in Costa Rican protected areas under different hunting restrictions. *Conservation Biology*, 14(6): 1580–1591.

Cervantes, F. & Yépez, L., 1995. Species richness of mammals from the vecinity of Salina Cruz, Coastal Oaxaca, Mexico. *Anales del Instituto de Biología. Serie Zoología*, 66(1): 113–122.

Collins, S. L., Gleen, S. M. & Gibson, D. J., 1995. Experimental analysis of intermediate disturbance and initial floristic composition: Decoupling cause and effect. *Ecology*, 76: 486–492.

Colwell, R. K., 2000. *EstimateS* version 3.2. Department of Ecology & Evolutionary Biology, University of Connecticut, Storrs, U.S.A.

Cruz–Lara, L., Lorenzo, C., Soto, L., Naranjo, E. & Ramírez–Marcial, N., 2004. Diversidad de mamíferos en cafetales y selva mediana de las cañadas de la selva Lacandona, Chiapas, México. *Acta Zoológica Mexicana*, 20(1): 63–81.

Cueva, X. A., Morales, N., Brown, M. & Peck, M, 2010. Macro y mesomamíferos de la Reserva Comunitaria Santa Lucía, Pichincha, Ecuador. *Boletín Técnico 9, Serie Zoológica*, 6: 98–110.

Gallina, S., Delfín, C., Mandujano, S., Escobedo, L. & González, R., 2007. Situación actual del venado cola blanca en la zona centro del estado de Veracruz, México. *Deer Specialist Group News. Newsletter*, 22: 29–33.

García, E., 1988. *Modificaciones al sistema de clasificación climática de Köppen.* Instituto de Geografía, Universidad Nacional Autónoma de México, México.

García–Trejo, E. A. & Navarro, A. G., 2004. Patrones biogeográficos de la riqueza de especies y el endemismo de la avifauna en el Oeste de México. *Acta Zoológica Mexicana (n.s.)*, 20(2): 167–185.

González, G., Briones–Salas, M. A. & Alfaro, A. M., 2004. Integración del conocimiento faunístico del estado. In: *Biodiversidad de Oaxaca*: 349–366 (A. J. García Mendoza, M. J. Ordóñez & M. Briones–Salas, Eds.). Instituto de Biología, UNAM–Fondo Oaxaqueño para la Conservación de la Naturaleza–World Wildlife Found, México.

Guzmán–Lenis, A. & Camargo–Sanabria, A., 2004. Importancia de los rastros para la caracterización del uso de hábitat de mamíferos medianos y grandes en el bosque Los Mangos (Puerto López, Meta, Colombia). *Acta Biológica Colombiana*, 9(1): 11–22.

Herrera–Flores, J. C., Fredericksen, T. S. & Rumíz, D., 2002. Evaluación rápida de mamíferos en base a huellas para observar los impactos del manejo forestal. *Ecología en Bolivia*, 37(1): 3–13.

Jost, L., 2006. Entropy and diversity. *Oikos*, 113:

363–375.

Laurance, W. F., Croes, B. M., Tchignoumba, L., Lahm, S., Alonso, A., Lee, M. E., Campbell, P. & Ondzeano, C., 2006. Impacts of roads and hunting en Central Africa Rainforest mammals. *Conservation Biology*, 20(4): 1251–1261.

Lavariega, M. C., Briones–Salas, M. A. & Gómez–Ugalde, R. M., 2012. Mamíferos medianos y grandes de la Sierra de Villa Alta, Oaxaca, México. *Mastozoología Neotropical*, 19(2): 163–178.

Litvaitis, J. A., Titus, K. & Anderson, E. M., 1994. Measuring vertebrate use of terrestrial habitats and foods. In: *Research and Management Techniques for Wildlife and Habitats*: 254–274 (Th. A. Bookhout, Ed.). The Wildlife Society Bethesda, Maryland.

Luna, M. D., 2005. Distribución, abundancia y conservación de carnívoros en Santiago Comaltepec, Sierra Madre de Oaxaca, México. Tesis de Licenciatura, Instituto Tecnológico Agropecuario de Oaxaca.

Magurran, A. E., 1988. *Ecological diversity and its measurement*. Princeton University Press, New Jersey.

Martorell, C. & Peters, E., 2005. The measurement of chronic disturbance and its effects on the threatened cactus *Mammilaria pectinifera*. *Biological Conservation*, 124: 199–207.

Michalski, F. & Peres, C. A., 2005. Anthropogenic determinants of primate and carnivore local extinctions in a fragmented forest landscape of southern Amazonia. *Biological Conservation*, 124: 383–396.

Monroy–Vilchis, O., Zarco–González, M. M., Rodríguez–Soto, C., Soria–Díaz, L. & Urios, V., 2011. Fototrampeo de mamíferos, en la Sierra Nanchititla, México: abundancia relativa y patrón de actividad. *Revista de Biología Tropical*, 59(1): 373–383.

Moreno, C. E., Barragán, F., Pineda, E. & Pavón, N. P., 2011. Reanálisis de la diversidad alfa: alternativas para interpretar y comparar información sobre comunidades ecológicas. *Revista Mexicana de Biodiversidad*, 82: 1249–1261.

Murcia, C., 1995. Edge effects in fragmented forests: implications for conservation. *Trends in Ecology and Evolution*, 10(2): 58–62.

Navarro, E., 2005. Abundancia relativa y distribución de los indicios de las especies de mamíferos medianos en dos coberturas vegetales en el santuario de flora y fauna Otún Quimbaya, Pereira, Colombia. Tesis de Licenciatura, Pontificia Universidad Javeriana.

Noss, R. F., 1987. Corridors in real landscapes: a reply to Simberloff and Cox. *Conservation Biology*, 1: 159–164.

Ojasti, J., 2000. *Manejo de fauna silvestre neotropical*. SIMAB Series N° 5. Smithsonian Institution / MAB Program. Washington, D.C., U.S.A.

Orjuela, O. J. & Jiménez, G., 2004. Estudio de la abundancia relativa para mamíferos en diferentes tipos de coberturas y carretera, Finca Hacienda Cristales, área Cerritos–La Virginia, Municipio de Pereira, Departamento de Risalda, Colombia. *Universitas, Revista de la Facultad de Ciencias*, 9: 87–96.

Ortega, D., Sánchez, G., Solano, C., Huerta, M. A., Meza, V., Romero, J., Cruz, L., Palacios, T.,

Montes, E. & Galindo–Leal, C., 2010. *Áreas de Conservación Certificadas en el Estado de Oaxaca*. WWF–CONANP Oaxaca, México.

Ortiz– Martínez, T., Rico–Gray, V. & Martínez–Meyer, E., 2008. Predicted and verified distributions of *Ateles geoffroyi* and *Alouatta palliata* in Oaxaca, Mexico. *Primates*, 49: 186–194.

Pérez–Irineo, G. & Santos–Moreno, A., 2012. Diversidad de mamíferos de talla grande y media de una selva subcaducifolia del noreste de Oaxaca, México. *Revista Mexicana de Biodiversidad*, 83: 164–169.

Peters, E. & Martorell, C., 2000. *Conocimiento y conservación de las mamilarias endémicas del Valle de Tehuacán–Cuicatlán*. Reporte final del proyecto R166–CONABIO, México, D.F.

Peterson, A. T., Soberón, J. & Sánchez–Cordero, V., 1999. Conservatism of ecological niches in evolutionary time. *Science*, 285: 1265–1267.

Ramírez, J., Arroyo, J. & Castro, A., 2005. Estado actual y relación nomenclatural de los mamíferos terrestres de México. *Acta Zoológica Mexicana (nueva serie)*, 21(1): 21–82.

Robinson, J. G. & Redford, L. H., 1986. Body size, diet, and population density of Neotropical forest mammals. *The American Naturalist*, 128: 665–680.

Rocha, E. C. & Dalponte, J., 2006. Composição e caracterização da fauna de mamíferos de médio e grande porte em uma pequena reserva de Cerrado em Mato Grosso. Brasil. *Revista árvore*, 30(4): 669–678.

Roemer, G. W., Gompper, E. & Van Valkenburgh, B., 2009. The ecological role of the mammalian mesocarnivore. *BioScience*, 59(2): 165–173.

Sánchez, O. & López, G., 1988. A theoretical analysis of some indices of similarity as applied to biogeography. *Folia Entomologica Mexicana*, 75: 119–145.

Santos–Moreno, A. & Ruíz–Velásquez, E., 2011. Diversidad de mamíferos de la región de Nizanda, Juchitán, Oaxaca, México. *Therya*, 2(2): 155–168.

Simonetti, J. A. & Huareco, I., 1999. Uso de huellas para estimar diversidad y abundancia relativa de los mamíferos de la Reserva de la Biosfera– Estación Biológica del Beni, Bolivia. Nota técnica. *Mastozoología Neotropical*, 6(1): 139–144.

Soto, G. & Herrera–Flores, J. C., 2003. *Respuestas de mamíferos y aves terrestres a las diferentes intensidades de aprovechamiento forestal en la época húmeda y seca*. Documento Técnico No. 132. Proyecto BOLFOR, Santa Cruz, Bolivia.

Soulé, M. E., Alberts, A. C. & Bolger, D. T., 1992. The effects of habitat fragmentation on chaparral plants and vertebrates. *Oikos*, 63: 39–47.

Sutherland, W. J., Ed., 1996. *Ecological census techniques: a handbook*. Cambridge University Press, Cambridge, U.K.

Treves, A. & Karanth, K. U., 2003. Human–carnivore conflict and perspectives on carnivore management worldwide. *Conservation Biology*, 17(6): 1491–1499.

Whittaker, R. H., 1972. Evolution and measurement of species diversity. *Taxon*, 21(2/3): 213–251.

Wilson, M. V. & Shmida, A., 1984. Measuring beta diversity with presence–absence data. *Journal of*

Ecology, 72: 1055–1064.

Zar, J. H., 1999. *Biostatistics.* Prentice Hall, Englewoods Cliffs.

Zarco, M., 2007. Distribución y abundancia de mamíferos medianos y grandes en la Sierra Nanchititla. Tesis de Licenciatura, Universidad Autónoma del Estado de México.

Zarza, H., 2001. Estructura de la comunidad de pequeños mamíferos en diversos hábitats en La Selva Lacandona, Chiapas, México. Tesis de Licenciatura, Universidad Nacional Autónoma de México, México, D.F.

18

The impact of
an invasive exotic bush
on the stopover ecology
of migrant passerines

J. Arizaga, E. Unamuno, O. Clarabuch & A. Azkona

Arizaga, J., Unamuno, E., Clarabuch, O. & Azkona, A., 2013. The impact of an invasive exotic bush on the stopover ecology of migrant passerines. *Animal Biodiversity and Conservation*, 36.1: 1–11.

Abstract

The impact of an invasive exotic bush on the stopover ecology of migrant passerines.— Migration is highly energy–demanding and birds often need to accumulate large fuel loads during this period. However, original habitat at stopover sites could be affected by invasive exotic plants outcompeting native vegetation. The impact of exotic plants on the stopover behavior of migrant bird species is poorly understood. As a general hypothesis, it can be supposed that habitat change due to the presence of exotic plants will affect migrants, having a negative impact on bird abundance, on avian community assemblage, and/or on fuel deposition rate. To test these predictions, we used data obtained in August 2011 at a ringing station in a coastal wetland in northern Iberia which contained both unaltered reedbeds (*Phragmites* spp.) and areas where the reedbeds had been largely replaced by the invasive saltbush (*Baccharis halimifolia*). Passerines associated with reedbeds during the migration period were used as model species, with a particular focus on sedge warblers (*Acrocephalus schoenobaenus*). The saltbush promoted a noticeable change on bird assemblage, which became enriched by species typical of woodland habitats. Sedge warblers departed with a higher fuel load, showed a higher fuel deposition rate, and stayed for longer in the control zone than in the invaded zone. Invasive plants, such as saltbush, can impose radical changes on habitat, having a direct effect on the stopover strategies of migrants. The substitution of reedbeds by saltbushes in several coastal marshes in Atlantic Europe should be regarded as a problem with potential negative consequences for the conservation of migrant bird species associated with this habitat.

Key words: *Acrocephalus* spp., Biological conservation, Biological invasion, Coastal marshes, Fuel deposition rate, Saltbush (*Baccharis halimifolia*).

Resumen

Impacto de un arbusto exótico invasor en la ecología de los puntos de parada de los paseriformes migradores.— La migración requiere un elevado gasto de energía y las aves suelen necesitar acumular grandes cantidades de grasa durante este período. Sin embargo, el hábitat original de los puntos de parada podría verse afectado por plantas exóticas invasoras que compiten con la vegetación autóctona. Se conocen poco los efectos de las plantas exóticas en el comportamiento de las especies de aves migradoras en cuanto a los puntos de parada. Como hipótesis general, puede suponerse que el cambio del hábitat debido a la presencia de plantas exóticas afectará a las aves migradoras e influirá negativamente en su abundancia, la composición de la comunidad de aves y el índice de deposición de grasa. Para comprobar estas predicciones, utilizamos los datos obtenidos en agosto de 2011 en una estación de anillamiento situada en los humedales costeros del norte de la península ibérica en los que había carrizos inalterados (*Phragmites* spp.) y en zonas en las que los carrizos habían sido sustituidos en gran parte por el bácaris invasor (*Baccharis halimifolia*). Se utilizaron como modelo a los paseriformes asociados a los carrizales durante el período de migración y se prestó especial atención al carricerín común (*Acrocephalus schoenobaenus*). El bácaris propició un cambio notable en la composición avícola, que se enriqueció con especies típicas de hábitats forestales. Los carricerines partieron con una cantidad de grasa superior, mostraron un índice de deposición de grasa más elevado y permanecieron más tiempo en la zona de control que en la zona invadida. Las plantas invasoras, como el bácaris, pueden forzar cambios radicales en el hábitat y tener un efecto directo en las estrategias de parada de las aves migradoras. La sustitución de los carrizales por bácaris en diversas marismas de la costa atlántica de Europa debería considerarse un problema con posibles consecuencias negativas para la conservación de las especies de aves migradoras asociadas a este hábitat.

Palabras clave: *Acrocephalus* spp., Conservación de la diversidad biológica, Invasión biológica, Marismas costeras, Índice de deposición de grasa, Bácaris (*Baccharis halimifolia*).

Juan Arizaga, Edorta Unamuno & Ainara Azkona, Dept. of Ornithology, Aranzadi Sciences Society, Urdaibai Bird Center, Orueta 7, E–48314 Gautegiz–Arteaga, Bizkaia, España (Spain).– Oriol Clarabuch, Catalan Institute of Ornithology, Museu de Ciències Naturals de Barcelona, Passeig Picasso s/n., E–08003 Barcelona, Espanya (Spain).

Corresponding author: J. Arizaga. E–mail: jarizaga@aranzadi–zientziak.org

Introduction

Migration is considered to have carry–over effects on several parameters in the life cycle of avian migrants (Newton, 2004). Suboptimal stopover places may hamper an adequate fuel deposition rate and compromise not only survival, but also future life history aspects such as mating or breeding success (Sandberg & Moore, 1996; Smith & Moore, 2003).

Generally, migrant birds cannot gain enough fuel at a single site to reach their destination areas in a single uninterrupted flight or in several flights without refuelling at en route stopovers. They therefore need to stop over periodically to accumulate sufficiently high energy stores to accomplish the next flight bout successfully. The rate of fuel accumulation at stopover sites influences speed of migration, and it has been considered to be an indicator of habitat quality at these stopover sites (Alerstam & Lindström, 1990). Accordingly, any factor that affects fuel deposition rate, such as food availability or predator disturbance, can be crucial for migrant bird species in terms of migration success or survival.

Exotic plants can displace native vegetation, causing habitat changes which are often linked to changes in biodiversity (Vitousek et al., 1997). This phenomenon also affects bird migrants when they land to refuel in an altered habitat (e.g., Cerasale & Guglielmo, 2010). The impact of exotic plants on stopover behavior of migrant bird species is, however, poorly understood (Cerasale & Guglielmo, 2010). This issue is worth taking into consideration, especially if we consider that several bird species, both at population and individual levels, tend to use the same stopover sites year after year (Newton, 2008). Habitat changes in these areas can thus have negative consequences, even though alternative stopover sites may be available.

As a global hypothesis, it can be stated that native habitat change (i.e., deterioration) due to the presence of exotic plants will have a negative impact, either in relation to bird assemblage or stopover behavior (fuel deposition rate, stopover duration, etc.), on individual migrants originally associated with native vegetation.

The predictions tested in this study were: (1) A decrease in the area covered by native vegetation will have a negative impact on bird abundance at a local scale, because these species are adapted to use their particular native habitats. Consequently, these species would not use areas heavily invaded by exotic plants. From a structural standpoint, the community may change toward species better adapted to the new conditions (e.g., Sol et al., 2002). (2) If a zone affected by an invasive exotic plant species turns into a suboptimal area (e.g., offering worse fuelling opportunities), migrants should be expected to move from this to better nearby areas with native vegetation. Thus, we expect a higher number of within–season recaptures from the affected, a priori suboptimal site (affected by exotic plants), to the unaffected site, especially if migrants are able to look for better sites at a micro–scale level within the same stopover area (Delingat & Dierschke, 2000). (3) A decrease in the

rate of fuel deposition of migrants in a zone affected by exotic vegetation as compared to a zone with only native vegetation is also expected (Cerasale & Guglielmo, 2010), either because migrants associated with native vegetation are not good foragers in a foreign habitat (in suboptimal habitats, migrants should be expected to forage less efficiently; e.g., Jenni–Eiermann et al., 2011), or because local insects (i.e., birds' food supply) cannot feed on exotic plants and, therefore, food availability is lower than that found in areas of native vegetation. Consequently, migrants departing from a stopover site affected by exotic plants should have lower fuel loads than migrants departing from an unaffected site.

The aim of this study was to evaluate the impact of exotic plants on the community structure and stopover ecology of migrant birds, particularly their effect on fuel load and fuel deposition rate. We used data obtained at a coastal wetland in southwestern Europe that had both unaltered reedbeds (Phragmites spp.) and areas where the reedbeds had been largely replaced by the invasive saltbush (Baccharis halimifolia). The study was carried out at two levels: the first level focused on bird assemblages, while the second level focused on the sedge warblers (Acrocephalus schoenobaenus) as an avian model species typical of reedbeds to compare body mass, fuel deposition rate and stopover duration in invaded and control areas.

Material and methods

Study system

Reedbeds in Europe play a relevant role as stopover sites for marsh–associated birds during migratory periods (Schaub et al., 2001; Arizaga et al., 2006). This habitat, however, has suffered a notable decline in several areas due to the saltbush (Sanz et al., 2004), a shrub originally found along the coast of eastern North America (Cronquist, 1980). Typical of plains within coastal marshes, it is currently widespread worldwide as an exotic, invasive plant that occupies wetlands with slight to moderate levels of brackish waters. The saltbush has become a problem of primary importance in many wetlands in Europe, Asia and Australia, and it has been the target of numerous (and usually costly) management plans (e.g., Palmer et al., 1993). Its eradication has thus been a priority in several projects (e.g., Life projects such as LIFE08NAT/E/000055) orientated to preserve habitats of interest in Europe.

The reed–associated warblers (Acrocephalus spp.) are a group of closely–related species of small insectivorous birds that are normally adapted to exploit vertically–structured vegetation in Europe, Africa, Asia and some parts of Oceania (Cramp, 1992; Leisler & Schulze–Hagen, 2011). A paradigmatic case of this specialization is birds such as sedge warblers, which mainly forage on aphids (Hyalopterus spp.) found in reedbeds (Bibby & Green, 1981). Sedge warblers breed in humid habitats from the west of Europe to Central Siberia (90° E), between the July isotherms of 12 and 30ºC (Cramp, 1992). They overwinter in tropical/southern Africa or southern Asia (Cramp, 1992). During migration (mainly from mid–July to September during the autumn migration period), they preferably occupy reedbeds, both along the coast and inland (Cramp, 1992). Sedge warblers have a direct dependence on certain aphid species on which they feed to accumulate large fuel loads before crossing the Sahara Desert in autumn (Bibby & Green, 1981). Stopover normally takes place in strategic areas with good foraging conditions (Bibby & Green, 1981; Bensch & Nielsen, 1999), so habitat changes at stopover areas could be particularly damaging to the conservation of this species.

The sedge warbler is present in northern Iberia only during the migration period (Tellería et al., 1999), when they are common in many wetlands. During the autumn migration period, the species accounts for ca. 20% of the captures obtained at a constant–effort ringing station in Urdaibai, a coastal wetland in Northern Iberia (Unamuno & Arizaga, unpubl. data). The sedge warbler is a good avian model to test the effect of invasive plants on stopover behavior of migrants associated with reedbeds, since all birds captured in Urdaibai are non–breeding, true migrants.

Study area and data collection

Data were obtained at Urdaibai, a Ramsar coastal estuary in the southeastern Bay of Biscay, northern Iberia. The wetland spreads over an area of 945 ha and has a relatively high richness of habitats, determined by tide regimens and the degree of salinity. Such habitats range from beach and dunes (situated within the lowest part of Urdaibai) to tidal flats of limes and plant species adapted to tidal flooding in the lower marsh and reedbeds and freshwater–associated vegetation in the upper marsh and nearby polders.

Birds were captured with mist nets of 16 mm mesh (144 linear m/zone) at two zones which had a different degree of invasion by saltbushes (table 1). Originally, both zones were occupied by reedbeds (as dominant vegetation) together with *Juncus* spp. and *Aster* spp., but currently the vegetation in one zone has been almost completely replaced by saltbushes. Hereafter, we will respectively call these zones 'control' and 'invaded'. Both zones are found in the upper marsh and are subject to daily tidal flooding. From July to September the area is used by warblers coming from the British Isles and Western–Central Europe (Cantos, 1998) en route to their wintering areas in tropical Africa (Cramp, 1992).

Mist nets at both zones were open daily during August 2011 for a period of 4 h starting at dawn. Since the sampling was carried out in parallel with another project, focused on finding key stopovers for the aquatic warbler (*A. paludicola*) in the bay of Biscay, we also used tape lures with the song of a male aquatic warbler (one lure per 36 linear m of mist nets) (Julliard et al., 2006). Since the sampling effort with both mist nets and lures was constant in these two zones, using tape lures would not be expected to create any bias

for the comparison of stopover behavior between the two trapping sites.

Once captured, each bird was ringed and its age (first year or adult bird) was determined according to Svensson (1996). Additionally, we measured body mass (± 0.1 g) and the length of the P3 primary feather (± 0.5 mm; numbered from outermost to innermost). Following measurements, the birds were released. No bird was retained for longer than 1 h.

Statistical analyses

We compared community richness (number of birds), as well as a diversity index (H') using a bootstrap procedure as calculated by the software PAST 2.1 (Hammer et al., 2001). H' stands for the Shannon index, which ranges from 0 (communities with only a single taxon) to high values in communities with many taxa, each with few individuals. The bootstrap procedure consisted in generating a 95% confidence interval (CI) by taking 1,000 random sub–samples from the total pooled data. The number of captures per day at each zone was compared with a t–test. Number of captures was log–transformed to normalize the data.

Contingency tables were used to compare the proportion of recaptures between zones, and to test whether cross–recaptures from the control zone into the invaded zone were more common than vice versa.

As body mass in excess of structural mass in migrants is mainly stored as fat and, to a lesser extent, proteins (Jenni & Jenni–Eiermann, 1998), we used body mass controlled for body size (including P3 as a covariate) as a surrogate for fuel load (e.g., Arizaga et al., 2010, 2011a). Analyses of fuel load and fuel deposition rate were run only for sedge warblers.

Captures from ordinary trapping sessions at a stopover site are subject to certain constraints. In particular, the first and last captures of each individual are not always obtained on the exact dates of arrival or departure (Schaub et al., 2001). It is thus impossible to estimate daily fuel load var iation for the entire stopover period of each individual. However, trapping sessions at ringing stations suffice to estimate fuel load of stopping–over migrants (Schaub & Jenni, 2000a, 2000b; Arizaga et al., 2010, 2011b). To obtain a closer estimation of fuel load both on arrival at the site and on departure, we selected the ten lightest and heaviest migrants in each zone (Ellegren & Fransson, 1992). Since migrants are expected to gain fuel whilst stopping–over at a site, the lightest birds are those most likely to have just arrived, whereas the heaviest birds are those most likely about to depart. We compared fuel loads between zones using an ANCOVA on body mass with zone and fuel category (lightest/heaviest) as factors, and P3 as a covariate that controlled for body size. Although body mass can differ between age classes (Grandío, 1999), we did not consider age as an additional factor due to the relatively low sample size, particularly in one of the zones (the invaded one). Body mass was normally distributed (K–S test: P > 0.05).

Table 1. Main (mean ± SE) vegetation characteristics at the two trapping stations. Statistics are calculated for a sample size of 12 nets at each trapping site. The habitat was studied within an area of 240 m² from each side of each net (*i.e.*, 480 m² around each net). The distance (straight line) from one zone to another was 300 m: * We show here the first and second dominant herbaceous plants.

*Tabla 1. Características principales (media ± EE) de la vegetación en las dos estaciones de trampeo. Los estadísticos se calcularon para un tamaño de muestra de 12 redes en cada punto de trampeo. El hábitat se estudió en una superficie de 240 m² a cada lado de las redes (es decir, 480 m² alrededor de cada red). La distancia (en línea recta) entre las dos zonas era de 300 m: * Aquí mostramos la primera y la segunda plantas herbáceas dominantes.*

	Affected by saltbush (Invaded zone) 43° 20' 43.44" N 02° 39' 44.30" W	Non–affected by saltbush (Control zone) 43° 20' 52.90" N 02° 39' 44.04" W
Vegetation cover (%)		
No vegetation (mud flats and water)	0.4 ± 0.4%	4.6 ± 1.9%
Tree	–	3.3 ± 2.2%
Bush	88.8 ± 4.7%	5.6 ± 2.8%
Herbaceous	10.8 ± 4.5%	86.5 ± 3.8%
Dominant vegetation		
Tree	–	*Tamarix*
Bush	*Baccharis*	*Baccharis*
Herbaceus*	*Juncus–Phragmites*	*Phragmites–Aster*
Vegetation height (over 12 nets)		
< 1 m	–	2/12
1–2 m	–	3/12
2–3 m	6/12	7/12
> 3 m	6/12	–

Contingency tables were also used to see whether the proportions of the first–year birds and adults differed between the control and invaded zones. This analysis was done only for sedge warblers.

The importance of a stopover site cannot be determined only in relation to how many birds a given site hosts. Consideration must also be given to sites which allow migrants to gain fuel (Alerstam & Lindström, 1990). Indeed, the fuel deposition rate is considered to reflect the habitat quality of a stopover site (Newton, 2008). To estimate fuel deposition rate, we considered the difference of body mass between the last and first capture event of each bird, divided by the number of days elapsed between these two events. To compare the fuel deposition rate between zones we used an ANCOVA on the fuel deposition rate with zone as a factor, and P3 and number of days between first and last capture as covariates. Since the estimation of the fuel deposition rate using recaptures of ringed migrants can be subject to bias due to a handling effect, especially when recaptures are obtained shortly after the first capture (Schwilch

& Jenni, 2001), we repeated the analysis removing the migrants recaptured the day after their first capture event. Intermediate recaptures (in migrants recaptured more than once) were not considered in any of the analyses.

Finally, we also calculated the minimum stopover duration as: $(t_i - t_0) + 1$, where t_i and t_0 were respectively the date of the last and first capture event of each bird, respectively. The '+1' was added because sedge warblers are nocturnal migrants. We did not use a more accurate estimation of stopover duration (*e.g.* by means of the use Cormack–Jolly–Seber models, which allow separation of the survival and recapture probabilities) because our sample was too small to allow us to separately estimate survival (here, probability of a bird remaining at the site from one day to the next) and the recapture rate (Lebreton et al., 1992). We used a *t*–test assuming heteroscedasticity to test if the stopover duration differed between the control and invaded zones.

For statistical procedures, we used the software PAST 2.1. (Hammer et al., 2001), and SPSS 18.0. All means are given ± SE.

Results

Assemblage characteristics

We captured a total of 30 bird species in the invaded zone and 28 of these were passerines. In the control zone we captured 28 species and 26 were passerines. The number of captures at each zone was 529 and 558, respectively (table 2).

Bird assemblage did not differ in terms of richness, but it did differ in terms of diversity (table 3), which was lower in the control zone. Another notable aspect was that bird assemblage in the control zone was richer in species typical of reedbeds and wetlands, while in the invaded zone assemblage was richer in woodland–related species (table 3). In the control zone, six out of the ten most abundant species were clearly associated with reedbeds and wetlands (only classified as 'reed' in table 2), while conversely, in the invaded zone, six species were clearly associated with woodlands (classified as 'wood' in table 2; fig. 1). The aquatic warbler, the only species linked to wetlands and globally threatened, only appeared in the control zone ($n = 7$).

Capture rates were similar in both zones (invaded zone: 18.0 ± 2.7 cap./day, $n = 29$; control zone: 18.2 ± 3.0 cap./day, $n = 31$; $t_{58} = 0.25$, $P = 0.80$). However, the difference was significant when only captures of species clearly related to reedbeds/wetlands were considered (invaded zone: 6.2 ± 0.7 cap./day, $n = 29$; control zone: 10.0 ± 1.0 cap./day, $n = 31$; $t_{58} = 3.12$, $P = 0.003$).

Recaptures

We found 26% of the birds captured in the control zone were recaptured (at any of the study zones), while only 15% of the birds captured in the invaded zone were recaptured (at any of the study zones). This difference was significant ($\chi_1^2 = 13.04$, $P < 0.001$). Focusing on species related to reedbeds, the proportion was 38% and 25%, respectively ($\chi_1^2 = 4.71$, $P = 0.03$).

Moreover, 6% of the birds first captured in the control zone were thereafter recaptured in the invaded zone, while 24% of the birds first captured in the invaded zone were thereafter recaptured in the control zone. This difference being significant ($\chi_1^2 = 9.78$, $P = 0.002$). Focusing on species typical of reedbeds, the difference was even higher (6% *versus* 32%; $\chi_1^2 = 12.97$, $P < 0.001$), even for sedge warblers (3% *versus* 58%; $\chi_1^2 = 11.49$, $P = 0.005$), indicating that species from reedbeds tended to move to the control zone much more often than the other way around.

Body mass, fuel deposition rate, age ratios and stopover duration in sedge warblers

Sedge warblers captured in the invaded and control zone did not differ in body mass when considering only the ten lightest migrants in each zone, but there was a difference at the other end of the scale (fig. 2; ANCOVA: zone, $F_{1,39} = 20.73$, $P < 0.001$; fuel category, $F_{1,39} = 385.25$, $p < 0.001$; zone × fuel category, $F_{1,39} = 35.18$, $P < 0.001$; P3, $F_{1,39} = 0.001$, $P = 0.98$).

The body mass (controlled for body size) of the ten heaviest migrants in the control zone was higher than the body mass of the ten heaviest migrants in the invaded zone (fig. 2).

Sedge warblers last recaptured in the control zone showed significantly higher fuel deposition rates than those recaptured in the invaded zone (table 4).

The proportion of first–year birds and adults did not differ between the invaded (48.6%) and control zones (57.1%; $\chi_1^2 = 0.80$, $P = 0.37$).

Minimum stopover duration in the control zone (6.4 ± 0.9 days, $n = 32$) was longer than in the invaded zone (3.0 ± 0.6 days, $n = 3$; $t_{17.228} = 3.20$, $P = 0.005$).

Discussion

The saltbush is an exotic shrub that invades humid habitats in Europe and displaces the native vegetation (*e.g.*, reedbeds). In this work, we show that the saltbush may have a negative impact on both the abundance and stopover characteristics (fuel management and habitat use) of migrants associated with reedbeds.

Bird assemblage characteristics

The diversity index was found to reach higher values in the invaded zone than in the control zone, mainly due to the fact that in the control zone there were three passerines accounting for 70% of the captures, whilst in the invaded zone only two species were clearly dominant, making up 50% of the captures. Reedbeds constitute particular habitats where vertical vegetation is dominant. Exploitation of this type of vegetation requires high adaptation/specialization, so reedbeds constitute an adequate habitat for relatively few species (Poulin et al., 2000, 2002; Leisler & Schulze–Hagen, 2011) compared to other habitats where horizontal vegetation is dominant (*e.g.*, Arizaga et al., 2009). It is of note that in species that are typical of wooded areas and occupy reedbeds during the non–breeding period, such as robins (*Erithacus rubecula*), it is the juvenile fraction which is detected in reedbeds and wetlands (suboptimal habitats), whilst adults monopolize more suitable habitats such as forests (Figuerola et al., 2001). Thus, the captures in the control zone were mostly of species typical of reedbeds and wetlands, whereas the species detected in the invaded zone were, to a larger extent, typical of forested habitats. Moreover, reedbed–associated species were more abundant in the control zone than in the invaded zone. This result highlights that the saltbush not only had a negative impact on the abundance of reedbed–associated species, but also affected bird assemblage by facilitating the presence of woodland passerines.

Stopover behavior

The finding that the proportion of recaptures was higher for the control zone than for the invaded zone suggests that migrants tended to stay longer in the control zone. We cannot overlook the possibility that the recapture rate was site–dependent and this may have imposed some

Table 2. Number of captures/recaptures of species caught with mist nets in the invaded and control zones. Species were assigned to the specific habitats that they normally occupy during the non–breeding period, either when foraging or roosting (Cramp, 1988, 1992; Cramp & Perrins, 1994): Reed. Reedbeds and wetlands; Wood. Woodland, including shrubs, hedgerows, forested areas and parks with trees/shrubs; Others. Urban areas, open habitats, crops. Recaptures refer to birds recaptured ≥ 1 days after the first capture event, to sites where birds were first captured (i.e., a bird captured in the control zone and subsequently recaptured in the invaded area is included in the column of recaptures for the control zone), and only one recapture per bird is considered: Cap. Captures; Rcap. Recaptures.

Tabla 2. Número de capturas/recapturas de las especies atrapadas en redes japonesas en las zonas invadida y de control. Las especies se asignaron a los hábitats que suelen ocupar durante el período no reproductivo, bien mientras forrajeaban, bien mientras reposaban (Cramp, 1988, 1992; Cramp & Perrins, 1994): Reed. Carrizales y humedales; Wood. Tierras arboladas, incluidos arbustos, setos, áreas boscosas y parques con árboles y arbustos; Others. Áreas urbanas, hábitats abiertos y cultivos. Las recapturas hacen referencia a las aves que se volvieron a capturar uno o más días después de la primera captura y a los sitios en que se había capturado a las aves la primera vez (esto es, un ave capturada en la zona de control y posteriormente vuelta a capturar en la zona invadida figura en la columna de las recapturas de la zona de control); solo se tuvo en cuenta una recaptura por ave: Cap. Capturas; Rcap. Recapturas.

Specific name	Code	Main habitats	Invaded zone		Control zone	
			Cap.	Rcap.	Cap.	Rcap.
A. arundinaceus	ACRARU	Reed	0	0	1	1
A. paludicola	ACROLA	Reed	0	0	7	2
A. schoenobaenus	ACRSCH	Reed	37	7	106	40
A. scirpaceus	ACRSCI	Reed	106	22	145	56
A. caudatus	AEGCAU	Wood	17	1	0	0
A. atthis	ALCATT	Reed	9	5	7	3
C. brachydactyla	CERBRA	Wood	1	0	0	0
C. cetti	CETCET	Reed	23	13	12	4
C. juncidis	CISJUN	Others + Reed	2	0	38	9
E. rubecula	ERIRUB	Wood	21	9	5	4
F. hypoleuca	FICHYP	Wood	3	0	1	0
F. coellebs	FRICOE	Wood + Others	3	0	1	0
G. glandarius	GARGLA	Wood	1	0	0	0
H. pollyglotta	HIPPOL	Wood	6	0	7	0
J. torquilla	JYNTOR	Wood	1	0	0	0
L. collurio	LANCOL	Others + Wood	4	0	4	0
L. luscinioides	LOCLUS	Reed	0	0	1	1
L. naevia	LOCNAE	Wood + Reed	6	0	3	0
L. megarhynchos	LUSMEG	Wood	2	0	2	0
L. svecica	LUSSVE	Reed	6	0	30	13
P. caeruleus	PARCAE	Wood	10	2	8	3
P. cristatus	PARCRI	Wood	2	0	0	0
P. major	PARMAJ	Wood	5	0	6	0
P. domesticus	PASDOM	Others + Reed	3	0	1	0
P. montanus	PASMON	Wood + Reed	0	0	2	0
P. ibericus	PHYIBE	Wood	23	4	4	1
P. trochylus	PHYLUS	Wood + Reed	155	7	145	5
R. aquaticus	RALAQU	Reed	0	0	1	0

Table 2. (Cont.)

| | | | Invaded zone | | Control zone | |
Specific name	Code	Main habitats	Cap.	Rcap.	Cap.	Rcap.
R. ignicapillus	REGIGN	Wood	2	1	1	0
S. torquata	SAXTOR	Wood	1	0	0	0
S. vulgaris	STUVUL	Others + Reed	0	0	1	0
S. atricapilla	SYLATR	Wood	25	1	0	0
S. borin	SYLBOR	Wood	7	0	1	0
S. communis	SYLCOM	Wood	10	0	14	0
T. troglodytes	TROTRO	Wood	11	3	4	2
T. merula	TURMER	Wood	27	4	0	0

bias in relation to the estimation of stopover duration (Schaub et al., 2001). Unfortunately, our sample was too small to separately estimate survival (here meaning the probability of a bird remaining in the site from one day to the next) and the recapture rate (Lebreton et al., 1992).

Mean fuel deposition rate in the control zone (+ 0.2 g/d) was even higher than the rate reported in another coastal reedbed in northern Iberia (+ 0.1 g/d; Grandío, 1998). In optimal stopover habitat, the species has been reported to reach mean rates of > + 0.3 g/d (Schaub & Jenni, 2000a). Thus, the rate of fuel accumulation at Urdaibai seemed high, and hence the control zone can well be considered an optimal habitat for the species. The fact that the proportion of birds that moved to the control zone from the invaded zone was higher than the opposite scenario supports the hypothesis that the saltbush area was suboptimal for migrants as compared to reedbeds. Also supporting the hypothesis that the saltbush did not provide a proper fuelling chance for migrants was the finding that the fuel deposition rate in the control zone was much higher than in the invaded zone. Together with the previous result, this is in accordance with the idea that migrants quickly depart from a site when experiencing a very low fuel accumulation rate (Alerstam & Lindström, 1990). From an evolutionary standpoint, this response allows migrants to look for better stopover places and thus have a second chance

to gain fuel during migration period. However, this flexible behavior depends on current fuel load, and birds with small fuel loads will therefore be hampered in finding a better stopover site if this is associated with long displacements. In this scenario, invasive plants imposing radical habitat changes, like the saltbush, constitute a severe problem from a fuelling standpoint. This is particularly applicable if native vegetation is replaced by exotic plants across very large areas. The substitution of reedbeds by saltbushes in several coastal marshes in the Bay of Biscay, including Urdaibai, must be regarded as a problem with unknown consequences for the conservation of migrant bird species associated with reedbeds.

The body mass of the ten heaviest and the ten lightest sedge warblers from each zone was used as a surrogate of body mass at departure and arrival, respectively (Ellegren & Fransson, 1992). The analysis of these data indicate that sedge warblers arrived at both zones with a similar fuel load, but departed with more fuel from the control zone, a result that is in accordance with the control zone favoring fuel accumulation. This result cannot be considered to be caused by a possible bias between age classes (body mass can vary between age classes; Grandío, 1999) and zones, since the proportion of each age class was constant for the invaded and control zones. Our results also support the hypothesis that a number of

Table 3. Diversity–related statistics between the invaded and control zones.

Tabla 3. Estadísticos relacionados con la diversidad entre las zonas invadida y de control.

	Invaded zone	Control zone	Bootstrap (*P*–values)
Taxa (richness)	30	28	0.556
Shannon diversity index (*H'*)	2.485	2.124	< 0.001

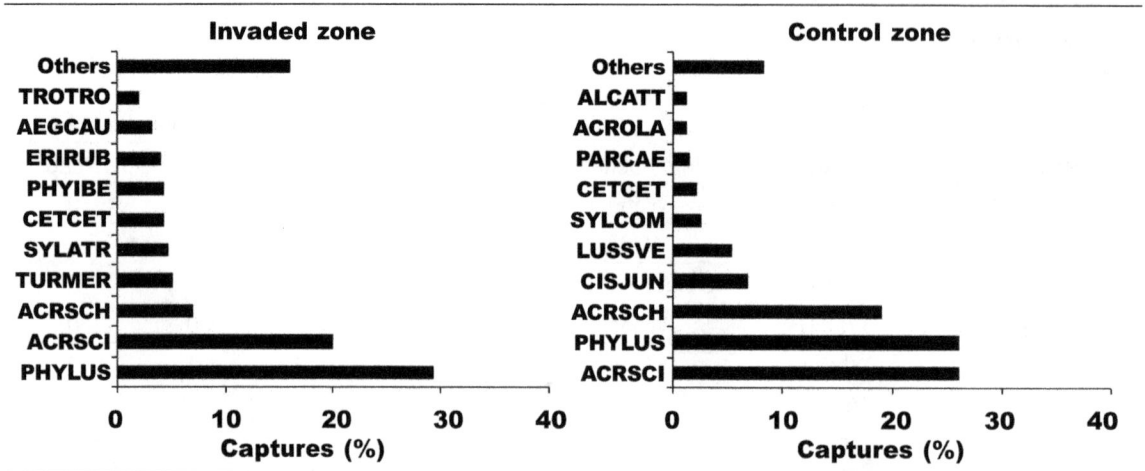

Fig. 1. Relative number of captures (first capture event) of the ten most frequent species at each sampling zone.

Fig. 1. Número relativo de capturas (primera captura) de las diez especies más frecuentes en cada zona de muestreo.

warblers in the saltbush zone may have been able to compensate for their low fuel accumulation rate/fuel load on departure by moving to a better site (the control zone) with higher fuelling chance.

Conclusion and perspectives

The saltbush had a negative impact on several species closely associated with reedbeds, and promoted

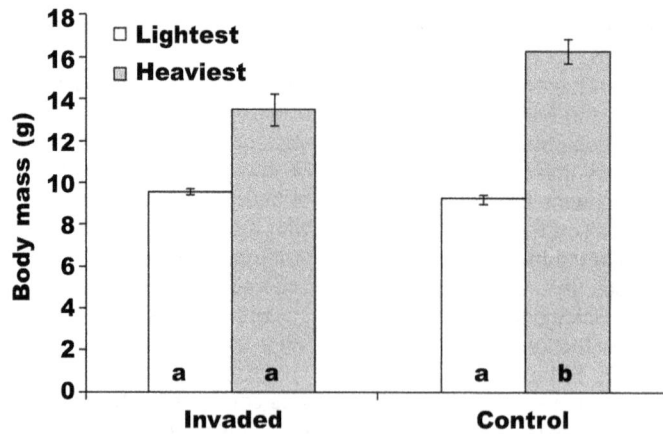

Fig. 2. Mean (± SE) body mass of sedge warblers first captured at the invaded and control zones. For each zone, we considered the ten lightest and heaviest captures of sedge warblers in each zone (ages pooled). Also within each zone, average values bearing the same letters represent non–significant differences in body mass between fuel categories (lightest/heaviest). Differences in body size were controlled including P3 as a covariate.

Fig. 2. Media (± EE) del peso corporal de los carricerines que se capturaron por primera vez en las zonas invadida y de control. Para cada zona, se tuvieron en cuenta las capturas de los diez carricerines de mayor y menor peso (agrupadas por edad). También para cada zona, los valores medios asociados a las mismas letras representan diferencias no significativas en cuanto al peso corporal entre las categorías de cantidad de grasa (más ligero/más pesado). Las diferencias de tamaño corporal se controlaron con la inclusión de la P3 como covariable.

Table 4. Mean (± SE) fuel deposition rates for sedge warblers captured in the invaded and control zones. Sample size in brackets. Data on both zones were compared with an ANCOVA on fuel deposition rate with zone as factor and P3 and number of days elapsed between the first and last captures as covariates; we only show F–values used to test for the effect of zone. The number of recaptures differed from table 1 since not all birds had the P3 measured.

Tabla 4. Media (± EE) de los índices de deposición de grasa para los carricerines capturados en las zonas invadida y de control. El tamaño de la muestra entre paréntesis. Los datos de ambas zonas se compararon con una ANCOVA del índice de deposición de grasa con la zona como factor y la P3 y el número de días transcurridos entre la primera captura y la última como covariables; mostramos únicamente los valores de F utilizados para comprobar los efectos de la zona. El número de recapturas difirió de las de la tabla 1 porque no se midió la P3 a todas las aves.

Days	Invaded zone	Control zone	F	P
All	− 0.1 ± 0.1	+ 0.2 ± 0.1	0.062	0.812
	(4)	(41)		
> 1	− 0.2 ± 0.2	+ 0.1 ± 0.1	4.316	0.016
	(3)	(31)		

a noticeable change on bird assemblage, which was found to be enriched by species typical of woodland habitats. A migrant species strongly associated with reedbeds during migration period, sedge warblers captured at the control zone departed with a higher fuel load, showed a higher fuel deposition rate, and remained for longer than those that stayed at a nearby site occupied by saltbushes (an exotic invasive bush from America). Our results suggest that saltbushes reduce the habitat quality for migratory sedge warblers. Unfortunately, data comparing bird migration in the control and invaded areas before saltbushes invasion are not available. Consequently, although unlikely, it might be possible that both areas already differed in their quality before invasion. Future research with replicates will be necessary to confirm these results. Further studies are also needed to accurately quantify how the wetlands (coastal marshes) from southern Europe have been affected by saltbushes, and to better understand the ecology of migrants so as to to be able to properly evaluate the real impact of saltbushes on the stopover behavior and migratory performance of bird migrants associated with reedbeds. From a management standpoint, restoration of reedbeds must be regarded as a priority tool to preserve the optimal stopover habitat for passerines associated with this type of vegetation.

Acknowledgements

This research was funded by the Basque Government, the Bizkaia Council and BBK. Ringing activities were authorized by the Bizkaia Council. V. Salewski, D. Serrano and an anonymous editor provided very valuable comments that helped us to improve an earlier version of this work.

References

Alerstam, T. & Lindström, Å., 1990. Optimal bird migration: the relative importance of time, energy and safety. In: *Bird migration: the physiology and ecophysiology:* 331–351(E. Gwiner, Ed.). Springer–Verlag Heidelberg, Berlin.

Arizaga, J., Alcalde, J. T., Alonso, D., Bidegain, I., G., B., Deán, J. I., Escala, M. C., Galicia, D., Gosá, A., Ibáñez, R., Itoiz, U., Mendiburu, A., Sarassola, V. & Vilches, A., 2009. *La laguna de Loza: flora y fauna de vertebrados. Munibe* (Supl.), 30.

Arizaga, J., Alonso, D., Campos, F., Unamuno, J. M., Monteagudo, A., Fernandez, G., Carregal, X. M. & Barba, E., 2006. ¿Muestra el pechiazul *Luscinia svecica* en España una segregación geográfica en el paso posnupcial a nivel de subespecie? *Ardeola*, 53: 285–291.

Arizaga, J., Arroyo, J. L., Rodríguez, R., Martínez, A., San–Martín, I. & Sallent, Á., 2011a. Do Blackcaps *Sylvia atricapilla* stopping over at a locality from Southern Iberia refuel for crossing the Sahara? *Ardeola*, 58: 71–85.

Arizaga, J., Barba, E., Alonso, D. & Vilches, A., 2010. Stopover of bluethroats (*Luscinia svecica cyanecula*) in northern Iberia during the autumn migration period. *Ardeola*, 57: 69–85.

Arizaga, J., Sánchez, J. M., Díez, E., Cuadrado, J. E., Asenjo, I., Mendiburu, A., Jauregi, J. I., Herrero, A., Elosegi, Z., Aranguren, I., Andueza, M. & Alonso, D., 2011b. Fuel load and potential flight ranges of passerine birds migrating through the western edge of the Pyrenees. *Acta Ornithologica*, 46: 19–28.

Bensch, S. & Nielsen, B., 1999. Autumn migration speed of juvenile Reed and Sedge Warblers in relation to date and fat loads. *Condor*, 101: 153–156.

Bibby, C. J. & Green, R. E., 1981. Migration strategies of reed and sedge warblers. *Ornis Scandinavica*, 12: 1–12.

Cantos, F. J., 1998. Patrones geográficos de los movimientos de sílvidos transaharianos a través de la Península Ibérica. *Ecología*, 12: 407–411.

Cerasale, D. J. & Guglielmo, C. G., 2010. An integrativa assessment of the effects of tamarisk on stopover ecology of a long–distance migrant along the San Pedro river, Arizona. *Auk*, 127: 636–646.

Cramp, S., 1988. *Handbook of the Birds of Europe, the Middle East and North Africa. Vol. 5.* Oxford Univ. Press, Oxford.

– 1992. *Handbook of the Birds of Europe, the Middle East and North Africa. Vol. 6.* Oxford Univ. Press, Oxford.

Cramp, S. & Perrins, C. M., 1994. *Handbook of the Birds of Europe, the Middle East and North Africa. Vol. 8.* Oxford Univ. Press, Oxford.

Cronquist, A., 1980. *Vascular flora of the Southeastern United States.* Univ. of North Carolina Press, Chapel Hill, North Carolina.

Delingat, J. & Dierschke, V., 2000. Habitat utilization by Northern Wheatears (*Oenanthe oenanthe*) stopping over on an offshore island during migration. *Vogelwarte*, 40: 271–278.

Ellegren, H. & Fransson, T., 1992. Fat loads and estimated flight–ranges in four *Sylvia* species analysed during autumn migration at Gorland, South–East Sweden. *Ringing and Migration*, 13: 1–12.

Figuerola, J., Jovani, R. & Sol, D., 2001. Age–related habitat segregation by Robins *Erithacus rubecula* during the winter. *Bird Study*, 48: 252–255.

Grandío, J. M., 1998. Comparación del peso y su incremento, tiempo de estancia y de la abundancia del carricerín común (*Acrocephalus schoenobaenus*) entre dos zonas de la marisma de Txingudi (N de España). *Ardeola*, 45: 137–142.

– 1999. Migración postnupcial diferencial del carricerín común (*Acrocephalus schoenobaenus*) en la marisma de Txingudi (N de España). *Ardeola*, 46: 171–178.

Hammer, Ø., Harper, D. A. T. & Ryan, P. D., 2001. PAST: Palaeontological Statistics software package for education and data analysis. *Palaentologia Electronica*, 4.

Jenni, L. & Jenni–Eiermann, S., 1998. Fuel Supply and Metabolic Constraints in Migrating Birds. *Journal of Avian Biology*, 29: 521–528.

Jenni–Eiermann, S., Almasi, B., Maggini, I., Salewski, V., Bruderer, B., Liechti, F. & Jenni, L., 2011. Numbers, foraging and refuelling of passerine migrants at a stopover site in the western Sahara: diverse strategies to cross a desert. *Journal of Ornithology*, 152 (Suppl. 1): S113–S128.

Julliard, R., Bargain, B., Dubos, A. & Jiguet, F., 2006. Identifying autumn migration routes for the globally threatened Aquatic Warbler *Acrocephalus paludicola*. *Ibis*, 148: 735–743.

Lebreton, J. D., Burnham, K. P., Clobert, J. & Anderson, D. R., 1992. Modelling survival and testing biological hypothesis using marked animals: a unified approach with case studies. *Ecological Monographs*, 62: 67–118.

Leisler, B. & Schulze–Hagen, K., 2011. *The Reed Warblers*. KNNV Publishing, Zeist.

Newton, I., 2004. Population limitation in migrants. *Ibis*, 146: 197–226.

– 2008. *The migration ecology of birds.* Academic Press, London.

Palmer, W. A., Diatloff, G. & Melksham, J., 1993. The host specificity of *Rhopalomyia california* felt (Diptera: Cecidomyiidae) and its importation into Australia as a biological control agent for *Baccharis halimifolia* L. *Enthomological Society of Washington*, 95: 1–6.

Poulin, B., Lefebvre, G. & Mauchamp, A., 2002. Habitat requirements of passerines and reedbed management in southern France. *Biological Conservation*, 107: 315–325.

Poulin, B., Lefebvre, G. T. & Pilard, P., 2000. Quantifying the breeding assamblage of reedbed passerines with mist–net and point–count surveys. *Journal of Field Ornithology*, 71: 443–454.

Sandberg, R. & Moore, F. R., 1996. Fat stores and arrival on the breeding grounds: Reproductive consequences for passerine migrants. *Oikos*, 77: 577–581.

Sanz, M., Sobrino, E. & Dana, E. D., 2004. *Atlas de las plantas alóctonas invasoras de España.* Organismo Autónomo de Paques Nacionales, Madrid.

Schaub, M. & Jenni, L., 2000a. Fuel deposition of three passerine bird species along the migration route. *Oecologia*, 122: 306–317.

– 2000b. Body mass of six long–distance migrant passerine species along the autumn migration route. *Journal of Ornithology*, 141: 441–460.

Schaub, M., Pradel, R., Jenni, L. & Lebreton, J. D., 2001. Migrating birds stop over longer than usually thought: An improved capture–recapture analysis. *Ecology*, 82: 852–859.

Schwilch, R. & Jenni, L., 2001. Low initial refuelling rate at stopover sites: a methodological approach? *Auk*, 118: 698–708.

Smith, R. J. & Moore, F. R., 2003. Arrival fat and reproductive performance in a long–distance passerine migrant. *Oecologia*, 134: 325–331.

Sol, D., Timmermans, S. & Lefebvre, L., 2002. Behavioural flexibility and invasion success in birds. *Animal Behaviour*, 63: 495–502.

Svensson, L., 1996. *Guía para la identificación de los paseriformes europeos.* Sociedad Española de Ornitología, Madrid.

Tellería, J. L., Asensio, B. & Díaz, M., 1999. *Aves Ibéricas. II. Paseriformes* (J. M. Reyero, Ed.). Madrid.

Vitousek, P. M., D'Antonio, C. M., Loope, L. L., Rejmanek, M. & Westbrooks, R., 1997. Introduced species: A significant component of human–caused global change. *New Zealand Journal of Ecology*, 21: 1–16.

Permissions

All chapters in this book were first published in ABC, by Natural Science Museum of Barcelona; hereby published with permission under the Creative Commons Attribution License or equivalent. Every chapter published in this book has been scrutinized by our experts. Their significance has been extensively debated. The topics covered herein carry significant findings which will fuel the growth of the discipline. They may even be implemented as practical applications or may be referred to as a beginning point for another development.

The contributors of this book come from diverse backgrounds, making this book a truly international effort. This book will bring forth new frontiers with its revolutionizing research information and detailed analysis of the nascent developments around the world.

We would like to thank all the contributing authors for lending their expertise to make the book truly unique. They have played a crucial role in the development of this book. Without their invaluable contributions this book wouldn't have been possible. They have made vital efforts to compile up to date information on the varied aspects of this subject to make this book a valuable addition to the collection of many professionals and students.

This book was conceptualized with the vision of imparting up-to-date information and advanced data in this field. To ensure the same, a matchless editorial board was set up. Every individual on the board went through rigorous rounds of assessment to prove their worth. After which they invested a large part of their time researching and compiling the most relevant data for our readers.

The editorial board has been involved in producing this book since its inception. They have spent rigorous hours researching and exploring the diverse topics which have resulted in the successful publishing of this book. They have passed on their knowledge of decades through this book. To expedite this challenging task, the publisher supported the team at every step. A small team of assistant editors was also appointed to further simplify the editing procedure and attain best results for the readers.

Apart from the editorial board, the designing team has also invested a significant amount of their time in understanding the subject and creating the most relevant covers. They scrutinized every image to scout for the most suitable representation of the subject and create an appropriate cover for the book.

The publishing team has been an ardent support to the editorial, designing and production team. Their endless efforts to recruit the best for this project, has resulted in the accomplishment of this book. They are a veteran in the field of academics and their pool of knowledge is as vast as their experience in printing. Their expertise and guidance has proved useful at every step. Their uncompromising quality standards have made this book an exceptional effort. Their encouragement from time to time has been an inspiration for everyone.

The publisher and the editorial board hope that this book will prove to be a valuable piece of knowledge for researchers, students, practitioners and scholars across the globe.

List of Contributors

M. J. T. Assunção–Albuquerque
Dept. of Ecology, Univ. of Alcalá, 28871 Alcalá de Henares, Madrid, España (Spain)

J. M. Rey Benayas
Dept. of Ecology, Univ. of Alcalá, 28871 Alcalá de Henares, Madrid, España (Spain)

M. Á. Rodríguez
Dept. of Ecology, Univ. of Alcalá, 28871 Alcalá de Henares, Madrid, España (Spain)

F. S. Albuquerque
Grupo de Ecología Terrestre, Centro Andaluz de Medio Ambiente, Univ. Granada–Junta de Andalucía, Av. del Mediterráneo s/n., 18006 Granada, España (Spain)

Jorge Lozano
Dept. de Ecología, Univ. Autónoma de Madrid, c/ Darwin 2, Edificio de Biología, E–28049 Cantoblanco, Madrid, España (Spain)
Dept. de Ecología, Univ. Complutense de Madrid, c/ José Antonio Novais 12, Ciudad Universitaria, E–28040 Madrid, España (Spain)

Jorge G. Casanovas
Dept. de Ecología, Univ. Complutense de Madrid, c/ José Antonio Novais 12, Ciudad Universitaria, E–28040 Madrid, España (Spain)

Juan M. Zorrilla
Dept. de Ecología, Univ. Complutense de Madrid, c/ José Antonio Novais 12, Ciudad Universitaria, E–28040 Madrid, España (Spain)

Emilio Virgós
Dept. de Biología y Geología, Univ. Rey Juan Carlos, c/ Tulipán s/n., E–28933 Móstoles, Madrid, España (Spain)

Louri Klemann Júnior
Ecology and Conservation Post–Graduation Program, Univ. Federal do Paraná, P. O. Box 19020, Curitiba, Paraná, Brazil

Juliana S. Vieira
Entomology Post–Graduation Program, Univ. Federal do Paraná, P. O. Box 19020, Curitiba, Paraná, Brazil

Jorge Ortega
Lab. de Ictiología y Limnología, Depto. de Zoología, Escuela Nacional de Ciencias Biológicas, Inst. Politécnico Nacional, Prolongación de Carpio y Plan de Ayala s/n., Col. Sto. Tomas, 11340, México, D. F.

Daya Navarrete
Lab. de Ictiología y Limnología, Depto. de Zoología, Escuela Nacional de Ciencias Biológicas, Inst. Politécnico Nacional, Prolongación de Carpio y Plan de Ayala s/n., Col. Sto. Tomas, 11340, México, D. F.

Jesús E. Maldonado
Center for Conservation and Evolutionary Genetics, Smithsonian Conservation Biology Inst., National Zoological Park, Washington, DC 20008, USA

Gregorio Rocha
Dept. of Agro–forestry Engineering, Univ. of Extremadura, Avda. Virgen del Puerto 2, 10600 Plasencia, Cáceres, Spain

Petra Quillfeldt
Inst. für Tierökologie und Spezielle Zoologie, Justus Liebig Univ. Gießen, Heinrich–Buff–Ring 38, 35392 Gießen (Germany)

José Guerrero–Casado
Dept of Zoology, Univ. of Córdoba, Campus de Rabanales, E–14071, Córdoba, España (Spain)
Inst. for Terrestrial and Aquatic Wildlife Research, Univ. of Veterinary Medicine Hannover, Bischofsholer Damm 15, 30173 Hannover (Germany)

Francisco S. Tortosa
Dept of Zoology, Univ. of Córdoba, Campus de Rabanales, E–14071, Córdoba, España (Spain)

Jérôme Letty
Office National de la Chasse et de la Faune Sauvage, Direction des Etudes et de la Recherche, 147 route de Lodève, F–34990, Juvignac (France)

Carlos A. Mancina
Inst. de Ecología y Sistemática, carretera de Varona km 3 ½ Capdevila, Boyeros, A. P. 8029, C. P. 10800, La Habana, Cuba

Daysi Rodríguez Batista
Inst. de Ecología y Sistemática, carretera de Varona km 3 ½ Capdevila, Boyeros, A. P. 8029, C. P. 10800, La Habana, Cuba

Edwin Ruiz Rojas
Centro de Estudios y Servicios Ambientales, Villa Clara (Cuba)

Miroslav Kulfan
Dept. of Ecology, Fac. of Natural Sciences, Comenius Univ., Mlynská dolina B–1, SK–84215 Bratislava, Slovakia

Pavel Beracko
Dept. of Ecology, Fac. of Natural Sciences, Comenius Univ., Mlynská dolina B-1, SK-84215 Bratislava, Slovakia

Milada Holecová
Dept. of Zoology, Fac. of Natural Sciences, Comenius Univ., Mlynská dolina B-1, SK-84215 Bratislava, Slovakia

Milena Matos
Dept. of Biology & CESAM, Univ. of Aveiro, 3810-193 Aveiro, Portugal

Michelle Alves
Dept. of Biology & CESAM, Univ. of Aveiro, 3810-193 Aveiro, Portugal

Maria João Ramos Pereira
Dept. of Biology & CESAM, Univ. of Aveiro, 3810-193 Aveiro, Portugal

Inês Torres
Dept. of Biology & CESAM, Univ. of Aveiro, 3810-193 Aveiro, Portugal

Sara Marques
Dept. of Biology & CESAM, Univ. of Aveiro, 3810-193 Aveiro, Portugal

Carlos Fonseca
Dept. of Biology & CESAM, Univ. of Aveiro, 3810-193 Aveiro, Portugal

Iván de la Hera
Depto. de Zoología y Biología Celular Animal, Fac. de Farmacia, Univ. del País Vasco (UPV/ EHU), Paseo de la Universidad 7, 01006 Vitoria-Gasteiz, España (Spain)

Juan Arizaga
Urdaibai Bird Center. Sociedad de Ciencias Aranzadi, Orueta 7, 48314 Gautegiz-Arteaga, Bizkaia, España (Spain)

Aitor Galarza
Depto. de Agricultura, Diputación Foral de Bizkaia, 48014 Bilbao, España (Spain)

Mathieu Sarasa
Grupo Biología de las Especies Cinegéticas y Plagas (RNM-118), España (Spain)

Juan-Antonio Sarasa
Grupo de Caza Mayor de Urdués, España (Spain)

Cédric De Danieli
Fédération Départementale des Chasseurs de la Haute-Savoie, Impasse des Glaises 74350 Villy-le-Pelloux

Mathieu Sarasa
Fédération Nationale des Chasseurs, 13 rue du Général Leclerc 92136, Issy les Moulineaux, France

José Carlos Báez
Inst. Español de Oceanografia (IEO), Centro Oceanográfico de Málaga, Puerto pesquero de Fuengirola s/n., 29640 Fuengirola, Málaga, España (Spain);
investigador asociado de la Fac. de Ciencias de la Salud, Univ. Autónoma de Chile, Chile

David Macías
Inst. Español de Oceanografia (IEO), Centro Oceanográfico de Málaga, Puerto pesquero de Fuengirola s/n., 29640 Fuengirola, Málaga, España (Spain)

Maite de Castro
Fac. de Ciencias de Ourense, Univ. de Vigo, Ourense, España (Spain)

Moncho Gómez-Gesteira
Fac. de Ciencias de Ourense, Univ. de Vigo, Ourense, España (Spain)

Luis Gimeno
Ephyslab Fac. de Ciencias de Ourense, Univ. de Vigo, Ourense, España (Spain)

Raimundo Real
Depto. de Biología Animal, Fac. de Ciencias, Univ. de Málaga, 29071 Málaga, España (Spain)

Luis M. García-Feria
Red de Biología y Conservación de Vertebrados, Inst. de Ecología A. C., carretera Antigua a Coatepec 351, El Haya, CP 91070, Xalapa, Veracruz, México

Cinthya A. Ureña-Aranda
Lab. de Sistemática Filogenética, Red de Biología Evolutiva, Inst. de Ecología A.C., carretera antigua a Coatepec 351, El Haya, CP 91070, Xalapa, Veracruz, México

Alejandro Espinosa de los Monteros
Lab. de Sistemática Filogenética, Red de Biología Evolutiva, Inst. de Ecología A.C., carretera antigua a Coatepec 351, El Haya, CP 91070, Xalapa, Veracruz, México

Bianca P. Vieira
Post-graduate Research Program, Inst. of Biodiversity, Animal Health and Comparative Medicine, Univ. of Glasgow, G12 8QQ, Glasgow, U. K.

Carlos Fonseca
Depto. de Biologia & CESAM, Univ. de Aveiro, Campus Santiago, 3810-193, Aveiro, Portugal

Rita G. Rocha
Depto. de Biologia & CESAM, Univ. de Aveiro, Campus Santiago, 3810–193, Aveiro, Portugal

Mohd. Shahnawaz Khan
WWF India, 172 B Lodi Estate, New Delhi, 110 003 (India)

Asghar Nawab
WWF India, 172 B Lodi Estate, New Delhi, 110 003 (India)

Nand Kishor Dimri
WII, 18, Chandrabani, Dehradun, Uttarakhand, (India)

Orus Ilyas
AMU, Aligarh, Uttar Pradesh 202 002 (India)

Parikshit Gautam
FES, NDDB House PB. 4906 Safdarjung Enclave, New Delhi, 110 029 (India)

Juan Arizaga
Dept. of Ornithology, Aranzadi Sciences Society, Urdaibai Bird Center, Orueta 7, E–48314 Gautegiz–Arteaga, Bizkaia, España (Spain)

Edorta Unamuno
Dept. of Ornithology, Aranzadi Sciences Society, Urdaibai Bird Center, Orueta 7, E–48314 Gautegiz–Arteaga, Bizkaia, España (Spain)

Ainara Azkona
Dept. of Ornithology, Aranzadi Sciences Society, Urdaibai Bird Center, Orueta 7, E–48314 Gautegiz–Arteaga, Bizkaia, España (Spain)

Oriol Clarabuch
Catalan Institute of Ornithology, Museu de Ciències Naturals de Barcelona, Passeig Picasso s/n., E–08003 Barcelona, Espanya (Spain)